EXS
Experientia Supplementum
Vol. 51

Birkhäuser Verlag
Basel · Boston

Advances in Aerobiology

Proceedings of the 3rd International Conference on Aerobiology, August 6–9, 1986, Basel, Switzerland

Edited by

G. Boehm
R. M. Leuschner

1987

Birkhäuser Verlag
Basel · Boston

Addresses of the Editors:

Ruth M. Leuschner, PhD
Department of Research
Div. Dermatology/Allergology
Kantonsspital
CH–4031 Basel, Switzerland

Prof. G. Boehm
Herrengrabenweg 51
CH–4054 Basel, Switzerland

Library of Congress Cataloging in Publication Data

International Conference on Aerobiology (3rd : 1986 :
    Basel, Switzerland)
    Advances in aerobiology.

    (Experientia. Supplementum ; v. 51)
    Bibliography: p.
    Includes index.
    1. Air – – Microbiology – – Congresses.   2. Pollen – – Dispersal
– – Congresses.   3. Airborne infection – – Congresses.
4. Air quality – – Congresses.   I. Leuschner, R. M.
(Ruth M.), 1922–   .   II. Boehm, G. (Gundo E.),
1904–   .   III. Title.   IV. Series.
QR101.I4   1986   574.909'612   87-11672

CIP-Kurztitelaufnahme der Deutschen Bibliothek

Advances in aerobiology : proceedings of the
3rd Internat. Conference on Aerobiology, August 6–9,
Basel, Switzerland / ed. by
G. Boehm ; R. M. Leuschner. – Basel ; Boston ;
Birkhäuser, 1987.
    (Experientia : Supplementum ; Vol. 51)
    ISBN 978-3-0348-7493-9   ISBN 978-3-0348-7491-5 (eBook)
    DOI 10.1007/978-3-0348-7491-5
NE: Leuschner, Ruth M. [Hrsg.]; International
Conference on Aerobiology ⟨03, 1986, Basel⟩;
Experientia / Supplementum

© 1987 Birkhäuser Verlag Basel
Softcover reprint of the hardcover 1st edition 1987

ISBN 978-3-0348-7493-9

FOREWORD

The keynote of this volume of Proceedings is 'Aerobiology today'. The 3rd International Conference on Aerobiology in Basel
brought together a very large number of contributions on this
main theme, in the form of lectures and posters, from many
different areas. The scope of the volume of abstracts shows
the great variety of the presentations. It seemed appropriate
that we should also publish the Proceedings, to prevent much
of what was discussed in Basel being forgotten. However, since
the cost had to be considered, some selection was necessary.
As the reader will see, the contribution have been arranged,
as far as possible, according to a number of main themes. As
a whole - as the list of contents shows - the volume provides
a general survey of current endeavour in the field of Aerobiology.

There is an introductory article about the origin of the expression 'Aeroplankton' and, in particular, about the creator
of the word 'Aerobiology'.

The editiors hope that the systematic arrangement of the contents, and above all the detailed keyword index, will make it
easy for the reader to find his way through the volume.

       The editiors:

       Prof. G. Boehm   Dr.R.M.Leuschner

Basel, spring 1987

Contents

I Introduction

Boehm, G.: Aerobiology – its past and its future 3

Frinking, H. D.: Philipp Herries Gregory, FRS, DSc (1907-1986) 9

II Airborne pollen and biometeorology

1. Lectures of general importance

Benninghoff, W. S.: Environmental influences on deposition of airborne
particles 13

Leuschner, R. M.; Boehm, G. and Brombacher, Chr.: Influence of inversion
layers on the daily pollen count and on the allergic attacks of patients in
Basel (Switzerland) 19

Wedler, E.; Fegeler, U.; Moyzes, R. and Eberhard, K.: Influences of pollution
and weather on obstructive respiratory tract diseases of children in Berlin
(West) 25

Coetzee, J. A.: Comparison of recent and quarternary pollen data 31

Mandrioli, P.: Biometeorology and its relation to pollen count 37

Zerboni, R.; Manfredi, M.; Campi, P.; Arrigoni, P. V.: Correlation between
aerobiological and phytogeographical investigations in the Florence area
(Italy) 43

2. Airborne pollen in different countries

Hurtado, I. and Riegler-Goihman, M.: Air sampling studies in a tropical area.
Four year results 49

El-Ghazaly, G. and Fawzy, M.: A study of airborne pollen grains of
Alexandria (Egypt) 55

Singh, A. B.: Air borne pollen types of allergenic significance in India. . . . 61

Dhorranintra, B.; Bunnag, Ch. and Limsuvan, S.: Survey of atmospheric
pollens in various provinces of Thailand. 65

III Allergology and airborne particles

1. Lectures of general importance: diagnosis and therapy

Fuchs, E.: Inhalative allergens. 71

Kneist, W.; Düngemann, H.; Gehrken, H. and Borelli, S.: Relationship of airborne pollen and spores to symptoms on the skin and mucous membranes of patients in the high altitude climate in Davos (Switzerland) . 81

Boehm, G. and Leuschner, R.M.: Experiences with the «Individual Pollen Collector» developed by G. Boehm 87

Fasani, F. and Gorini, M.: Airborne pollens and symptoms in allergic patients undergoing immunotherapy 89

2. Special allergens: Pollen

Macchia, L.,; Aliani, M.; Caiaffa, M. F.; Carbonara, A. M.; Gatti, E.; Iacobelli, A.; Strada, S.; Casella, G. and Tursi, A.: Monitoring of atmospheric conditions and forecast of Olive pollen season (Bari, Italy) 95

O'Rourke, M.K. and Buchmann, St. L.: Pollen yield of two cvs. of Olea europaea L. («Manzanillo» and «Swan hill») (Tucson, Arizona) 101

Galán, C.; Ruiz De Clavijo, E.; Infante, F. and Gallego, G.: Annual and daily variation of pollen from Olea europea L. in the atmosphere of Cordoba (Spain) along two year of sampling 109

Sutra, J. P.; Ickovic, M.-R.; De Luca, H.; Peltre, G. and David, B.: Chestnut pollen counts related to patients pollinosis in Paris 113

Déchamp, C. and Cour, P.: Pollen counts of ragweed and mugwort (Cour Collector) in 1984 measured at 12 meteorological centers in the Rhône Bassin and surrounding regions (France) 119

Wedner, H. J.: Zenger, V. E. and Lewis, W. H.: Allergic reactivity to Parthenium hysterophorus pollen: An unrecognized type I allergen. . . . 125

3. Different aeroallergens

Goldstein, I. F.; Reed, C. E.; Swanson, M. C. and Jacobson, J. S.: Aeroallergens in New York inner-city apartments of asthmatics 133

4. Special allergens: Fungus spores

Gallup, J.; Kozak, P.; Cummins, L. and Gillman, S.: Indoor mold spore exposure: Characteristics of 127 homes in southern California with endogenous mold problems. 139

Burge, H. A.; Simmons, E. G.; Muilenberg, M.; Hoyer, M.; Gallup, J. and Solomon, W.: Intrinsic variability in airborne fungi: Implications for allergen standardization 143

Frankland, A. W.: Clinical cross-reactions between various moulds with special reference to Fusarium 147

Schultze-Werninghaus, G.; Lévy, J.; Bergmann, E. M.; Kappos, A. D. and Meier-Sydow, J.: Clinical significance of airborne Alternaria tenuis-spores: Seasonal symptoms positive skin and bronchial challenge tests with Alternaria in subjects with asthma and rhinitis 153

Infante, F.; Ruiz De Clavijo, E.; Galán, C. and Gallego, G.: Occurrence of Alternaria Nees ex Fr. in indoor and outdoor habitats in Cordoba (Spain). . 157

Spieksma, F. Th. M.; Nolard, N.; Beaumont, F. and Vooren, P. H.: Concentrations of airborne Botrytis conidia, and frequency of allergic sensitization to Botrytis extract 165

5.   Special allergens: Algae

Tiberg, E.: Microalgae as aeroplankton and allergens 171

6.   Special allergens: Insects and mites

Rudolph, E.; Stresemann, E.; Stresemann, B. and Haupthof, M.: Sensitizations against Tribolium confusum Du Val in patients with occupational and non-occupational exposure 177

Fuchs, E.; Bossert, J.; Honomichl, K.; Maasch, H.-J.; Risler, H. and Wahl, R.: Gibbium psylloides («mite beetle») as an inhalative allergen 183

Bischoff, E. and Schirmacher, W.: Investigations of allergen-containing dust samples from the interior of the house 189

Menz, G.; Petri, E.; Lind, P. and Virchow, Chr.: House dust mite in different altitudes of Grisons (Switzerland) 197

Pauli, G.; Tenabene, A.; Bessot, J. C. and Hoyet, C.: Guanine dosage in house dust samples and quantification of mite allergens. 203

IV   Air, our environment

1.   Lectures of general importance

Doll, Sir Richard: The quantitative significance of asbestos fibres in the ambient air 213

Guillemin, M.; Litzistorf, G.; Madelaine, P.; Iselin, F. and Buffat. Ph.: The indoor asbestos problem: Facts and questions 221

Bylin, G.; Lindvall, T.; Rehn, T. and Sundin, B.: Effects of short-term exposure
to ambient nitrogen dioxide concentrations on human bronchial reactivity
and lung function 227

Seemayer, N. H.; Hadnagy, W. and Tomingas, R.: Mutagenic and carcinogenic
effects of airborne particulate matter from polluted areas on human and
rodent tissue cultures. 231

Jakob, G. J.: Modulation of pulmonary defense mechanisms by acute
exposures to Nitrogen Dioxide. 235

Orsi, E. V.; Bavlsik, M.; Viera, R.; Petersheim, M. and Baturay, O. F.:
Lysosome response and cytoskeleton alteration in cell cultures exposed to
airborne lead 243

2.   Outdoor air

Stalder, K. and Kundel, M.: Health risks from respirable dusts produced
during the operation of big harvesters 251

Marty, H.: The influence of meteorological and air pollution factors on acute
diseases of the airways in children as illustrated by the Biel region
(Switzerland. 255

Brown, H. M. and Jackson, F. A.: Airborne crystals of anhydrous Calcium
Sulphate – a new air pollutant 261

Schrader, G.: Coniometric dust measurements during a sea voyage from the
northern to the southern hemisphere, along the Antarctic coast, and at a
coast station in Antarctica 267

3.   Indoor air

Macher, J. M.: Inquiries received by the California indoor air quality program
on biological contaminants in buildings. 275

Gravesen, S.: Microbial and dust pollution in non-industrial work places . . . 279

Elixmann, J. H.; Jorde, W. and Linskens, H. F.: Filters of an airconditioning
installation as disseminators of fungal spores 283

Kroeling, P.: The building illness syndrome: Comparative studies in air
conditioned and conventional heated buildings 287

Pitkänen, E.; E. Pellikka, M.; Kalliokoski, P. and Jantunen, M. J.: Bioaerosols
and office building ventilation system 297

Burkart, W.: Radon and its decay products in the indoor environment:
Radiation exposure and risk estimation 303

V Phytopathology

Kramer, Chs. L. and Eversmeyer, M. G.: Uniformity of airspora concentrations 313

Edmonds, R. L. and Basabe, F. A.: Ozone and acidic inputs into forests in Northwest USA. 319

Masuch, G.; Paul, V. H. and Mallant, R. K. A. M.: Effects of acidic fog containing H_2O_2 on the sensitivity of agricultural crops to important fungal diseases 325

Limpert, E.: Spread of Barley mildew by wind and its significance for Phytopathology, Aerobiology and for Barley cultivation in Europe 331

Azuma, A.; Onda, Y. and Ichikawa, R.-h.: Falling behaviour of sterilized and slept fruitefly and melonfly 337

VI Microbiology

Sellers, R. F.: Airborne spread of microorganisms affecting animals. 347

Donaldson, A. I.; Lee, M. and Gibson, C. F.: Improvement of mathematical models for predicting the airborne spread of foot-and-mouth disease . . . 351

Fitzgeorge, R. B.; Baskerville, A. and Featherstone, A. S. R.: Fine particle aerosols in experimental Legionnaires' disease: Their role in infection and treatment. 357

Willeke, K. and Baron, P. A.: The size distribution of whirlpool-generated droplets, their ability to contain bacteria and their disposition potential in the human respiratory tract 361

Boutin, P.; Torre, M.; Moline, J. and Boissonot, E.: Bacterial atmospheric contamination in wastewater treatment plants 365

Crook, B.; Higgins, S. and Lacey, J.: Airborne Gram negative bacteria associated with the handling of domestic waste 371

VII Methods in aerobiology

Boehm, G.: Special applications of the Burkard-Sampler for collecting airborne pollen and spores

I. Semi-quantitative continuous determination of dust-particles, soot, flue ash etc. which may activate diseases of the respiratory tract 377

Boehm, G.: Special applications of the Burkard-Sampler for collecting airborne pollen and spores

II. Record of radioactive dust (after Chernobyl) in the air of Basel (Switzerland) 381

Rantio-Lehtimäki, A.; Kauppinen, E.; Koivikko, A.: Efficiency of a new sampler in sampling Betula pollen for antigen analyses. 383

Henningson, E. and Fängmark, I.: Collection efficiency of two samplers for microbiological aerosols 391

Waldner-Sander, S. and Botzenhart, K.: Microorganisme as biological indicators of air pollution 395

Amman, K.; Herzig, R.; Liebendörfer, L. and Urech, M.: Statistical correlation of deposition data on 8 air pollutants to lichen data in a small town in Switzerland 401

Buttler, A. and Girard, M.: Methods for yearly comparisons of aeropalynological data 407

Agashe, S. N. and Chatterjee, M.: Aircraft sampling of the upper airspora. . . 411

Kettrup, A. and Schmidt, P. R.: Methods to reduce allergic effects: Elimination of allergens (pollen, mites, indoor dust, bacteria, gases etc.) from indoor air 415

1 Introduction

AEROBIOLOGY - ITS PAST AND ITS FUTURE

Introductory lecture given at the opening of the Third International Conference on Aerobiology, held in Basel, Switzerland from Aug. 6 - 9, 1986.

G. Boehm, Basel

Dear friends and colleagues,

I should like to open this conference with a brief look at the past, and a glance into the future.

I asked myself, " Who was actually the originator of the term 'aeroplankton'?" Pasteur and other investigators of his time and later, certainly knew that there are things like bacteria and fungal spores in the air around us. The expression 'aeroplankton' was apparently coined in 1912 by the botanist Molisch in Vienna. Hans Molisch (Fig.1)(1.) was a much-travelled man, and one of the greatest botanists of his time. He published numerous papers, and several important books. In 1916, in a popular lecture, he discussed the 'aeroplankton' in rather more detail (2.).

Under the heading 'aeroplankton', Molisch included not only inanimate suspended particles, like dust of all kinds, but also particles of living matter, or ones with an organised structure, like pollen grains, fungal spores, algae, bacteria, and the fine hairs in 'plane-tree dust', which can cause what has sometimes been called 'plane-tree cough'. He was a man of broad vision, and already at that time he recognised how important more exact studies of the aeroplankton would be. That would also apply to studies of pollen with reference to allergic reactions; very little was known about these at that time.

Molisch himself suffered from 'hay-fever' in the season when wild grasses and cereal crops were flowering. In this connection, he had given much thought to the question of wether other kinds of pollen might trigger off the unpleasant symptoms he suffered from in spring, before the grasses came into flower. He considered it very important that more detailled investigations

of the pollen involved should be made. For health resorts, too,
it would be important to keep a constant check on the animate
and inanimate suspended particles in the air. (Injurious gases,

Fig. 1

which are the subject of so much research today, are not mentioned by Molisch in his article.)

The term 'aerobiology' which is more comprehensive than
'aeroplankton', was coined in the thirties by the American plant
pathologist Fred Campbell Meier(Fig.2)(3.).A parallel to the term
'hydrobiology',which was already well known, was created.

F.C.Meier worked for the U.S.Departement of Agriculture
for many years in various capacities. He was interested, among
other things, in plant diseases which were distributed by airborne fungal spores. In order to investigate the atmosphere at va-
rous levels, in different areas and over considerable distances

he worked with the famous aviator Charles A. Lindbergh (the first
man to fly solo across the Atlantic). At the request of F. C.
Meier, Lindbergh undertook a number of reconnaissance flights,

FRED CAMPBELL MEIER
1893 - 1938

Fig. 2

for example between Maine and Denmark - a route which included
Arctic areas - to collect 'aerobiological' specimens such as fungus spores, fragments of insects'wings etc. A special piece of
equipment called a 'sky-hook' was used to collect suspended particles. At first, microscopic slides covered with a layer of adhesive were attached to the apparatus; later Petri dishes containing agar were used.

On flights over the Eastern United States, F.C.Meier
was able to collect numerous spores which still had been viable.
Examples from 17 genera were identified at heights of ca. 150 m
(500 ft.) and 5.500 m (18.000 ft.). Spores collected in the Ca-
ribean area, between 800 and 1.200 km from the next coast, were

also capable of germinating, and so were spores of a number of different genera which were collected at a height of ca.11.000 m (36.000 ft) during a flight into the stratosphere commissioned by F.C.Meier.

As I mentioned above, all this work was undertaken as part of a study of the spreading of plant diseases. However, on the basis of his results, F.C.Meier was able to convince the National Research Council that it was of great importance to extend investigations of this kind to other living particles in the air, for example to pollen as the cause of hay-fever. As everybody who knew F.C.Meier realised, human welfare was always close to his heart.(4.)

It was clear to him that further advances in such studies of the atmosphere could only be made along interdisciplinary lines. With the 'overwhelming enthusiasm' which his colleagues knew so well, he aroused the interest of botanists, meteorologists, zoologists, bacteriologists, plant pathologists and medical experts in further work of this topic. The group included his friend medical adviser, E.B.McKinley, who was also a founder member of what was later called the 'Committee on Aerobiology'. The committee met for the first time on November 12th 1937, in Washington. Support for their projects came from a number of sources, including government departements. Collaborators included influential people from the U.S.Army, the Navy, the Coast Guards and the National Research Council. They also optained the co-operation of Pan-American Airlines.

After these first important achievements, F.C.Meier's life ended in tragedy. He set out on a flight from California in the direction of Manila, to carry out further studies of the atmosphere over the Pacific, in the Pan-American Airway' 'Hawaii Clipper' - a very powerful machine for those days. The aeroplane disappeared on July 29th 1983. The last audible radio message from the pilot gave their exact position and an altitude of 9.100 ft. (ca.2.750 m), and spoke of problems with a rainstorm ('rain static'). In spite of the most intensive searches, no trace was ever found of the aeroplane, its crew of nine, or the six passengers, who included the above-mentioned doctor, E.B.McKinley.

Today, Fred Campbell Meier is almost forgotten. But his ideas and inspiration have survived. What he certainly hoped for has become reality. From 1938 to the present day, under the heading 'Aerobiology' - the term he coined - scientists from a wide variety of disciplines have been continuing the work. Evidence for their activities is given by the nunerous papers and monographs published, and the symposia and conferences held - of which the most recent is the 'Third International Conference on Aerobiology', which begins here in Basel today, the 6th of August 1986, with representatives from a wide variety of disciplines.

People who know little about this topic might ask what the goal of this research is. The answer that the specialist gives will depend on the area of his particularly interest. But for all of us, the objective is to protect people, animals and plants - and buildings too - from damage resulting from the wide variety of pollutants in the atmosphere, and to clean up the endangered ocean of air around us as much as possible, using all the means at our disposal. In some centres of population that has already been partially achieved.

When we look at the many urgent problems that must still be solved - for example that of dying forests - we might say that the solutions are very far away. But we should set to work with a will, inspired by what has already been achieved, and inspired by the old Latin call, 'Sursum corda!' - 'Lift up your hearts!'

Bibliography:

1. From: H.MOLISCH:'Erinnerungen und Welteindrücke eines Natur-
                  forschers'.Emil Haim & Co.,Wien u.Leipzig,
                  1934.
2. H.MOLISCH:'Populäre biologische Vorträge'.Gustav Fischer,Jena
                  2.Aufl.1922,pp 209 - 226:'Biologie des atmo-
                  sphärischen Staubs - Aeroplankton - .
3. From: Phytopathology, 29, 1939, pp 292.

4. Obituary: Royat J. HASKELL, Howard P.BAKES: 'Fred Campbell
                Meier, 1883 - 1938'. Phytopathology 29, 1939,
                pp 293 - 301 (see ibid. p.398).
5. The pioneering achievements of Fred Campbell Meier are descri-
                bed in detail in:
                P.H.GREGORY:'The microbiology of the atmosphe-
                re'.(3rd ed. 1973) p.14 and 183 - 184.

Authors's address:

Prof. Dr. med. G. Boehm, Herrengrabenweg 51, CH-4054 Basel,
                                        Switzerland.

PHILIP HERRIES GREGORY, FRS, DSc

(1907 – 1986)

Dr. Philip H. Gregory is no longer among us. He passed away on 9 February 1986, at the age of 78 years after a fall at his home in Harpenden, England.
Philip Gregory was an outstanding phytopathologist with special interests in the epidemiological aspects of phytopathology. He had an important impact on theories concerning the spread of plant diseases, to the extent that he is deserving to be called the pioneer of modern aerobiology. His speciality in modern aerobiology gave him the opportunity to link many scientific disciplines. Philip Gregory was fascinated by the possibilities of this inter-scientific cooperation. He promoted it, and it is obvious that his satisfaction was like that of a proud father at the birth and sometimes laborious development of the *International Association of Aerobiology*. He was given the distinction of Honorary Member of the Association at the first General Assembly meeting in August 1974 at The Hague.

During the 3rd International Conference on Aerobiology in Basel he was commemorated by Dr. A. William Frankland who, though being a scientist from a different discipline, for some time had worked with him in the field of aerobiology, and in this way had the good fortune to personally known him well. Unfortunately this opportunity will not exist for most younger aerobiologists, but for them Philip Gregory will always be the author of *Microbiology of the Atmosphere'*, the standard work on the application of aerobiology. All those who have met him personally have been impressed, not only by his great knowledge and experience, but also by his personality, from which radiated kindness and warmth. Personally I experienced his genuine interest in the scientific work of others, when I had the opportunity to speak with him for the last time, at the Seattle International Airport. We were both waiting for our return transport after the 2nd International Conference on Aerobiology in Seattle in 1982, where he was the guest of honour.
All, who have known Philip Gregory will remember him with great respect. We will miss him.

Ir. H. D. Frinking, PD
Lab. Phytopathology, Agr. Univ.
NL-6709 PD Wageningen

II Airborne pollen and biometeorology

I. Lectures of general importance

EXS 51:
Advances in Aerobiology
©1987 Birkhäuser Verlag Basel

ENVIRONMENTAL INFLUENCES ON DEPOSITION OF AIRBORNE PARTICLES

W. S. Benninghoff

The University of Michigan, Department of Biology, Ann Arbor, Michigan, USA

Introduction

The most elementary model of an aerobiological process recognizes launching of particles, their transport in the atmosphere, and their deposition. I wish to comment on some of the relevant environmental influences coming to light, especially as they affect the transport, deposition, and viability of the airborne particles. Suspended particulate matter refers to solid or liquid particles in the air. The particles and the gases in which they are suspended make up the aerosol. We are concerned mostly with primary particles, those emitted into the atmosphere directly as particles, in contrast to secondary particles, which are formed in the atmosphere by gas-to-particle transformation. It is common practice among air quality technicians to classify particles by size into the nuclei mode (Aitken nuclei) 0.005 to 0.1 um, accumulation mode 0.1 to 2 um, and coarse mode 2 to 100 um (COMMITTEE ON PARTICULATE CONTROL TECHNOLOGY, 1980). The nuclei are most numerous but the smallest in volume. The accumulation mode particles have the most surface and a large part of the mass. This discussion will relate primarily to the coarse particulate matter in size from 5 to 100 um.

General considerations

Residence time in the troposphere depends upon the particle's density, "sail" area or effective section, the movement of the enclosing air mass, the humidity of the air mass, and probably the sign and strength of the electric potential gradient of the atmosphere. The more important characteristics of the deposition or lodgment surface are the roughness (or irregularities), the scales of roughness, the polarity of the particle charge and of the possible lodgment surface, and the condition of the surface (wet, sticky, waxy, etc.). The roughness must be considered, for example, at the scale of forest in contrast to grassland, or at scales of mossy hummocks, or hairy leaves. Obviously, roughness affords opportunities for particles to fall into air that is moving more slowly, allowing sedimentation. Forests collect more particles and a wider range of sizes than grassland. However, hairy leaves do not, in all instances, collect more particles than glabrous leaves.

Most aerosol particles carry some electric charge; some may be highly charged (HINDS, 1982). The most important electrostatic effect in aerosol mechanics is the force exerted on a charged particle in an electrostatic field. For highly charged particles the electrostatic force can be thousands of times greater than the force of gravity. However, one must remember that the gravitational force on a particle of very small mass is relatively minute. An electric field exists in space around a charged object and causes a charged particle in this space to experience a force, the electrostatic force. The late Dr. Philip H. GREGORY's (1958) classic experiment with positive and negative charges on parallel plates beneath a Ganoderma bracket fungus demonstrated that the spores when shed carried positive charges and were impacted on the negatively charged plate. Not all spores thus released are positively charged; some are negative and a few have been found to have no charge. At least a large proportion of the coarse airborne particles do carry a charge, commonly positive, in the positive fair weather potential gradient of the atmosphere.

The solid earth (i.e. soils, bedrock, and water bodies) is polarized negatively, or grounded (earthed). Usually rooted plants are polarized negatively, although the plant body is a relatively poor conductor and the upper portions are less strongly polarized. In one of the few experiments to measure these effects, CORBET et al. (1982) demonstrated that pollen grains of Brassica rapa were drawn to receptive stigmas of the same species through distances of several millimeters. By polarized collectors such as Tauber sedimentation traps with conductive covers the charged conditions of at least a large proportion of coarse airborne particles has been demonstrated (BENNINGHOFF & BENNINGHOFF, 1982). It is reasonably certain that the electrostatic charging of airborne particles and of possible deposition surfaces are important factors in the lodgment of the particles.

The atmospheric environment

Airborne particles of a wide range of sizes, but especially those in the Aitken nuclei range, serve as nuclei for condensation of water vapor. In the higher levels of the troposphere ice crystals form around the nuclei; in the lower, warmer levels water droplets form. If the droplet or ice crystal accretes to sufficient mass, it will fall as a "hydrometeor" (raindrop or snowflake). Only about 10% of all clouds actually precipitate; the remainder age and dissipate, leaving behind a residue of aerosol particles after evaporation of the cloud droplets. "Thus, nonprecipitating clouds can act as 'reaction vessels' for reactive gases and a 'pumps' that transfer material from the boundary layer through the inversion into the middle and upper troposphere" (WORKSHOP ON TROPOSPHERIC TRANSPORT OF POLLUTANTS TO THE OCEAN, 1978, p. 7). Convective systems of thunderstorms and tall cumulus clouds that lift particles to higher levels are confined to the tropics, subtropics, and temperate latitude warm seasons. These same systems, and precipitation from clouds at lower levels, are effective at scavenging airborne particles by formation or accretion with raindrops or snowflakes.

The most common aerosol nucleating agent in industrial areas is SO_2. The SO_2 may become oxidized within cloud droplets, and when the water evaporates, the sulfate is released as solid particulate matter (SAXENA AND HENDLER, 1983). What would be the effect on a living spore or pollen grain in such a droplet? It is possible air pollutants such as SO_x, NO_x, and others may be hazards to the survival of airborne organisms.

The terrestrial environment

The so-called "dry deposition" is dependent upon wind force, roughness, humidity, the electrostatic environment, and blowing dust or snow. The potential gradient of the atmosphere in fair weather is positive and moderate (50 to 150 V/m), with low humidity and blowing dust or snow stronger positive, and with falling rain or snow usually negative and relatively weaker. Evidence is accumulating that under conditions of blowing dust, sand, or snow and low humidity, ozone is generated near the ground. Although soil, water, and snow are sinks for ozone, continuous generation could result in increased concentrations. Natural levels of ozone are believed to be in the 25 to 30 ppb range, on the average, with the principal source being the stratosphere.

However, concentrations 10 to 20 times the natural level have been observed in industrialized areas because of photochemical transformations of NO_x and gaseous hydrocarbons (WORKSHOP ON TROPOSPHERIC TRANSPORT...1978). It is possible that ozone near ground level may be a hazard to the viability of at least some airborne organisms. I doubt that ozone tolerance levels are known for any spores or pollen grains.

In recent years it has been found that vegetation and at least some soil surfaces undergo change in electric charge and polarity under different weather conditions. The intense electric fields beneath thunderstorms commonly give rise to electrical discharges (coronae) from the tips of objects such as tree tops, shrubs, and other pointed objects attached to the earth, causing strong influences on the electric field at the ground (STANDLER AND WINN, 1979). Without the presence of thunderstorms, there are small ionic discharges from tips of trees, shrubs, and other pointed objects attached to the ground and local changes in electrostatic charge occur on surfaces of soils and foliage as they are moistened or dried. Charles LEACH (1976, 1982) has found that leaf surfaces of maize and bean become charged during periods of high humidity to 200 to 4,000 V/cm potential gradient at 10 mm distance. Lowering the relative humidity causes the charge to become static and, as the leaf and the superficial conidia of a phytopathic fungus both have the same charge, there is a repulsion between spore and sporophore, the juncture breaks and the electrostatic force propels the spore into the air. Electrostatic deposition of pesticide sprays has been under study for several decades. Edward LAW (1981, 1983) found that with maize and cotton plants, the pointed leaves as targets restricted depositon of charged spray droplets. Gaseous discharges between sharp leaf tips and incoming charged spray clouds limited the effectiveness of the method. This is experimental proof that foliar morphology is a factor in the distribution of electrostatic charges on plants, and therefore, most probably on the deposition of charged airborne particles on foliage.

The aquatic environment

Deposition of airborne particles on water is very different from that on land. Deposition on the sea or a lake is under the influence of the relationships of the negative polarity of the water body, the wind and wave conditions which may generate a positive gradient stronger than the common fair weather gradient in the meter or so above the surface, and the increased probability of "capture" of airborne particles by wind-spray droplets. The fair weather potential gradient over a freshwater pond where I have sampled is usually about 60 V/m over the lee edge as opposed to 100 V/m over the windward edge. Sea water presents even a greater difference from land conditions because of the dissolved salts and generally greater wind waves and blown spume. The bursting bubbles of breaking waves are associated with an elevated potential gradient which is especially strong on coasts with heavy surf (BLANCHARD, 1963). Crystals of sea salts left as the water evaporates are sometimes blown far inland, and perhaps this holds also for airborne particles that have been captured by sea spray.

Conclusions

The above discussion points out some, but by no means all, of the potential environmental influences on the deposition of airborne particles. Measurements of the rates and efficiencies of the different processes and their influences on each other are needed. For example, there is the possibility of predicting the mast fruiting of trees of the Fagaceae from the abundance of pollen shed in spring, providing we can interpret the environmental conditions attending anthesis and fertilization. Similarly, it should be possible to improve the warning systems for phytopathic infections of crops with better knowledge of the environmental influences on the processes of spore transport and deposition. Baseline measurements are necessary for calibration of the environmental forcing functions and their effects in order to improve our models and to detect and measure change.

References

Benninghoff, W S and Benninghoff, A S. (1985) Wind transport of electrostatically charged particles and minute organisms in Antarctica. In: Antarctic Cycles and Food Webs, ed. by W R Siegfried, P R Condy, and R M Laws. Springer, Berlin, pp. 592-596.

Blanchard, D C. (1963) The electrification of the atmosphere by particles from bubbles in the sea. In: Sears, M. (ed.) Progress in Oceanography, vol. 1. Macmillan Pergamon, New York. pp. 71-202.

Committee on Particulate Control Technology. (1980) Controlling Airborne Particles. Environmental Studies Board, National Research Council, National Academy of Sciences, Washington, D.C. 114 p.

Corbet, S, Beament, A J and Eisikowitch, D. (1982) Are electrostatic forces involved in pollen transfer? Plant, Cell and Environment, $\underline{5}$, 126–129.

Gregory, P H. (1957) Electrostatic charges on spores of fungi in air. Nature, London, $\underline{180}$, 330.

Hinds, W C. (1982) Aerosol Technology: Properties, behavior, and measurement of airborne particles. Wiley-Interscience, John Wiley & Sons, New York. 424 p.

Law, S E and Lane, M D. (1981) Electrostatic deposition of pesticide spray onto foliar targets of varying morphology. Trans. Amer. Soc. Agric. Engineers $\underline{24}$(6), 1441–1445, 1448.

Law, S E. (1983) Electrostatic pesticide spraying: concepts and practice. IEEE Transactions on Industry Applications, $\underline{1A}$–$\underline{19}$, no. 2, pp. 160–168.

Leach, C M. (1976) An electrostatic theory to explain violent spore liberation by Drechslera turcica and other fungi. Mycologia, $\underline{68}$(1), 63–86.

Leach, C M. (1982) Leaf surface electrostatics related to sporulation and active spore liberation (abstract). British Mycological Society Symposium, Glasgow, U.K., April 13–15, 1982.

Saxena, V K and Handler, A H. (1983) In-cloud scavenging and resuspension of cloud active aerosols during winter storms over Lake Michigan. In: Precipitation Scavenging, Dry Deposition, and Resuspension, ed. by H R Pruppacher, R G Semonin, and W G N Slinn. Elsevier, New York, vol. 1, pp. 91–101.

Standler, R B and Winn, W P. (1979) Effects of coronae on electric fields beneath thunderstorms. Quart. J. R. Met. Soc., $\underline{105}$ (443), 285–302.

Workshop on Tropospheric Transport of Pollutants to the Ocean. (1978) The Tropospheric Transport of Pollutants and Other Substances to the Oceans. National Research Council, National Academy of Sciences, Washington, D.C. 243 pp.

Professor William S. Benninghoff, Department of Biology, The University of Michigan, Ann Arbor, Michigan 48109-1048, USA.

INFLUENCE OF INVERSION LAYERS ON THE DAILY POLLEN COUNT AND ON THE ALLERGIC ATTACKS OF PATIENTS

Ruth M. Leuschner [1], G. Boehm [2], Chr. Brombacher [3]

(1) Dept.Forschung,Kantonsspital,CH-4031 Basel,Switzerland
(2) Herrengrabenweg 51, CH-4054 Basel, Switzerland
(3) Botanisches Institut der Universität, CH-4056 Basel

Introduction:

The influences of special weather situations on human beings - also of inversion layers - have been studied earlier(De RUDDER 1952). The research reported here concerns the influence of temperature inversions on man with the aid of special measurements. Severe inversions pose a serious threat to both sick and healthy people. The smog which occurred in winter during recent years in Athens, Madrid and also in Berlin should be mentioned here. In summer such inversions usually occur during the night. we have recorded them with an acoustic echo sounder. (LEUSCHNER und BOEHM 1980 und 1981.)

Methods: 1. Acoustic Echo Sounder (= AES = Sodar)

This instrument is shown and explained in two figures (fig. 1 and 2). It serves to measure the inversion layers and is placed on the area of the 'Hardwasser AG' in Pratteln near the city of Basel. Measurements are not possible in the city itself owing to the constant level of noise. The AES works similarly to radar. Short acoustic impulses (frequencies 1-2 kHz) are sent into the atmosphere and if they meet some discontinuity they are reflected as echos and registered by a sensitive receiver. This acoustic reflection can be brought by a special method onto paper for registration. On these papers we see two kinds of echos. For our investigation only the layer echo is relevant. It marks regions with the inversion layers, that means warm layers of air lying over colder ones. For the graphs(fig.3) this layer echos were used. If an inversion is present, the normal temperature distribution is reversed which means - as already mentionned - that warmer air rests on the cooler layers(lake

of cool air). Such inversions hinder the rise of air into greater heights through convection. There are therefore accumulations
of particles and gases in the lower air.
2. The accumulation of sulphur dioxide - measured as mg/m^3 air
was continously registered by a special apparatus.
3. For trapping the pollen we used the Burkard pollen-and-spore-
trap, which works volumetrically. Pollen, fugal spores and sometimes particles of air pollutants (c.f. BOEHM and LEUSCHNER 1975
and 1979) are caught on a transparent strip smeared with vaseline, and is moved continously by a clockwork day an night. The
method is described in LEUSCHNER 1974.
4. Allergic attacks of the patients in the allergological out-
patient clinic of the dermatological university clinic (Dir.
Prof..R.Schuppli, head of the out-patient clinic Dr.F.Wortmann)
were recorded daily by the patients on special forms. The attacks are divided into complaints on the eyes, the nose and the
bronchi.
5. The precipitation is recorded at the meteorological station
in the surroundings of the city.
6. The results of an anemometer - placed near the pollen-and-
spore trap -are not used in this case, as they are not relevant
for this investigation.

<u>Discussion and results:</u>

Looking at the graphs (fig.3), in which the pollen counts for
night and day are give separately, we can see that there is an
accumulation of pollen in the lower parts of air when there was
an inversion for some hours at night; there was at this time also a greater amount of sulphur dioxide because the inversion hindered the rise of it into greater heights. An accumulation of
this kind of gas increases often the allergic attacks of patients.
When - always looking at fig.3 there is no inversion then there
is less pollen, the sulphur dioxide registration is lower and
the attacks of the patients are not so severe. (C.f. LEUSCHNER
and BOEHM 1980 and 1981).

Fig. 1

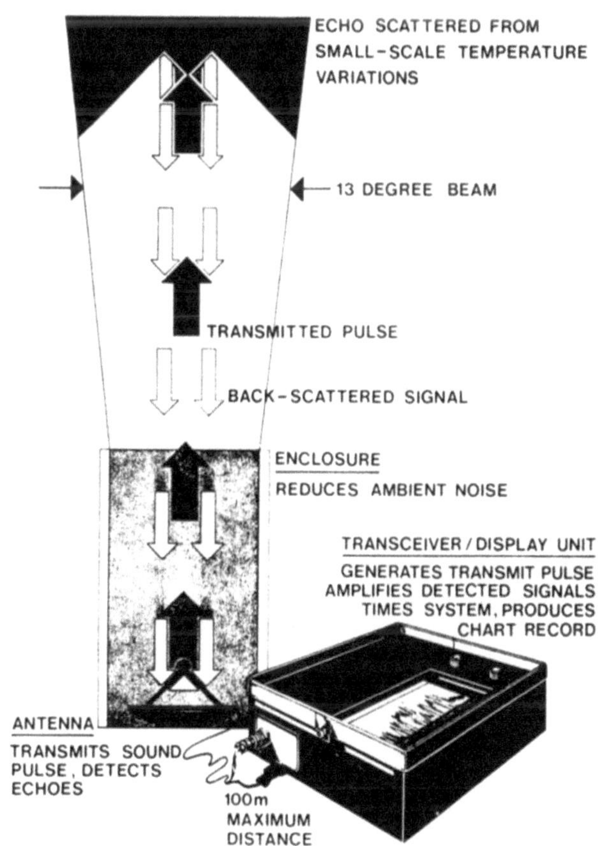

Fig. 1 and 2 are taken from a prospectus of AEROVIRONMENT INC. Vista Ave. 145, Pasadena, Ca. 91107 USA (Unfortunately the author got no answer when asking them.)

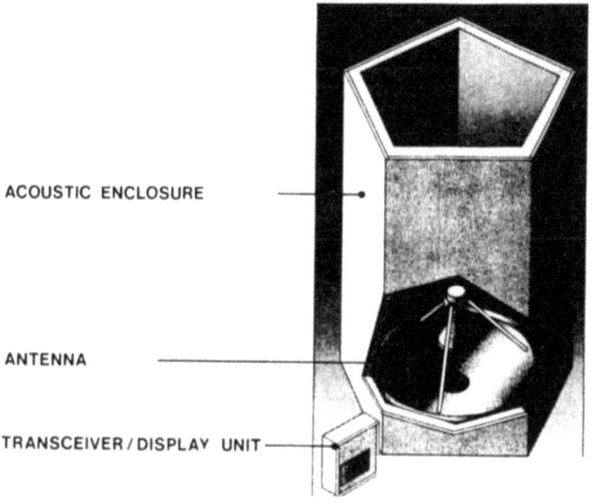

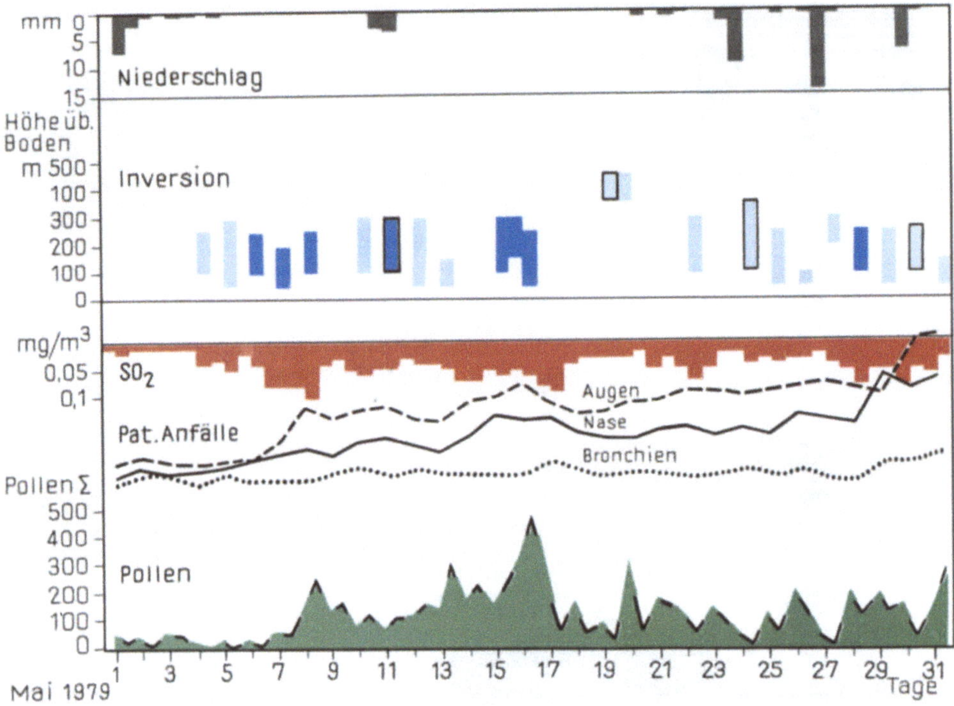

Fig.3: Inversion layers in May 1969(31 days)

Niederschlag = precipitation (brown)
Höhe üb.Boden = hight above ground level
Inversions: light blue: inversion not heavy
            dark blue: inversion heavy
            black margin around a blue field
            means: inversion lasts more than
            six hours
SO₂ = sulphur dioxide (red)

Pat.Anfälle = hay-fever attacks of patients,
            patients scores
            Augen = eyes, Nase = nose, Bron-
            chien = bronchi

The short black lines above the pollen curve (green)
mark the pollen content at night

Mai = May   Tag = day

Literature:

BOEHM,G. und LEUSCHNER,Ruth M.:'Beobachtungen über unbelebte üb-
liche und besondere Schwebeteilchen der Aussenluft.'
Schweiz.Ges.f.Reinraumtechnik, Arb'tag. April 1975,
Chem.Rundschau, Juni 1975.

BOEHM,G. and LEUSCHNER,Ruth M.:'A simple procedure for the pho-
tometric evaluation of incidental findings on the recording strips of the Burkard pollen-and-spore trap'.
Experientia 35, 1415 - 1416, 1979.

DE RUDDER,B.:'Grundriss einer Meteorobiologie des Menschen.Wet-
ter- und Jahreszeiteinflüsse. 3.Aufl. Berlin, Göttingen
Heidelberg, 1952, 303pp, see p 107ff and 116.

LEUSCHNER,Ruth M.und BOEHM,G.:'Beobachtungen an Luftpollen in
Verbindung mit besonderen Wetterlagen und im Hinblick
auf Pollinosis-Beschwerden'. Abstract.XI Congress of
the Europ.Acad.of Allerg.and.Clin.Immunol.Vienna,Oct.
6-10,1980.In:Allergologia et Immunopathologia 8/4,
1980,p.4.

LEUSCHNER,Ruth M.and BOEHM,G.:'Influence of inversion layers on
pollen concentrations in the air and attacks of pollinosis'. Abstract Volume 9th Int.Congress of Biometeorology.(Eds.D.Overdieck,J.Mueller,H.Lieth.) Osnabrück/
Stuttgart 23.Sept.-3.Okt.1981.Biometeorology 8, 1981,
p 225.

Aknowledgement:

This investigation has been partly supported (Dr. Ruth M.
Leuschner and cand. phil. II Chr. Brombacher) by the Swiss
National Foundation for Scientific Research (Schweizerischer
Nationalfonds zur Förderung wissenschaftlicher Forschung).

Author's address:

Ruth M.Leuschner, Dr.phil.II, Dept.Forschung, Abt.Dermatologie/
Allergie, Hebelstrasse 20, CH-4031 Basel, Switzerland.

Home address: Missionsstrasse 17 A, CH-4055 Basel, Switzerland

INFLUENCES OF POLLUTION AND WEATHER ON OBSTRUCTIVE RESPIRATORY TRACT DISEASES OF CHILDREN IN BERLIN(WEST)

E. WEDLER[1], U. FEGELER[2], R. MOYZES[1], K. EBERHARD[3]

1 Institut für Meteorologie der Freien Universität Berlin
2 Rittberg Kinderklinik vom Deutschen Roten Kreuz Berlin
3 Fachhochschule für Sozialarbeit und Sozialpädagogik Berlin

During the period July 1979 to June 1982 the frequency of children living in districts of Berlin (West) with different load of air pollution and suffering from obstructive diseases of the airways were investigated (FEGELER, MOYZES, WEDLER, EBERHARD 1985). The data basis was the patients' rate of those children living in the 75% supply areas (SAH) of the Neukölln children's hospital (NK) and the Rittberg children's hospital (RK) (Fig. 1). The air pollution load of 75% SAH NK was average higher than that of RK. Air pollution was registered by 31 measuring points representing the Berlin air quality measuring network (BLUME). The values of each measuring point were representative for a nearly 16 km^2 area (BLUME-area).

Since 1984 the influence of allergenic strain caused by pollen during the vegetation period is investigated (WEDLER et al. 1986).

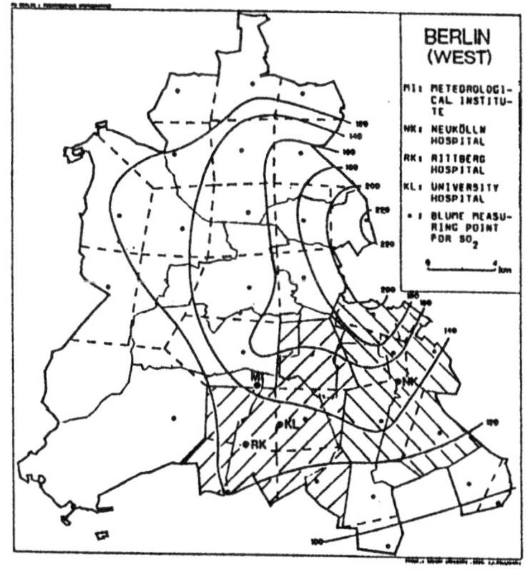

Fig. 1 Spatial distribution of mean SO_2 concentration winter period (Oct-Apr) 1975-80; BLUME-areas; 75% SAH of NK, RK hospital

Data Basis, Parameters of Investigation

1. Independent variables: The values of SO_2 concentration obtained by BLUME were used as an indicator of gaseous and dusty air pollution (Fig. 1). The corresponding meteorological parameters are: the equivalent temperature as an indicator for the cooling load of the mucosa of the respiratory tract (MRT) while staying outdoors and the relative humidity of indoor climate (RHIC) indicating the desiccation load of MRT indoors (FEGELER et al. 1985). These data were provided by the measuring devices of the Meteorological Institute of the Free University Berlin, Berlin-Dahlem.

2. Dependent variables: The dependent variable "patients' rate" (PR) was defined as the daily number of children under medical treatment provided by either first aid care or taking up in one of the hospitals (hospital care). In the course of the examination period the children's department of NK gave medical care to 1.085, the RK to 297 patients suffering from obstructive respiratory tract diseases (ORTD). 83% belonged to the age group 0-6 years. The average distribution of age shows its maximum at the first (NK) or second (RK) year of life. Boys fell ill 1.8 times more often than girls.

Methods of Evaluation

Regarding the time series of daily PR, a remarkable interdiurnal variation is noticeable within the weekly rhythm with peaks on Saturdays, Sundays and Wednesdays corresponding to the closing hours of pediatrician-practices. These weekly rhythms which did not depend on air pollution or weather conditions were eliminated. Independent and dependent variables were correlated for the same time and with regard to the influence of incubation periods with time lag (linear and univariate correlations). Furthermore, partial multivariate correlations were computed.

Meteorological and Air Quality Characteristics of the Examination Period

The maximum of SO_2 concentration in the winter 1979/80 was the highest ever registered by BLUME since the beginning of SO_2 measurements in 1975 while it was lowest in winter 1980/81. The maximum of January 1982 showed average values. As mean SO_2 concentration for the 75% SAH of NK 156 $\mu g/m^3$ were registered for the invernal periods (Nov to Feb) 1979/80 to 1981/82 while the corresponding value for RK amounted to 128 $\mu g/m^3$.

Several cold periods occurred during the winter periods 1979/80 and 1981/82 due to weather conditions with poor vertical and horizontal exchange. During these periods the values of air pollution were remarkably above the average. On the other hand, in the winter period 1980/81 thermal conditions generally were on an average. Resulting from strong ventilation at that time rather low amounts or air pollution substances were registered (Fig. 2).

RESULTS - TIME SERIES COMPARISONS

1. Time Scale: Months

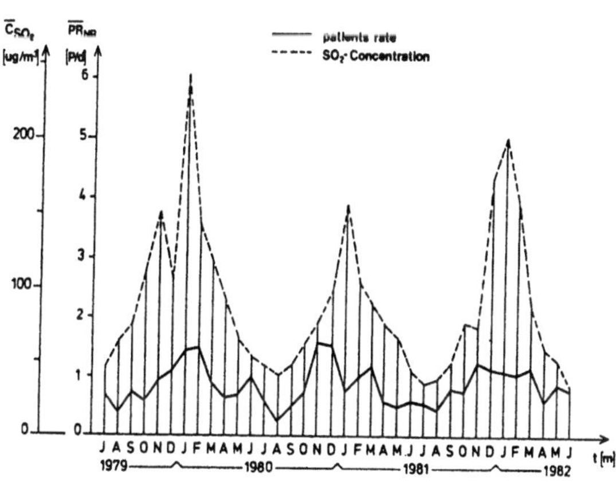

Fig. 2 Annual variations of the PR and SO_2 concentration (SAH NK + RK)

The annual PR rhythms of ORTD indicate peaks in winter periods (Nov to Feb) and minima in summer (Jul/Aug). Secondary maxima recorded in May, June or July (Fig. 2) were possibly due to pollinosis. During the research period the annual rhythms of both SO_2 concentration and PR were in phase. But primarily this is only a formal correlation. The main increase to the high winter level generally occurred in the course of the October/November.

In order to relate the differences in time and space for SAH of NK and RK a standardized quotient has been calculated relating mean winter PR of ORTD to mean summer PR of both SAH. Contrary to the mean PR - which showed only small unimportant differences during winter periods - the interannual variations of this SAH-quotient showed clear differences: The increase from mean summer PR to mean winter PR was higher for the SAH of NK relatively to the SAH of RK by +40% in 1979/80, +93% in 1980/81 and +34% in 1981/82. I.e. the maximum difference between the both SAH was found in winter 1980/81 which had air pollution values below the average. Compared to the other two winter periods in question the winter 1980/81 showed rather strong ventilation and relatively frequent changes of relatively warm and cool weather periods, thus indicating a considerable influence of meteorological environmental parameters on the PR level registered at NK.

Dates of the epidemiology of infectious agents were available only for the winter season (Sep to Feb) 1981/82. For this period the influences of infectious agents (Mykoplasma pneunomiae) on the PR of ORTD registered at NK cannot be excluded.

2. Time Scale: Weeks

Linear correlation methods have been applied on weekly means of the PR of the ORTD. The correlation coefficients exceeded |0.5| in only a few cases. Regarding the partial correlation, i.e. determining the influence of the single independent variables equivalent temperature, relative indoor humidity and SO_2 concentration show a strong mutual correlation with the PR of the ORTD. That means at first approximation the PR of ORTD with children is equally effected by both thermal conditions and environmental air pollution influences.

3. Time Scale: Days

3.1 Patients Rate Median of the ORTD in Relation to SO_2 High Concentration Periods

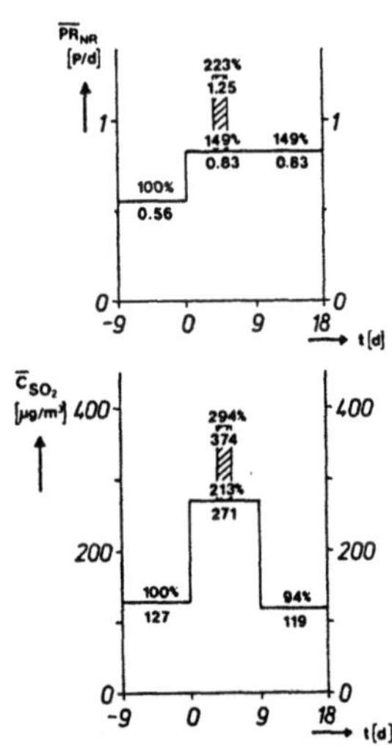

Five equally long periods before, during, and after epochs of high pollution concentration (epoch length: 5-13 days with median values of SO_2-concentration of 271 µg/m³) were investigated. During these periods the median value of PR rose from 100% before the high pollution epoch to 149% during the epoch itself and the following period (Fig. 3). The median value of PR on days with first stage of smog alert was 223% and therefore exceeded the one registered for epochs of high pollution for about 74%. This indicates an immediate influence of air pollution on the ORTD.

Fig. 3 PR before, during, and after periods of high SO_2-concentration; ▨: smog alert days

3.2 Patients Rate Median of the ORTD in Relation to Daily SO_2 Concentration

The daily median SO_2 values coming up to 400 µg/m³ for 75% SAH of NK – thus exceeding the MIK-value (= maximum concentration of air pollution; VDI-Richtlinien) – correspond to the remarkable PR increase on the very day of load (+85% with regard to the average of the first four classes) and on the third day after the main SO_2 concentration (+71%; Fig. 4). These results point to a direct (first PR maximum) and to an indirect (second PR maximum) influence of air pollution, respectively.

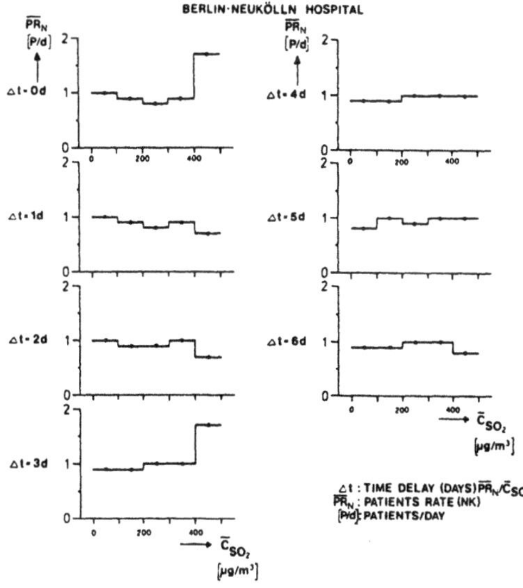

Fig. 4 Daily PR of ORTD (NK) due to daily SO_2 concentration 1979-82

3.3 Patients Rate of the ORTD in Relation to Concentrations of BETULA Pollen

Since 1984 registrations of pollen concentration are carried out in Berlin (W). During the 1984 and 1986 periods when dusting of BETULA pollen was most intense - during third decade of April (maximum) until first or second decade of May - the mean PR of ORTD at NK and RK hospital exceeded the mean values of the vegetation periods (Apr-Sep) in 1979/82 by factor 1.5 (1984) or 2.6 (1986), respectively. During the flowering period 1985 the maximum pollen concentration was 10 to 50 times lower than 1984 and 1986. Therefore the mean PR of ORTD in 1985 showed nearly average values. Relating the daily PR to daily values of BETULA pollen concentrations during 1984 and 1986 flowering period a saturation effect was observed when the concentration exceeded 250 pollen/m³. The saturated PR level was 150% (1984) or 300% (1986) above the average during the vegetation periods, respectively.

RESULTS - COMPARISON OF THE SUPPLY AREAS

<u>1. Comparison between Mean Values of Patients Rate of the ORTD and SO_2 Concentration at the 75% SAH of NK and RK Hospital:</u> During the winter periods 75% of BLUME areas of the both 75% SAH with mean SO_2 concentrations exceeding the SAH average respectively coincided with relative morbidity above the mean value.

<u>2. Correlation between Patients Rate of the ORTD and SO_2 Concentration at Single BLUME Areas:</u> A significant result was obtained by correlating the monthly mean values of PR of the ORTD in the winter periods 1979/80, 1980/81 and 1981/82 and the corresponding air pollution concentration at single BLUME areas of the 75% SAH of NK and RK: r = 0.69 (p < 0.05).

3. Mean Relative Increase of Patients Rate of the ORTD of the Supply Area NK in Relation to the Supply Area RK from Summer to Winter Level (SAH Quotient):

The mean SAH quotient of the three-year-examination period indicates a relative increase of PR of the ORTD at NK amounting to 53% compared to the rate at RK in the course from summer to winter level. The same tendency can be observed regarding the three single years (see above), thus making the mean result significant (p < 0.04).

The three methods of comparing NK and RK SAH show results with the same tendency: During the winter season (Nov to Feb) the SAH of NK hospital showing air pollution values above the average registrated a higher percentage of children suffering from ORTD than the SAH of RK hospital where pollution reached only normal values or even values below the average.

CONCLUSIONS

Epidemiological interdependence of air pollution concentrations and respiratory diseases is too complex to be derived completely by univariate, linear correlation analysis. The relation of atmospheric environmental parameter and obstructive respiratory tract diseases (ORTD) apparently has a multivariate non-linear character.

It can be seen by our study that air pollution as well as meteorological factors of the environment can influence the patients rate (PR) of the ORTD. No matter whether they occur isolated or combined, in either case they seem to effect the mucosa of the respiratory tract (MRT), the symptoms of illness being recognizable either immediately or with time lag. The underlying mechanism of these effects could be that during the winter seasons the MRT is continuously stressed by both outdoor environmental parameters (high air pollution concentration, low equivalent temperature) and indoor factors (low relative humidity due to nearly constant heating). They may effect MRT damages causing symptoms without delay or open the way for infectious agents (symptoms with delay).

REFERENCES

Fegeler U, Moyzes R, Wedler E, Eberhard K
Immissions- und Wettereinflüsse auf Erkrankungen der oberen und unteren Luftwege von Kindern in Berlin (West) 1979-1982   Berlin 1985

Fegeler U, Moyzes R, Wedler E, Eberhard K
Immissions- und Wettereinflüsse auf Atemwegserkrankungen von Kindern in Berlin (1979-1982) Methodik und Ergebnisüberblick
Ann. Meteorol. (NF) 22 (1985) 72-74

VDI-Richtlinien
Maximale Immissionskonzentrationen für Schwefeldioxid
VDI 2310, Bl. 11 Berlin 1984

Wedler E, Klaschka F, Schick B, Schlaak P, Silbermann M
Interdisciplinary Pollen Working Group Berlin - Methods and First Results
3rd Int. Conf. Aerobiology, Basel 1986

Address

Dipl.ing. E. Wedler, Institut für Meteorologie, Freie Universität Berlin, Dietrich-Schaefer-Weg 6 - 10, D-1000 Berlin 41, FRG.

EXS 51:
Advances in Aerobiology
©1987 Birkhäuser Verlag Basel

COMPARISON OF RECENT AND QUATERNARY POLLEN DATA

J A Coetzee, Botany Department, University of the OFS, Bloemfontein, South Africa.

Introduction

The study of fossil pollen and spores has become one of the most important techniques for the reconstruction of palaeoenvironments, especially during the Quaternary. The interpretation of fossil pollen assemblages in terms of vegetation can, however, be seriously affected because of over- or under representation of pollen taxa. It is, therefore, of paramount importance that accurate assessments are made of recent pollen production, dispersal and deposition in the wide surrounding vegetation of a study site.

For the evaluation of pollen spectra the qualitative analogue method has been widely applied by palynologists. In this case modern plant communities are used as analogues of past vegetation types and certain indicator species with narrow ecological amplitudes are of great value.

In recent years certain quantitative methods have been proposed for more objective evaluations of pollen assemblages. Among these are the R-values or correcting factors for each pollen type which compensate for over- and under representation. The limitations of this approach have been discussed by BIRKS and BIRKS (1980). Another statistical model that provides adjustment indices is based on linear regression analyses (WEBB et al, 1981) but because of the lack of sets of data, this method is also not widely applied.

For the study of modern pollen deposition in certain vegetation types, surface soil samples, moss cushions and surface layers of lake sediments are mostly analysed, but a number of problems influence both qualitative and quantitative assessments (BIRKS & BIRKS, 1980). Pollen trapping devices like the Durham, Burkhard and Tauber traps have also been used but some practical aspects, especially those concerned with long-term operations, often hamper this approach. More recently COUR (1974) devised the filter method which has great advantages especially over the use of the surface soil samples as it gives a representative random sample from a fairly large surface. Gauze filters impregnated with silicone oil are mounted on the back of a vehicle to collect dust stirred up while driving according to certain specifications through a vegetation type.

In the present discussion the importance of the accurate knowledge of pollen deposition for the interpretation of pollen diagrams is illustrated by studies in the Sahara and Namib deserts and in the East African mountains. In the two hyper-arid areas the Cour method was applied while in the high altitudinal regions of East Africa surface samples were analysed.

The Sahara Desert
Recent and fossil pollen was studied by COUR and DUZER (1976) in the sebkha of Taoudenni in the Tanezrouft of northern Mali (Fig. 1). In this hyper-arid depression in the Sahara where salt has been excavated from time immemorial, salt and clay horizons with a depth of more than 4.50 m indicate the former presence of an extensive lake. The salt layers represent hyper-arid periods, while during more genial times aquifers may have replenished the basin. Surface inflow was responsible for the deposition of the clay horizons. The fossil pollen of these sediments with an age of some 8000 years (the ^{14}C date at 3.20 m depth is 5650 $\pm$150 years BP.) gives the impression that the surroundings were always covered with a grassy savanna of Sahelian origin. The total of Poaceae and Cyperaceae pollen reached 85 and even 93% while Compositae plus Cruciferae attained 1.7-5.3%.

To verify these findings pollen deposition on the desert surface was studied with the filter method at intervals of one degree latitude between 34 and 20°N. The following numbers of pollen grains were collected:
    in the hyper-arid Tanezrouft: 45-50,000/m^2
    in the hamada du Guir: 350,000/m^2
    and in the alfa steppe of Algeria: 1,600.000/m^2
A typical gradient in the occurrence of pollen types showed that in the traverse from north to south the mediterranean pollen was replaced gradually by Sahelian and tropical sporomorphs. In the hyper-arid Tanezrouft in the region of the Tropic of Cancer where at present no vegetation occurs the pollen sediments give the picture of a complex vegetation cover of grass steppe and savanna.

These findings were compared with the data of the fossil deposits and COUR and DUZER (1976) established that the fossil pollen assemblages, just as those from the Tanezrouft were the result of long-distance dispersal. They came to the interesting conclusion that the surroundings of the sebkha of Taoudenni have been without vegetation during most of the Holocene, while concentrations of Chenopodiaceae occurred only round the sebkha margin during this time.

An accurate comparison of the pollen data also made it possible to infer a southward shift of some 150 km of the Sahelian zone of the Sahara in the early Holocene. This was followed by a movement in the opposite direction resulting in the shifting of the Sahelian Sahara boundary some 350 km northward.

The Namib Desert
The Namib desert, well known for its remarkable endemic biota stretches as a narrow strip of less than 200 km wide for 2000 km along the southwestern coast of Africa. In its central part the southern erg consists of linear dunes 50-150 m high and more than 100 km long, running in a south-northerly direction. On the eastern side it is bordered by the great escarpment from which a number of ephemeral rivers like the Kuiseb, Tsondab, Tsauchab and Koichab flow into the desert (Fig. 2). In the past some of these rivers have been obstructed by newly formed dunes sothat silt is deposited in pans, vleis or floodplains by floodwaters coming from the escarpment. In these sediments fossil pollen has accumulated over long periods of time.

It is with regard to the history of the vegetation and the possible shifts in the climatic belts during the Late Cainozoic that E.M. VAN ZINDEREN BAKKER initiated palynological research in the southern erg.

In the erg itself the sparse vegetation varies according to the type of environment. The linear dunes are usually without vegetation except for the lower slopes, which in rare periods of more than normal rainfall, can be covered with the grass Stipagrostis, namaquensis. In the dry dune valleys the remarkable Acanthosicyos horridus is a dune builder. Along the ephemeral rivers Acacia crioloba dominates the woody species which diminish towards the desert. The floodplains and vleis may in the gravelly parts be covered with Stipagrostis fastigiata while in the sandy parts Salsola tuberculata traps windblown sand.

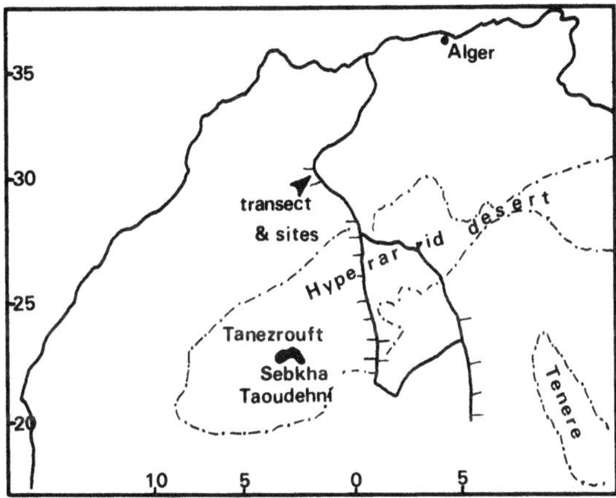

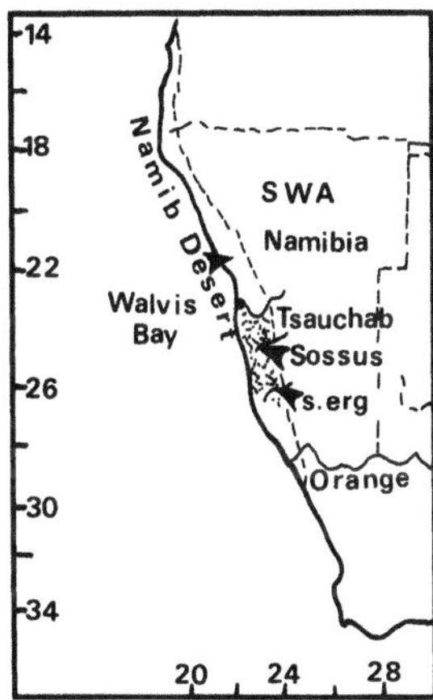

Fig. 1 Locality map: Sahara (from Cour & Duzer, 1976).

Fig. 2 Locality map: Namib desert.

The pollen results presented here were carried out by VAN ZINDEREN BAKKER and MULLER (in preparation) in the main southern erg. The Cour method was applied in the Tsauchab valley starting in the Naukluft mountains where the river arises and where there is still a remnant of a former woody vegetation. The pollen and vegetation survey followed the course of the 120 km ephemeral river to Sossus vlei and to the floodplains in the desert where in certain years water accumulates (Fig. 2).

The pollen spectra obtained by the filter method in the Tsauchab valley show a progressive decrease in riparian upland arboreal pollen and increasing aridity downstream. In the erg and serir five pollen spectra show high values of Poaceae (85%) indicating the sparseness of the vegetation in this hyper-arid region along the river.

In the surface pollen spectra of the floodplain and of Sossus vlei hardly any sporomorphs of the upland vegetation survived the erratic transport by the river and the exposure in the desert climate. The pollen assemblages here indicate the local vegetation which includes Acanthosicyos and Salsola. The fossil silt layers with an age of ca 2000 years contain pollen spectra throughout the sediments which are similar to the present surface pollen deposition. These results indicate that conditions have not changed during this time span. In this case long distance pollen cannot lead to erroneous conclusions as in the central Sahara.

These studies in the deserts clearly emphasise the need for accurate pollen deposition studies in which long distance dispersal can play such an important role in the problems of the reconstruction of former vegetation.

The East African Mountains

A third example to illustrate the use of recent studies for the interpretation of fossil pollen assemblages, is the high East African mountains where extensive palynological research has been conducted. (Summaries can be found in : COETZEE, 1967; HAMILTON, 1982). However, as a result of the difficult terrain the only types of samples that could be taken for the study of present day pollen deposition in these mountains were moss cushions, surface sediments of bogs or lakes, soil surface samples and airborne pollen collected in traps (HAMILTON and PERROTT, 1980). Analyses of these samples showed a clear distinction between three categories of pollen, one originating from the direct surroundings of the site, that produced by adjacent vegetation and the long distance component. The local pollen originates either from swamp plants or aquatics which are not very sensitive to climatic changes as they are largely dependent on edaphic conditions. The two other pollen components are of great interest to the pollen analyst since he wishes to relate vegetation change to climatic fluctuations. In this respect it is of great importance to separate "adjacent" and "long distance" elements.

Taking Mt Kenya, situated under the equator, as an example for these studies interesting data have been obtained on pollen productivity and transport in the vegetation belts, especially at and above the tree line which lies at about 3300 m. In the highest montane forest _Podocarpus milanjianus_ is an important element of the vegetation. Above the tree limit are the Ericaceous and Afroalpine belts with their respective low shrubs and grassland. This open vegetation with the characteristic and remarkable giant Senecios and Lobelias is exposed to the extreme diurnal climate, with frost every night and summer conditions every day.

It has been established from studies using traps and soil surface samples on the nearby Mt Elgon that the pollen production of the montane forest far exceeds that of the higher vegetation belts (HAMILTON and PERROTT, 1980). Much pollen is found to be exported to the higher altitudes by rising winds during the day. Most pollen taxa are, however, deposited in great numbers near to the source in the vegetation. This phenomenon is of great importance for the explanation of fossil pollen spectra.

Studies on recent pollen spectra in the East African mountains show that pollen taxa have different relative dispersal abilities. Four different categories outlined by HAMILTON (1972) indicate a range from relative high export ability to relative low export ability. From this characteristic the long distance transport of pollen types is assessed. This phenomenon, which is very important in connection with the reconstruction of vegetation, can be illustrated with an example of data obtained from studies of _Podocarpus_ pollen dispersal on Mt Kenya (COETZEE, 1967). The very high pollen production of Podocarpus occurring in the montane forest, has a considerable influence on the composition of the pollen spectra of the Ericaceous and Afroalpine belts where the pollen has been transported. At 1200 m above the forest limit Podocarpus pollen values reached 86.8%, nearly as high as in the uppermost forest. In the intermediate Ericaceous Belt only 19-31% Podocarpus pollen was recorded. This type of phenomenon is clearly reflected in the fossil pollen spectra of Mt Kenya (COETZEE, 1967; HAMILTON, 1982) an example of which will be described below. In the pollen sequence from Sacred Lake which has a total age of ca 34,000 years, pollen zone T (COETZEE, 1967) with an age of ca 26,000 to 14,050 ± 360 years shows conditions which are entirely different from the present. The lake, at an altitude of 2440 m, is at present surrounded by the upper moist montane forest. The pollen spectra of Zone T, however, show that the Ericaceous belt occupied the site at the time of the last glacial maximum. Five typical taxa of this belt are present in the spectra and trees of the montane forest belt with low pollen export are insufficiently represented to indicate the presence of forest there at this time. On the other hand trees with higher export dispersal values are relatively better represented. Podocarpus pollen values, for instance, were less than 15% which is in accordance with its values in the recent spectra of the Ericaceous belt. This important finding proves that the upper forest limit was lowered by at least 700 m, but more likely by 1000-1100 m during the last glacial maximum. This lowering represents a decrease in temperature of 5.1 to 8.8°C depending on the lapse rate used for the computation. Indications are that the climate was dry during this cold period.

The fossil pollen zone following this cold period shows a varying presence of trees with moderate and low export abilities indicating that they definitely grew round the lake. At the same time elements of the Ericaceous Belt were decreasing in number and rainfall was apparently high. This situation changed completely about 4000 years ago when Podocarpus pollen rose to 66% among the forest pollen taxa with low relative pollen export. This indicates that the Podocarpus-Juniperus forest type was growing round the lake in a drier climate much resembling present day conditions.

This example again shows the importance of the use of present day data for the explanation of fossil pollen spectra.

References

Birks, H.J.B. & Birks, H.H. (1980) Quaternary Palaeoecology (Edward Arnold, London).

Coetzee, J.A. (1967) Pollen analytical studies in East and Southern Africa. Palaeoecology of Africa 3, 1-146.

Cour, P. (1974) Nouvelles techniques de detection des flux et des retombées polliniques: etude de la sédimentation des pollens et des spores à la surface du sol. Pollen et Spores 16, 103-141.

Cour, P. & Duzer, D. (1976) Persistence d'un climat hyperaride au Sahara central et méridional au Cours de l'Holocène Rev. Géogr., Physique et Géol. Dynamique (2) 18, 175-198.

Hamilton, A.C. (1972) The interpretation of pollen diagrams from highland Uganda. Palaeoecology of Africa 7, 46-149.

Hamilton, A.C. (1982) Environmental history of East Africa. (Academic Press, London).

Hamilton, A.C. & Perrott, R.A. (1980) Modern pollen deposition on a tropical African mountain. Pollen et Spores 22, 437-466.

Van Zinderen Bakker, E.M. & Muller, M. Pollen studies on present and former vegetation in the Namib Desert (in preparation).

Webb, T.III, Howe, S.E., Bradshaw, R.H.W. & Heide, K.M. (1981) Estimating plant abundances from pollen percentrages: the use of regression analysis. Rev. Palaeobot. Palynol. 34, 269-300.

Prof J A Coetzee, Botany Department, Palynological Research, University of the OFS, P O Box 339, Bloemfontein 9300, South Africa.

EXS 51:
Advances in Aerobiology
©1987 Birkhäuser Verlag Basel

BIOMETEOROLOGY AND ITS RELATION TO POLLEN COUNT

Paolo Mandrioli

Istituto FISBAT. Consiglio Nazionale delle Ricerche, Bologna (I).

Introduction

Biometeorology is the study of direct and indirect effects of the physico-chemical micro- and macro-environments on physico-chemical system in general and on living organisms in particular. This is the definition given by the International Society of Biometeorology in 1970. which subdivides the biometeorology into six main groups: plant, zoological, human, cosmic, space and paleobiometeorology (1). Plant biometeorology studies the influence of weather and climate on the development and the distribution of healthy and diseaded plants for botanical, phytopatologycal and agricultural purposes, and also of the effects of both weather and climate on small plant organisms responsible for the development of plant, animal and human diseases. The study of vegetation in relation to soil conditions has been an active field of research for more than a century, on the contrary, the study of vegetation in relation to the atmospheric conditions has advanced more slowly. Actually in the past 30 years a remarkable effort has been made to achieve a better balance between weather and soil studies. In particular, about this argument, the micrometeorological studies are most important. The state of the atmosphere can be relatively simply defined and measured in terms of radiation and energy balance, temperature, water vapour and wind profiles. The state of the plant is much more difficult to be physiologically or biochemically defined and genetical variation adds to the difficulties of analysis. Many physiological processes determine the rates at which plants grow and develop. Growth rates are relatively easy to measure, for example, as an increase in dry weigth per unit time, but development rates are usually difficult to quantify and to measure, because the concept is not so clearly defined. The phenologycal studies, as merging of meteorological and biological sciences, may provide for this definition.

Micrometeorological influences on flowering process

The timing of pollen release and the quantity of pollen produced are principally results of genetically controlled adaptation, but both can be influenced by the climatic characteristics of the period before flowering. The variability in date of the first flower within a species is due, in part, to a variation in most of meteorological parameters and particularly it is a function of the accumulation of heat units above growth threshold temperature, usually near the freezing point, occuring after dormancy concludes (2). The relationship between the heat unit accumulation and the date of first flower, provide significant predictability, and it's possible to know the enhance or delay timing caused by light, temperature and moisture interactions (3)(4). At this point, when the value of heat unit has been met, the anthers are extended and only favourable micrometeorological conditions (warm and dry day) are required. The last phases of anthesis are different in trees and the weather conditions effect must be considered and studied for each case (5). Really, also the quantity of pollen available for spring flowering is determined by weather conditions of the previous summer, because at that moment the cells designated to become pollen grains are already present. In fact, when in the previous summer abnormally high temperature and low precipitation occur, it is frequent to notice an abundant spring pollen yield.

Micrometeorological influences on airborne pollen

The diurnal fluctuations of pollen concentration only partially reflects the pollen shedding rhythm. Once the anther walls have split, the pollen grains may be transferred in the turbulent air or intercepted by vegetational structures. The turbulent structure of the boundary layer is the most important characteristic of the atmosphere as regards the dispersion of large particles, and it is generated by a continuous fluctuation of wind speed and direction, called turbulence. The turbulent intensity, as a measure of eddies energy, depends on three factors: the roughness of the terrain, the change of wind speed with height, the temperature vertical profile. Near the ground, the amount of solar radiation and the wind speed determine principally the temperature profile and consequently the stability of the atmosphere or, in other words, its dispersive ability. Under unstability condition (the temperature decrease with height) the turbulent motions are easily induced and the atmospheric particles reinforce their vertical displacement. In this case the pollen has been dispersed: generally the concentration is uniform for a large area, but his value is low. A stable atmosphere, on the contrary, is not favourable to turbulence and consequently the values of pollen concentration near the source area have the possibility to increase under this conditions. In fact, frequently, in our measurements, it's possible to have the inversion layer in which the temperature increase with height. When the inversion layer is not adiacent to the ground, all particles are trapped and forced back down into the volume of air. In this layer, called mixing layer, the mixing processes are taking place. In the literature concerning the air pollution many methods and formulas for the calculation of height of the mixing layer are available (6)(7)(8),but often it's possible to have low correlations between mixing height and pollen concentration levels (9)(5). These results may be due to: a greater variation in the mixing height in time and space, an insufficient time for the particles to disperse into the mixing height and finally, the standard calculation of m.h. does not always give an accurate estimation of the height (10). Moreover, the big size and the shape of the pollen grains, if compared to other atmospheric particles, futher on influence pollen flight. In particular, under a convective condition, it's often insufficient to suppose that the particles are prevented from falling out (peak of concentration) only by a convective action. Recently the mechanism of "convective levitation" has been proposed (11) as an important process for keeping particles aloft. The suspicion the problem of the influence of meteorological factors to be complex has here been once again confirmed as well as the need to probe this field through a co-operation between biologists and physicists of the atmosphere.

References

(1) Tromp, S.W. (1980) Biometeorology, (Heiden, London).

(2) Wang, J.Y. (1960) A critique of the heat unit approach to plant response studies. Ecology 41, 785-790.

(3) Boyer, W.D. (1973) Air temperature, heat sums and pollen shedding phenology on long leaf pine. Ecology 54, 420-426.

(4) Taylor, F.G. (1974) Phenodynamics of production in a mesic deciduous forest. in Phenology and Seasonality Modeling (Springer, Berlin) pp 237-254.

(5) Kapylä, M. (1984) Diurnal variation of tree pollen in Finland. Grana 23, 167-176.

(6) Holzworth, G.C. (1967) Mixing depths, wind speeds and air pollution potential for selected locations in United States. J.appl.Met. 6, 1039-1044.

(7) Tennekes, H. (1973) A model for dynamics inversion above a convective boundary layer. J.Atmos.Sci. 30, 558-567.

(8) Garrett, A.J. (1981) Comparaison of observed mixed-layer depths to model estimates using observed temperatures and winds, and MOS forecasts. J.appl.Met. 11, 1277-1283.

(9) Spieksma, F.Th.M. (1983) Fluctuations in grass-pollen counts in relation to nightly inversion and air potential of the atmosphere. Int.J.Biometeor. 27, 107-116.

(10) Aron, R. (1983) Mixing height - an inconsistent indicator of potential air pollution concentrations. Atmospheric Environment 17, 2193-2197.

(11) Green, J.S.A. (1986) Levitation by convection, or does what goes up necessarily come down?. Weather 41, 110-114.

Dr. Paolo Mandrioli, Istituto FISBAT-C.N.R.,
Via Castagnoli 1, 40126 Bologna, ITALY.

EXS 51:
Advances in Aerobiology
©1987 Birkhäuser Verlag Basel

CORRELATION BETWEEN AEROBIOLOGICAL AND PHYTOGEOGRAPHICAL
INVESTIGATIONS IN THE FLORENCE AREA

°Zerboni R.,M.D.,°Manfredi M.,D.Sc.,°Campi P.,M.D.,°°Arrigoni
P.V.,Phytogeogr.Prof.

°Allergological Centre,Nuovo Osp.S.Giovanni di Dio,Florence,Italy
°°Laboratory of Phytogeography,Dept.of Veg.Biol.,Univ.of Florence

Introduction We present the preliminary data of our research
whose principal aims were: a)to check the correlation between
aerobiological and phytogeographical data in our area; b)to clarify certain discordances,for example between the presence of
some plants and the absence of their pollen at the trapping; c)to
know for each taxon: 1)presence and abundance in the area;
2)flowering period (by phenological checking) especially for
those families for which it is impossible to differentiate genera
and/or species by optical microscope; d)finally,to select the allergen panel for testing in our area.
          This research has been carried out by our Allergologi-
cal Centre since 1984 with the Institute of Botany at the University of Florence,Laboratory of Phytogeography,Dept.of Veg.Biol.

Methods and Materials A seven day recording volumetric pollen
trap was used;it is placed on the roof our hospital,near to a meteorological station.
          In Florence,in the warmer months,from April to Septem-
ber,when pollination is at its peak,the winds blow from west and
south-west,that is from flat areas intensively cultivated and
from the hills,both cultivated and woody.
          The phenological checking of the flowering period on
location was done for the most interesting species from an allergological point of view. Our observations were carried out every
10 days in 1986 through carefully planned routes in the area next
to the sampling station. MARCELLO scale has been used to express
the intensity of flowering. With regard to Cupressaceae and Oleaceae it is not easy to distinguish the pollen of one species from
that of the others,among each family. From the correlation
between flowering and pollination period it comes out that Cupressaceae show 2 peaks:the first refers to Cupressus sempervi-

rens and the second to Juniperus communis;similarly Oleaceae show 2 peaks:the former attributable to Fraxinus ornus and the later to Olea europaea.

Lay-out of the phytogeographical survey: a)definition of the area to be investigated; b)recognition of types of vegetation; c)survey of area and its assessment; d)collection and analysis of data from references; e)synthesis of data above mentioned. Three areas,surrounding our hospital,measuring 5,10 and 20 km in radius respectively have been used. Only with reference to plants either suspected or proved of being allergenic,we collected floristic data reported in the literature for the area investigated. Next we went through the circular areas,making observations and assessing the vegetation by collecting data on the abundance-dominance of the species (according to Braun-Blanquet). The data from references,those from our field survey and experience have been synthetized and evaluated according to a conventional scale of abundance-dominance. Allergenic plants have been considered in relation to 7 types of vegetation into which are grouped those plants present in the area investigated:xerophilous woods,mesophilous woods,conifers woods,parks and gardens,rural areas,meadows,border areas-weeds.

<u>Results</u> A good agreement has been found between incidence on total pollen count,presence of the plant in the vegetation and incidence of skin test positivity in relation to the pollen of the following families: Cupressaceae,Ulmaceae,Betulaceae,Salicaceae, Graminaceae,Oleaceae,Pinaceae,Urticaceae,Compositae,Chenopodiaceae and Plantaginaceae. The incidence of skin test positive to Fagaceae is scarce in relation to the fairly high pollen count: it perhaps depends on overlapping of pollination with Graminaceae:as Graminaceae are allergenically more important,very often Fagaceae have not been tested,but in the future we intend to test them routinely. In relation to some taxa,seldom or never found in pollen trap but fairly diffused in vegetation (Tilia sp.,Robinia pseudoacacia L,Medicago sativa L,Trifolium pratense L,Rumex sp, Polygonum sp,Erica arborea L,Erica scoparia L,Platanus sp,Aesculus hippocastanum L.) we attempted an explanation for each one. For example Platanus and Aesculus are trees,mostly located inside the town among high buildings;in the flowering period winds blow from our trap toward the town;there is a likely interference from pruning. Moreover for Platanus,we do not know how many male individuals are there and which is the biology of the hibrids.

<u>Conclusion</u> From the correlation between aerobiological and phyto-

geographical data has come out that: a)the principal allergological taxa observed in the area are well represented in our pollen
calendar; b)our pollen calendar does not include taxa which are
absent in the area investigated,20 km in radius; c)the observations in the area on the presence,number and flowering of the
different species under study are particularly useful in relation
to those families in which it is not easy to distinguish,at optical microscopy,the different genera and/or species; d)these data
will be useful in the selection of a panel of allergens for
testing in our area.

Reference Marcello,A.(1935) Nuovi criteri per osservazioni feno-
  logiche.Nuovo Giorn.Bot.Ital,N.S,42 (3):534-556.

Prof.Romano Zerboni,Centro Allergologico,Nuovo Osp. S.Giovanni di
Dio, Via di Torregalli 3. 50143 Firenze. Italy.

2. Pollen in different countries

EXS 51:
Advances in Aerobiology
©1987 Birkhäuser Verlag Basel

AIR SAMPLING STUDIES IN A TROPICAL AREA. FOUR YEAR RESULTS.

Inés Hurtado and Margalit Riegler-Goihman.

Instituto Venezolano de Investigaciones Científicas, Centro de
Microbiología y Biología Celular, Laboratorio de Alergia
Experimental, Caracas, Venezuela.

Our studies on the air-spora are intended as an aid
for the diagnosis and treatment of allergic conditions in the
tropics. Our air samplers operate at the Venezuelan Institute
for Scientific Research, in a suburban area of Caracas. Caracas,
which is the capital of Venezuela, spreads on a system of valleys
surrounded by mountains as high as 2,600 m. One of them the
Avila, separates the city from the Caribbean Sea. The climate
-at 10º N latitude- is mild and typically tropical. The mean
annual temperature is 21º C (range 19º to 22º) and the relative
humidity 80% (range 75 to 83). There are two seasons: rainy and
wet. The rainy season lasts from May to September, the dry
season from November to March. April and October are
intermediate months when it usually rains moderately.

We have used the gravity slide sampler (Durham) as
described in (1); the Rotorod sampler (Ted Brown Associates,
U.S.A.) and the Burkard spore trap (Burkard Mfg. Co., Great
Britain) as indicated by the manufacturers. Airborne particles
are counted on a 2.5 x 2.5 cm area of exposed microslides
(Durham) and on a 250 u path along exposed tapes of the Burkard
trap. Pollen grains are counted on 2 rods from a Rotorod sampler
set at 10% time; mold spores on 1 rod from a second Rotorod set
at 5% time. For the identification of the small and nondescript
conidia of some deuteromycetes we recurred on occasion to the
exposure of culture plates. Our sampling was started in 1981 and
continues to the present. Our findings, which have published in
part (2,3), are summarized in the following paragraphs.

In our area pollen grain and spores are airborne

continuously throughout the year. Approximately 60 different pollen, 4 fern spore and 250 fungus spore types compose the annual spectrum of airborne particles. Pollen of the following botanical families and genera have been recorded (the most frequent are marked with an asterisk): Amaranthaceae (Amaranthus, Pfaffia); Anacardiaceae; Casuarinaceae (Casuarina);Chenopodiaceae (Chenopodium); Compositae (several types); Cupressaceae (Cupressus *); Cyperaceae (more than one type); Gramineae * (many types); Euphorbiaceae; Leguminosae (Acacia, Calliandra, Mimosa, Enterolobium); Melastomataceae (Miconia); Moraceae * (Cecropia and other); Myricaceae (Myrica); Myrthaceae (Eucalyptus); Pinaceae (Pinus*); Plantaginaceae (Plantago); Rubiaceae; Polygalaceae (Bredemeyera); Tiliaceae (Heliocarpus).

Trees, grasses and weeds make unequal contributions to the pollen counts. The relative weight these different plant groups have on the counts, which was early established (2), has been observed to persist unaltered through successive years. Trees contribute from 60% to 70% of the pollen load and account for two major seasonal peaks. The tree peak is observed in April-May, during the transition from the dry to the rainy season. Grasses contribute 15% to 20% of the pollen load and account for the second major pollen peak which is observed in November at the beginning of the dry season. The pattern of deposition of individual pollen grains does not necessarily follows that of total pollen, i.e., a pattern of continuous fallout with two major peaks. Mulberry and grass pollens, although airborne throughout the year have one major peak, in April-May and November, respectively, when approximately 80% of their annual fallout is recorded. Yet other pollens fall only during certain months: Heliocarpus only during December and January, Plantago during February and March. Cypress pollen show no peak. The conifers (pinus and cypress)are of course imports from temperate zones. Cypress, with a share of approximately 40% of the annual counts, is the largest single pollen contributor; pinus is also imporant with a 9% share. Conifer pollen might be, however, of local importance only. It has not been found in areas distant from ours, in which case the total pollen load appears correspondingly decreased.

Fern spores are windborne throughout the year peaking in April-May, together with the trees.

From the large variety of fungus spores we continuously recover, we have visually identified the following : a) Deuteromycetes spores: Alternaria, Arthrinium, Asperisporium, Beltrania, Cercospora, Cladosporium, Cordana, Corynespora, Curvularia, Dictyoarthrinium, Dictyosporium, Diplococcium, Drechslera, Epicoccum, Exosporium, Fusarium, Gliomastix, Monodyctus, Nigrospora, Periconia, Pithomyces, Pullularia, Pteroconium, Sarcinella, Spegazzinia, Stemphylium, Tetraploa, Tetracocosporium, Tomenticula , Torula, Trichdochium ; b) Ascospores: Bombardia, Chaetomium, Chitonospora, Diplococcium Erysiphe, Leptosphaeria, Leptosphaerulina, Massaria, Melanomma, Paraphaesphaeria, Pleospora, Sordaria, Sporormiella, Triblidium, Venturia, Xylaria; c) Basiodispores: Agaricales (Agaricus, Coprinus); Aphyllophorales (Ganoderma); Uredinales (Puccinia); Ustilaginales. In addition we have recovered from exposed culture plates: Aspergillus, Aureobasidium, Penicillium and Scopulariopsis.

The counts of airborne fungus spores, in contrast to those of pollen and fern spores, do not show any definite seasonal peaks. The only trend observed so far is for the largest fallouts to occur at the beginning of the rainy season (April-May) and for the lowest fallouts at the end of the dry season (January-February). Most of the spores we recover are deuteromycetes spores which largely outnumber asco and basidiospores. The former, in turn, are usually more abundant than the latter. Spores from other fungus classes are virtually absent from our samplers.

Our yearly particle counts remain approximately same in successive years. This is shown by the gravimetric (Durham) data of table I. These data, however, would have been taken to indicate that more pollen grains than fungus spores are present in air. The use of the more efficient volumetric samplers indicates that, in fact, fungus spores largely outnumber pollen grains (Same table I, Rotorod counts; and table II).

Table I. Annual particle counts .

	Sampler	Unit	Pollen grains	Fern spores	Fungus spores
1982-83	Durham	cm^2	1,760	15	1,010
1983-84	"	"	1,776	108	995
1984-85	Rotorod	m^3	4,673	284	140,183

In our hands the efficiency of particle recovery increases for all class particles when going from the Rotorod to the Burkard sampler (table II).

Table II. Monthly particle counts (m^3)

	Sampler	Pollen grains	Fern spores	Fungus spores
April'86	Rotorod	376	29	7,560
May'86	"	593	66	14,785
April'86	Burkard	4,921	272	52,068
May'86	Burkard	8,803	584	231,910

The use of the volumetric samplers (Rotord and Burkard) resulted not only in a high recovery of total fungus spores, but also in a high recovery of Cladosporium spores. This was the most frequently and abundantly recorded fungus spore type. During 1985 10% to 15% of all fungus spores counted were Cladosporium spores. In contrast during the same year Cladosporium spores were recovered only 16 times on Durham slides. The percent contribution of this spore type to total counts was proportionally low.

When compared to that of temperate zones (4) the pollen deposition on our area seems rather light, while the fungus spore deposition appears average or high. Algae, insects, acari and various types of plant débris are not uncommonly recovered from our samplers.

Summary

Air sampling studies carried out in Caracas, Venezuela, indicate that in this tropical area pollen and fungus spores of allergenic potential are airborne throughout the year. The pollen grains of conifers, mulberry and grasses are the most frequently encountered. Many more fungus spores than pollen grains are recovered from the air; among the former, Cladosporium spores are the most frequent and abundant.

Acknowledgments

We thank Tibisay Cabrera-Maldonado and Henry Ramos for excellent technical assistance.

This work has been supported in part by a grant from Fundación Polar and by Conicit Travel Award S2-1078.

References

1 Ogden, E.C.; Raynor, G.S.; Hayes, J.V.; Lewis,D.M. & Haines,J. H. (1974) Manual for sampling airborne pollen (Hafner Press NY)

2 Hurtado, I & Riegler-Goihman, M. (1984) Air sampling studies in tropical America (Venezuela).Frequency and periodicity of pollen and spores. Allergol et Immunopathol 12, 449-454

3 Hurtado, I & Riegler-Goihman, M (1986) Air sampling studies in a tropical area. I Airborne pollen and fern spores. II Fungus spores Grana 25, 63-73.

4 1984 Statistical Report of the Pollen and Mold Committee. American Academy of Allergy and Applied Immunology. USA.

Correspondence to: Dr.Inés Hurtado, IVIC Ap 21827, Caracas 1020A, Venezuela.

EXS 51:
Advances in Aerobiology
©1987 Birkhäuser Verlag Basel

A STUDY OF AIRBORNE POLLEN GRAINS OF ALEXANDRIA, EGYPT

Gamal El-Ghazaly and M. Fawzy
Department of Botany, University of Alexandria, Egypt.

Introduction

There is only one report available on the aerobiology of Alexandria which was published over 28 years ago by Saad (1958). Since that time, the vegetation and cultivation in the city of Alexandria and its surroundings have changed, mainly due to industrialisation and urbanization after clearing of vegetation. Besides, the methods of trapping atmospheric pollen have been improved during the last few decades.

The present study includes a survey of atmospheric pollen grains in the city of Alexandria with the aim of providing information that may be useful in the treatment of aeroallergies. With increasing tourist activity in Alexandria, information of its aerobiology should be available to physicians in order to advise their travelling patients of potential allergic material in its environment.

Method

The trap was installed on the roof of the Botany Department, Faculty of Science, University of Alexandria (about 30 m above street level).

Results

Pollen grains belonging to 18 families were recorded in both 1981 and 1982 (Table 1). The beginning, duration, and end of pollen season were estimated, and the date of highest pollen concentration (number of pollen/m^3 air) was recorded. Pollen grains of nine families (Gramineae, Casuarinaceae, Chenopodiaceae, Compositae, Cruciferae, Leguminosae, Rosaceae, Urticaceae, Cupressaceae, and Pinaceae) were recorded in high frequency. The concentration of pollen grains of the families Gramineae, Casuarinaceae, Chenopodiaceae, and Urticaceae were represented as 7 day running means. The total number of pollen grains/day during the two years of investigation was represented as 3 day running means as their incidence was relatively not so high. Variation in pollen concentration (pollen grains/m^3 air) are presented in Table 1 for each of the dominant families encountered in both years of investigation.

The pollen season in 1981 generally started on February 11, two weeks later than that in 1982. However, they both ended in the same week (December 23). The total count of pollen grains trapped in 1982 (15 155 pollen/m^3 air was much higher than that in 1981 (5 122 pollen/m^3 air). The highest peaks in 1981 were on March 18 (91 pollen/m^3 air), April 1 (72 pollen/m^3 air), and June 10 (89 pollen/m^3 air). The highest peaks in 1982 were on April 1 and 15 (176 and 210 pollen/m^3 air respectively), June 3 (198 pollen/m^3 air) and July 1 (152 pollen/m^3 air). A relatively small peak (about 30 pollen m^3 air) was recorded on October 7 in both years. More than 75% of the total pollen count in both years were recorded in March, April, June and July. The minimum counts were recorded in January, February and December.

Table 1: Periods of prevalence of pollen grains of different families in the atmosphere of Alexandria, the date of highest pollen count for each family, and the pollen concentration (pollen/m³ air) recorded during that day, in 1981 & 1982.

Family	Beginning of the pollen season		Duration (days)		End of the pollen season		Date of highest pollen count		Concentr. /m³ air	
	1981	1982	1981	1982	1981	1982	1981	1982	1981	1982
Cyperaceae	23-03	06-07	16	6	30-06	06-10	14-05	28-06	4	6
Gramineae	11-02	28-01	146	161	23-12	23-12	09-06	06-06	36	70
Palmae	04-03	09-04	6	12	03-10	28-09	14-05	09-04	4	6
Caryophyllaceae	06-03	09-05	15	25	30-09	06-11	25-09	18-05	8	20
Casuarinaceae	04-03	18-02	59	49	23-12	23-12	18-03	01-04	456	740
Chenopodiaceae	25-02	01-03	65	83	21-10	21-10	04-10	09-04	30	94
Compositae	06-04	15-04	30	37	06-10	06-10	11-09	11-04	32	92
Cruciferae	04-03	01-04	46	76	23-07	07-08	23-05	29-05	50	620
Labiatae	08-04	02-04	4	12	07-07	02-07	16-06	02-04	2	12
Leguminosae	22-03	06-04	56	75	25-08	25-08	30-04	06-07	66	60
Myrtaceae	13-05	29-04	4	12	20-07	06-07	07-05	01-05	2	25
Polygonaceae	08-04	14-04	13	37	06-09	31-08	26-04	06-07	8	20
Rosaceae	19-03	07-03	46	43	08-11	15-10	16-04	10-03	50	360
Umbelliferae	02-05	14-04	7	12	16-09	22-08	08-06	26-06	8	8
Urticaceae	25-02	04-03	66	62	04-11	04-11	18-03	06-04	16	60
Ephedraceae	05-05	22-04	2	3	28-06	22-09	28-06	22-04	2	4
Cupressaceae	04-03	03-04	44	43	14-07	08-07	30-05	16-04	60	46
Pinaceae	31-03	31-03	49	60	25-08	25-08	04-06	07-06	170	116

<u>Discussion</u>

This study provides estimates of abundance of the atmospheric pollen in the air of Alexandria city for two years, 1981 and 1982. The striking feature of the results obtained is the much higher count in 1982 season than in 1981. These variations are explained by differences in flowering intensity and pollen productivity due to climatic variations. Anderson (1980) was able to evaluate the influence of climatic variations on annual pollen deposition on the basis of ten year records.

The duration of pollen-free atmosphere during this study was short and it occurred in January and February. In these months, the precipitation was at its maximum and the rain may have washed down the pollen content of the atmosphere. Saad (1958) observed a much longer pollen-free time of Alexandria atmosphere; he mentioned that pollen grains were nearly absent from the air during summer (July - September) and winter (December - February). Such long pollen-free period did not occur in both years of the present study. The main pollen season started from the first week of March and continued till the second week of July. During this period records included airborne pollen grains related to more than twenty species belonging to the families Casuarinaceae, Gramineae, Cruciferae, Pinaceae, Leguminosae, Chenopodiaceae, Urticaceae, Rosaceae, Cupressaceae, and Compositae. A short pollen season started from the middle of July till the end of December. During this short season, pollen grains of Casuarinaceae, Gramineae, Chenopodiaceae and Rosaceae were mainly present in the atmosphere of Alexandria. <u>Casuarina</u> seems to have two pollen cycles: a major one in March and April, and a minor one in October and November. The major cycle was in accordance with the main flowering season of <u>Casuarina</u>, while the minor one usually followed the early rainy period of October and November. Pollen peaks of grasses occurred on the first and the fifteenth of April, tenth of June, and in the first week of July, 1982. In 1981, several small peaks were recorded mainly in May and June. Early flowering grasses such as the annual <u>Poa annua</u> probably produced most of their pollen in April. In June and July, most grass pollen probably came from

<u>Zea</u> <u>mays</u>, <u>Poa</u> <u>pratensis</u>, <u>Dactylis</u> <u>glomerata</u>, <u>Agrostis</u> <u>alba</u> and <u>Agropyron</u> <u>repens</u>. A clear correlation between the flowering season and the high emission of pollen in the air was recorded in June for <u>Trifolium</u> <u>sativum</u>; this species is extensively cultivated in the fields around Alexandria, and its flowering season is in June.

References

Andersen, S.Th. (1980) Influence of climatic variation on pollen season severity in wind-pollinated trees and shrubs - Grana, <u>19</u>: 47-52.

Saad, S.I. (1958) Studies in atmospheric pollen grains and fungus spores at Alexandria. 1. A daily census of pollen - Egypt. J. Bot., <u>1</u> (1): 53-61.

Author's address

Gamal El-Ghazaly, Department of Botany, University of Qatar, P.O. Box 2713, Doha - Qatar

EXS 51:
Advances in Aerobiology

AIR BORNE POLLEN TYPES OF ALLERGENIC SIGNIFICANCE IN INDIA

A. B. Singh, CSIR Centre for Biochemicals, Delhi (India).

Introduction:

The magnitude of allergic problems in India is alarming. The morbidity surveys conducted at few places in India have revealed that more than 1% of the country's population (6.5 millions) suffer from bronchial asthma alone, and another 3-4% of the population are estimated to be afflicting from allergic rhinitis[1,2]. A rough estimate reveals that more than 10% suffer from major allergic ailments[3].

Airborne pollen and spores have been proved to be potent source of allergy and are major constituents of the atmosphere. India is a land of contrasts and is bestowed with a wide range of different biomass, ranging from desert to alpine thundra. Such a diversity in vegetation contributes enormous variations in the quality and quantity of aerial pollen from different ecozones of India. Accordingly the incidence and symptomatology of allergic patients from different ecozones are extremely variable.

AERIAL POLLEN DIVERSITY

<u>Source of Airborne Pollen</u>: Plant spores in a wider sense are found in all the divisions of the Plant Kingdom, but only those formed by the flowering plants, known as pollen grains, constitute significantly to the airspora. Wind pollinated (anemophilous) taxa are chief source of majority of inhalant pollen allergens (including grasses, chenopods, amaranths, and some trees). The flowers of these species possess a set of adaptive characteristics such as reduction in size and number of floral appendages, lack of nectar, and small, smooth, dry, simple and bouyant pollen. The amount of pollen production is also extremely high[4]. Many of the flowering taxa pollinating through vector mediation (entomophilous) are of less concern to respiratory allergic patients because of the limited exposure to their pollen, sufficient enough to sensitize them. Yet, many of the entomophilous taxa (such as <u>Carica papaya</u>, <u>Kigelia</u>, <u>Cocos</u>, <u>Phoenix</u> etc.) produce abundant pollen in the atmosphere[5,6]. Thus it is fallaceous to ignore all entomophilous species as offending allergens.

The airborne allergenic pollen are traditionally grouped as grass, weed or tree pollen. This is done merely because of convenience as they follow a generalised pattern in different seasons of the year. However, exceptions to the generalised flowering sequence does occur. For example <u>Prosopis juliflora</u> flowers erratically almost throughout the year, but with two peak pollination seasons - March-April and September-November[7]. However, differences in flowering pattern of some of the taxa is not uncommon for many species of different regions of India.

<u>Pollen Surveys:</u>

The atmospheric surveys conducted for airborne pollen and spores at various places in the country have been reviewed from time-to-time by different workers[8-11].

Recently an all-India co-ordinated project on aerobiology, involving more than 20 centres, was completed and the aerobiological informations have been compiled in the form of an edited book[12].

Some of the salient features of various pollen surveys reported from different regions of the country are:

1. In most of these surveys the sampling was based essentially on gravitational principle and Durham's gravity settling device has been used. However, in some of the recent surveys volumetric samplers have been employed.

2. Members of the families, Amaranthaceae, Cannabinaceae, Casuarinaceae, Chenopodiaceae, Compositae, Euphorbiaceae, Fabaceae, Moraceae, Palmae, Pinaceae and Poaceae are chief contributors to the airborne pollen.

3. Qualitative and quantitative variations in the pollen air spora of different ecozones. The variations observed are attributed to a variety of causes including climatic variations, vegetational cover and human interference with the nature. Study is required to evaluate the effect of each factor on the pollen airspora on a long term basis.

4. The maximum pollen were recorded in the air from anemophilous taxa, followed by amphiphilous and then entomophilous species. Similarly pollen from tree taxa constituted major part of the pollen air spora at various places compared to pollen from grasses and weeds.

5. The airborne pollen show diurnal, seasonal and annual variations. Two broad pollen seasons are: (1) February-April, and (ii) August-October. The former being dominated by pollen from tree species, while latter is predominantly the season for grasses and weeds.

6. Some of the dominant airborne pollen types reported from different parts of the country are: <u>Ailanthus</u>, <u>Artemisia</u>, <u>Azadirachta</u>, <u>Brassica</u>, <u>Cannabis</u>, <u>Chenopod/Amaranth</u>, <u>Dodonaea</u>, <u>Eucalyptus</u>, <u>Holoptelea</u>, <u>Morus</u>, <u>Prosopis</u>, <u>Putranjiva</u>, <u>Ricinus</u>, <u>Xanthium</u> and pollen from members of family Poaceae. Beside, pollen types confined to coastal areas are: <u>Borassus</u>, <u>Casuarina</u>, <u>Cocos</u>, <u>Peltophorum</u> and <u>Phoenix</u>.

Similarly pollen of <u>Alnus</u>, <u>Aesculus</u>, <u>Betula</u>, <u>Cedrus</u>, <u>Corylus</u>, <u>Quercus</u>, <u>Pinus</u>, <u>Prunus</u> and <u>Pyrus</u> are predominantly confined to Himalayan and sub Himalayan regions.

<u>Allergenic significance:</u>

In India, clinical investigations with pollen antigens have been carried out systematically only at V. P. Chest Institute, Delhi and was recently reviewed[13]. Work was also initiated at Jaipur about 3 decades ago but has now been discontinued for a long time. With the availability of allergenic extracts from CSIR Centre for Biochemicals, Delhi for diagnosis and immunotherapy, to various hospitals and physicians specially trained in the field of allergy and immunotherapy, a tremendous awareness has been created about the clinical significance of environmental allergens in the country in the last one decade. Efforts are also being made at different centres to confirm the positivity of skin reactions with immunological techniques for their specificity such as RAST and ELISA. As a result of various clinico immunological investigations from different parts of the country some of the offending pollen allergens recorded are: <u>Cenchrus ciliaris</u>, <u>Cynodon dactylon</u>, <u>Imperata cylindrica</u>, <u>Paspalum distichum</u>, <u>Pennisetum typhoides</u>, <u>Poa annua</u> and <u>Sorghum vulgare</u> from members of Poaceae (grasses); <u>Amaranthus spinosus</u>, <u>Artemisia scoparia</u>, <u>Brassica campestris</u>, <u>Cannabis sativa</u>, <u>Chenopodium album</u>, <u>Parthenium hysterophous</u>, <u>Ricinus communis</u> and <u>Xanthium strumarium</u> from weedy taxa; <u>Ailanthus excelsa</u>, <u>Anogeissus pendula</u>, <u>Cassia siamea</u>, <u>Cocos nucifera</u>, <u>Casuarina equisetifolia</u>, <u>Guazuma tomentosa</u>, <u>Holoptelea integrifolia</u>, <u>Prosopis juliflora</u> and <u>Salvadora persica</u> among tree species[14-19].

<u>References:</u>

1. Anonymous (1961) Morbidity survey of contributory health scheme beneficiaries, New Delhi, ICMR, New Delhi.

2. Vishwanathan, R. (1964) Definition, incidence, aetiology and natural history of asthma. Indian J. Chest Dis. <u>6</u> : 109-124.

3. Shivpuri, D. N. (1973) Unsolved problems in allergy. Aspects Allergy Appl. Immun. <u>6</u>: 1-7.

4. Singh, A. B. (1984) Pollen production in common allergenic plants of Delhi and their atmospheric prevalence. Science & Culture <u>50</u> : 65-66.

5. Singh, A. B. & Babu, C.R. (1982) Survey of atmospheric pollen allergens in Delhi: Seasonal periodicity. Ann. Allergy <u>48</u> : 115-122.

6. Ravindran, P. (1986) Studies in Aerobiology and Allergy in Kerala. ICMR Report, New Delhi.

7. Singh, A. B., Babu, C. R. & Shivpuri, D. N. (1979) Studies on atmospheric pollen of Delhi: Pollination calendar of plants of allergenic significance. Aspects Allergy Appl. Immun. 12 : 71-79.

8. Sreeramulu, T. (1967) Aerobiology in India, J. Sci. Ind. Res. 26 : 474-478.

9. Chanda, S. & Mandal, S. (1978) Aerobiology in India with reference to respiratory allergy. IAA News 8 : 7-15.

10. Singh, A. B. & Babu, C. R. (1983) Indian Aerobiological research in perspective. IAA News 19 : 14-19.

11. Singh, A. B. & Gangal, S. V. (1986) Sampling and distribution pattern of allergenic biopollutants in atmosphere. Biological Memoirs 12 : (In Press)

12. Nair, P. K. K., Joshi, A. P. & Gangal, S. V. (1986) Airborne pollen, spores and other plant materials of India - a survey. CSIR Centre for Biochemicals, Delhi.

13. Shivpuri, D. N. (1980) Clinically important pollen, fungal and insect allergens for nasobronchial allergy patients in India. Aspects Allergy Appl. Immun. 13 : 19-23.

14. Shivpuri, D. N. & Singh, K. (1971) studies in yet unknown allergenic pollen Delhi state: Clinical aspects. 59 : 1411-1420.

15. Agnihotri, M. S. & Singh, A. B. (1971). Observation on pollinosis in Lucknow with special reference to offending factors. Aspects Allergy Appl. Immun. 5 : 135-141.

16. Shivpuri, D. N., Singh, A. B. & Babu, C. R. (1979) New allergenic pollen of Delhi state, India and their clinical significance. Ann. Allergy 42 : 49-52.

17. Acharya, P. J. (1980) Skin test response to some inhalant allergens in patients of nasobronchial allergy from Andhra Pradesh. Aspects Allergy Appl. Immun. 13 : 14-18.

18. Nagendra Prasad, K. V. (1983) Dermal reactivity to various antigens in common allergic disorders in Bangalore. Aspects Allergy Appl. Immun. 16 : 64-69.

19. Agashe, S. N. & Anand, P. (1982) Immediate type hypersensitivity to common pollen and moulds in Bangalore city. Aspects Allergy Appl. Immun. 15 : 49-52.

Dr. A. B. Singh, CSIR Centre for Biochemicals,
V.P.Chest Institute Bldg., Delhi University, Delhi-110 007 (India)

EXS 51:
Advances in Aerobiology
©1987 Birkhäuser Verlag Basel

SURVEY OF ATMOSPHERIC POLLENS IN VARIOUS PROVINCES OF THAILAND

Boonchua Dhorranintra[*], Chaveevan Bunnag[**], Supatra Limsuvan[*]

[*]Division of Allergy and Immunopharmacology,
Department of Pharmacology
[**]Allergy clinic,Department of Otolaryngology,
Faculty of Medicine Siriraj Hospital, Bangkok, Thailand.

Thailand as well as other Southeast Asian Countries are situated in the tropical zone. They are evergreen agricultural countries with warm and humid climate. Therefore the atmospheric pollens and fungal spores occur abundantly throughout the year.

Introduction

The survey of aeroallergens have been done especially in Bangkok Metropolis by our Thai colleagus using both gravity and rotating arm methods[1,2,3] We are now expanding the aeroallergen survey to several provinces in every part of the country and keeping the continuous survey in Bangkok as a reference data.

Material and method

Since 1983, aeroallergens in four provinces,i.e., Uttaradith in the northern, Ubonratchthani in the northeastern, Bangkok in the central part and Cholburi on the east coast of the gulf of Thailand were surveyed. Rotorod Samplers were used for collecting airborne pollens and fungal spores. One hour collection was done at the same time of the day, three times a week for one year. The provincial meteorological reports were recorded and used for explanation of the aeroallergen finding.

<u>Result and discussion</u>

More than sixteen kinds of pollens could be identified in this
study. They were Cyperaceae, Fern, wild grass, cultivated grass
and Amaranthaceae. Among tree pollens were Lythraceae,Acacia,
Myrtaceae, Urticaceae, Pine and Palm.

The five common airborne pollens found in each province were

<u>Bangkok</u>	<u>Cholburi</u>	<u>Ubonratchthani</u>	<u>Uttaradith</u>
Cyperaceae	Cultivated grass	Cyperaceae	Wild grass
Wild grass	Fern	Wild grass	Amaranthaceae
Fern	Urticaceae	Mimosa	Cyperaceae
Cultivated grass	Wild grass	Urticaceae	Fern
Amaranthaceae	Cyperaceae	Cultivated grass	Urticaceae

The total count of the airborne pollen were high in the province
on the east coast and that on the northeastern part of the
country which are the agricultural areas. The total count was
low in the northern province where the high percentage of wild
grass and Amaranthaceae pollens were observed. The finding of
the central part was also lower than that of the previous two
provinces.

The high amount of airborne wild grass pollens which included
Bermuda grass, Para grass was observed in the northeastern
province especially during May and September, the finding was
comparatively lower in the dry season, October through March.
The findings in other three provinces were low especially
during March and October, and increased from October to
February.

In the east coast area, high amount of airborne cultivated grass
pollens could be detected and showed the abnormal high peaks in
May and November. In the northeastern province, the finding was

lower and fluctuated. In the northern province the finding was
low. In the central part the incidence was low from December to
August then gradually increased to the highest peak in November.
The high peaks of airborne cultivated grass pollens in all four
provinces seemed to be detected at the same month of November,
when cane sugar flowers produce high amount of pollens. And the
peak of Johnson grass pollination is in May.

Surprisingly, airborne fern spore could be detected in every
province. Very high incidences were found in central part and
east coast, where the land are low and damp promoting the
growth of various kinds of fern. The highest incidence were
found in June in the central part and September to November in
the east coast corresponding to the rainfall. On the high and
rather dry area like the northern and northeastern provinces
the finding of airborne fern spores were rather low throughout
the year.

We have already reported the significance of fern spores in Thai
allergic rhinitis patients.[4]

Cyperaceae grows abundantly in the areas of central and the
northeastern part. The pollen count were increased in the rainy
season during May to October in both provinces. The same
pollination character could be detected in other provinces but
in less amount.

Amaranthaceae grows in the wild and non-cultivated area
throughout the country. It grows very well in the dry
environment. The airborne pollens therefore could be detected
in moderate or low amount throughout the year. The highest
amount could be found in November in the central part and in
January, February in the east coast area.

Conclusion

According to the geography of the country. Many kinds of airborne pollens, wild grass and cultivated grass as well as tree pollens could be detected throughout the year. However, the amount of some pollens might increase in some particular months. Fern spore could be detected in the damp area and proved to be of clinical importance in its allergenicity. The survey of aeroallergens, therefore is very important for management of allergic diseases in the particular area.

References

1.Panichyakarn, P., Dhanamitta, S., Kraikitpanich, T. and Rungnirundornkul, D. (1974) Atmospheric pollens and molds survey in Bangkok. J. Med. Assn. Thailand. 57, 11-13.

2.Tuchinda, M. and Theptaranon, Y. (1976) Aeroallergens in Bangkok, Thailand. Ann. Allerg. 37, 47-54.

3.Bunnag, C. and Dhorranintra, B. (1974) A study of modern investigations and treatment of allergic rhinitis in Thai patients. Research final report to the China Medical Board of New York. Inc. p. 28.

4.Dhorranintra, B., Bunnag, C. and Klyprayong, S. (1986) Total and specific Immunoglobulin E (IgE) for the assessment of fern spore allergenicity. Presented at the III World conference on clinical pharmacology and therapeutics. Stockholm, July 27 - August 1, 1986.

Prof. Boonchua Dhorranintra, Division of Allergy and Immunopharmacology, Department of Pharmacology, Faculty of Medicine Siriraj Hospital, Mahidol University, Bangkok 10700, Thailand.

III Allergology and airborne particles

1. Lectures of general importance: Diagnosis and therapy

EXS 51:
Advances in Aerobiology
©1987 Birkhäuser Verlag Basel

INHALATIVE ALLERGENS

E. Fuchs

Deutsche Klinik für Diagnostik, Fachbereich Allergologie, Wiesbaden (D)

Introduction

Allergic pathogenesis is only <u>one</u> of the possible causes of developing asthma and its equivalents. However, unlike the primary forms of asthma due to chemical or toxic agents and physical irritation (such as heat, cold and inert dusts), allergic bronchial asthma shows a welldefined clinical picture and can be verified by adequate and - for the patient - acceptable diagnostic procedures, and can be treated causally.

In contrast to "spontaneous" allergic bronchial asthma caused by sensitization to allergens in our "natural" environment, occupational allergic bronchial asthma is characterized by features which are specific to particular professions. In respect of a possible sensitization by the occupational environment, occupational bronchial asthma differs essentially from that due to our "natural" environment in two important points (13):

1. the massive exposure to allergens characteristic of a special occupation, and

2. the aggressive potency of many occupational allergens which "enforce" sensitization.

It is <u>technology</u> itself which often creates conditions of massive exposure, which is followed by the development of an occupational allergic bronchial asthma. Such conditions include, for example, the use of spray guns for painting, the atomizing of materials by jets, as was, for example, previously common in the printing industry (11), and the massive dust exposure in various workung places, for example in the grinding shops of the timber industry (22), in the loading and storage halls, silos and carding rooms in the jute and hemp industry, and on farms, in mills and bakeries and many other industrial sites (12, 30).

Thus, conjunctivitis, rhinitis and bronchial asthma in coffee sorters, coffee blenders etc., are in most cases due to a monovalent sensitization to <u>raw</u> coffee-dust. Using a modified extracting procedure, we (15) could first show that the allergen in coffee is an aromatic substance which escapes during the roasting process and is destroyed by boiling for 3 minutes, as Bruun (6) was able to demonstrate. We were able to isolate the volatile component by a simple distilling process and to produce evidence of a sensitizing aromatic substance by these experiments, such as had been suspected by clinical observations. Reagin could be detected by direct and indirect intradermal experiments and also by bronchial challenge tests.

Catalogue of inhalative allergens

Gronemeyer and Fuchs (14) listed a catalogue of inhalative allergens according to the type and origin of allergen, including the diverse possibilities of professional exposure to the allergens in several occupational branches. Such a catalogue of allergens, of course, never can be complete and we will not try to list the individual items here. Fig. 1 gives a survey of the most important mixed allergens and some solitary allergens of special importance in certain industries. Mixed allergens and solitary allergens are arranged side by side. In the case of so-called mixed allergens, there are often hidden admixtures and inevitable impurities which can cause occupational sensitization themselves, and this fact should always be taken into account in diagnosis.

Moreover, the literature of the last few years has reported a number of new inhalant allergens, mainly drawn from case histories. We can only give attention to the most important ones.

1. Animal allergens:

1.1 Hair dust of lions and elephants (26, 27)

1.2 Dust of crickets (1)

Crickets are "relatives" of cockroaches and grasshoppers, which have been known for a long time to be potent sources of allergens.

1.3 Water-fleas (daphnids)

They are known as an occupational allergen in the production of aquarium fish food. While drying food by shaking in jigging sieves, a heavy exposure to dust occurs. This allergen is particularly aggressive, with a sensitization degree of 66 % in the exposed. The high sensitization degree is reflected by skin titration values of 10^{-5} to 10^{-6} (19). There may also be exposure in aquarium shops and in households in which aquarium fishes are kept.

1.4 Midges (chironomids)

The larvae - red-colored because of their haemoglobin content - are also often used to feed aquarium fishes. It is noteworthy that about one thired of the individuals exposed in fish food production develop inhalation allergies. Baur, working with Fruhmann in Munich, was able to isolate and identify specific molecules acting as antigens. They are monomer and dimer haemoglobins of the subspecies of Chironomus thummi which are very similar in structure. In sensitized persons it was possible to detect by RAST-technique a specific formation of immunoglobuline E against theses haemoglobins; in skin tests, corresponding immediate reactions were found. As many as four antigenic determinants could be identified within certain amino acid sequences of a haemoglobin. By bronchial provocation tests with low doses of about 0,2 µg of isolated haemoglobin dual bronchial reactions could be produces (3, 24).

1.5 Mite beetle (5)

1.6 Carmine

This is a dyestuff from a type of scale insect (Coccus cactus) which is used as colorant in cosmetics as well as in food stuffs like Campari and Ketchup. Clinically, allergic bronchial asthma may arise in persons working in the preparing industry, or a so-called cochineal cheilitis (dermatitis) may occur by contact (7).

1.7 Silk

The dust waste of wild silk also represents a very aggressive inhalant allergen. Wild silk is produced by wild living Antheraea (spp.) and is - because of its little weight - used as filling material for bed mattresses and rugs. The subject has been reported by Wüthrich (16, 32) and Forck recently.

Allergic diseases does not only occur in the producing industry. It has not yet been clearly worked out, whether we are dealing with a senitization to sericine (= silk glue), as I described 1955 (11, 12) in some cases with an allergic occupational silk worker's asthma (see also the new investigations by Dewair and coworkers (9)).

1.8 Silverfish (25)

2. Plant allergens:

2.1 Bromelain (from Ananas comosus)

2.2 Papain (from Carica papaya)

Proteases of bacterial origin, particularly from Bacterium subtilis in washing powders, have been known for a long time as occupational inhalant allergens.

Enzymes of plant origin (for example bromelain, papain) have been lately recognized as potent inhalant allergens, mainly in the production industry. This fact is also of general clinical importance, since there are other possibilities of exposure by the ingestion of enzyme-containing digestives and the consumption of meat prepared with so-called meat tenderizers. The addition of papain is supposed to make meat particularly tender and delicate. The first observations of rhinitis and asthma have been described in a population of cooks in restaurant kitchens, but it seems also to be possible, as experiments have shown, to elicit an allergic bronchial obstuction by oral administration. In brewing factories (but not in Bavaria) sulfhydril proteases are used to clarify beer.

Mention might also be made of the fact that proteases can activate the complement system and thus induce the formation of anaphylatoxin $C3_a$ and $C5_a$, followed by a release of mediators. Apart from their sensitizing capacity these proteases also have a chemically irritant effect and may destroy the tissue of the bronchial mucosa (31, 2).

3. Chemical allergens:

3.1 Chloramine-T

Recent investigations confirm the allergenicity of this potent desinfactant, which is used in butcher's shops, kitchens and also in operating theatres. These asthmatic reactions by chloramine-T are quite clearly IgE-mediated, while even patients with dual asthmatic reactions or with so-called asthmatic reactions of the late type do not show specific IgG in the serum (10).

3.2 Antibiotics

Spiramycin as an additive in chicken feed

3.3 Phthalic anhydride (18)

3.4 Piperazine dihydrochloride

3.5 Isocyanates

The discussion about the allergenic and/or toxic potency of iso-cyanates has been going on for more than 20 years and is still far from being resolved. It is under discussion that higher doses may have a toxic effect and smaller doses may possibly produce sensitization. The pathogenic vapours are generated either during production or by heating of end-products.

It is noteworthy that evidence has meanwhile been obtained of reaginic antibodies against tolylene diisocyanate (TDI) by PCA reactions in monkeys, by measuring complementbinding antibodies and also by inhibition of the asthmatic immediate reaction by the prophylactic inhalation of disodium cromoglycate. A test allergen for RAST has been developed by Karol et al., but with a different chemical isocyanate structure. About 15 to 18 % of clinical cases of TDI-asthma show at the same time these antibodies, detectable by RAST-technique with the new test allergen (p-tolyl-isocyanate, TMI) (8, 17).

Fruhmann's team (4) could also demonstrate isocyanate-specific IgE antibodies in 16 % of patients, but not in the sera of asymptomatic exposed subjects nor in the sera of unexposed controls. There also seems to be evidence that antibody carriers have a lower bronchial stimulus threshold for inhaled isocyanates than other people. Fruhmann pointed out that a specific formation of IgE against isocyanates in RAST, together with significant previous histories and clinical symptoms, should be an indication for an immediate change of working place. The investigation suggests that the maximal working place concentration of isocyanates, which is still now 0,02 ppm, should be lowered. We cannot go into further details here.

3.6 Metal salts

It has been well known for a long time that metal salts can produce allergic eczema, but it is not generally known that inhalation of the same salts of chromium, nickel and platinum may also provoke bronchial asthma, as may the inhalation of other simple chemical agents (23).

However, it is much more difficult to demonstrate an allergic reaction of the immediate type (type I), for example with platinum, than with full antigens containing proteins and polysaccharides (such as pollen etc.). The mode of action of these lowmolecular substances in such cases is not well understood. In the case of platinum salt in particular Schultze-

Werninghaus has paid attention to the frequently high degree of sensitization in occupational platinum salt asthma, and pointed out the problems that may therefore arise for <u>in vivo</u> skin testing (28).

The list of inhalants (dusts, gases, vapours and fumes) with an allergenic (and also sometimes simultaneous toxic irritant) action has not be brought up to date constantly as a result of industrial progress. Well analysed case reports are very helpful. It is noteworthy that a few substances, including some of low molecular weight, which have long been known as occupational allergens, can now also be studied by RAST-technique (Fig.2) and in some cases by histamine release from human leucocytes.

Cross-reactivity of herb pollen to spices and vegetables

There is no longer any doubt that food allergies to spices, herbs, and related vegetables in pollen-allergic individuals are increasing in West Germany, but obviously also in other European countries. Wüthrich and Hofer reported in 1984 (32) a "celery-mugwort-spice syndrome"; through personal communication we have learned that there may be an increase in this syndrome in Italy and France.

Although the patterns of sensitization to spices, herbs and vegetables may differ from country to country according to nutritional habits, there may well be a problem of common interest as regards the close relationship between sensitization to allergens of plant origin and pollen allergy.

Since 1980 we have seen more than 500 patients who are more or less sensitive to herbs, spices and related vegetables and after a follow-up of 5 years we may ofer some conclusions (29):

1. In case of sensitization to herb pollen a high incidence of spice and vegetable allergies can be expected. Mugwort pollen indeed seems a key allergen for detecting these patients. The specifity and presence of IgE antibodies for some allergens in question have been demonstrated by the Prausnitz-Küstner test, skin tests and RAST.

2. RAST inhibition tests have proven that allergen fractions exist in common for spices, herbs, vegetables <u>and</u> herb pollen.

3. The elimination of mixed and hot spices and industrial products ("ready meals") from the daily diet is usually followed by remission of several symptoms and also sometimes by a decrease in total IgE activity.

The nature of the allergen fractions shared in common by pollen and food still remains unclear and should be investigated. Skin tests have been performed with aqueous extracts and there is no evidence of sensitization to sesquiterpenlactones, which have been found to be responsible for contact allergies.

Early clinical recognition

Atopic subjects usually can be recognized by selective and comprehensive allergological anamnesis, together with repeated controls of the total IgE. However, to be certain, the diagnostic criteria of an immunological type of reaction must still be applied, and candidates for a special profession and also subjects potentially at risk have to be informed about the risk of sensitization by constant occupational exposure. Apart from atopic subjects, the development of an allergic obstructive lung disease is of course also a matter of fate - although specific working place-dependent factors are of crucial importance.

In many cases the development of the disease cannot be foreseen, and unfortunately we have to wait until the clinical manifestation takes place. But at the early stage of the "equivalents" (rhinitis and conjunctivitis), medical care should be initiated immediately.

Author's address

Prof. Dr. med. habil. Erich Fuchs, Stiftung Deutsche Klinik für Diagnostik GmbH, Fachbereich Allergologie, Aukammallee 33, D-6200 Wiesbaden (FR-Germany).

References

1) Bagenstose, A.H., Mathews, K.P., Homburger, H.A. and Saaveard-Delgado, A.P. (1980) Inhalant allergy due to crickets. J. Allergy Clin. Immunol. 65, 71-74.

2) Baur, X. (1979) Studies on the specificity of human IgE-antibodies to the plant proteases papain and bromelain. Clin. Allergy 9, 451-457.

3) Baur, X., Dewair, M., Fruhmann, G., Aschauer, H., Pfletschinger, J. and Braunitzer, G. (1982) Hypersensitivity to chironomids (non-biting midges): localization of the antigenic determinants within certain polypeptide sequences of hemoglobins (erythrocrourins) of Chironomus thummi thummi (Diptera). J. Allergy Clin. Immunol 69, 66-76.

4) Baur X., Dewair, M. and Fruhmann, G. (1984) Detection of immunologically sensitized isocyanate workers by RAST and intracutaneous skin tests.

J. Allergy Clin. Immunol. 73, 610-618.

5) Bossert, J., Fuchs, E., Honomichl, K., Maasch, H.J., Risler, H. and Wahl, R. (1983) Exogen-allergisches Asthma bronchiale durch den sogenannten "Buckelkäfer" oder "Kugelkäfer" - Gibbium psylloides. Allergologie 6, 255-259.

6) Bruun, E. (1957) Allergy to coffee. Acta Allerg. 11, 150.

7) Burge, P.S., O'Brien, I.M., Harries, M.G. and Pepys, J. (1979) Occupational asthma due to inhaled carmine. Clin. Allergy 9, 185-189.

8) Butcher, B.T., O'Neil, C.E., Reed, M.A. and Salvaggio, J.E. (1980) Radio-allergosorbent testing of toluene diisocyanate-reactive individuals using p-tolyl isocyanate antigen. J. Allergy Clin. Immunol. 66, 213-216.

9) Dewair, M., Baur, X. and Ziegler, K. (1985) Use of immunoblot technique for detection of human IgE and IgG antibodies to individual silk proteins. J. Allergy Clin. Immunol. 76, 537-542.

10) Dijkman, J.H., Vooren, P.H. and Kramps, J.A. (1981) Occupational asthma due to inhalation of chloramin-T. I. Clinical observations and inhalation-provocation studies. Int. Archs. Allergy appl. Immun. 64, 422-427.

11) Fuchs, E. (1973) Asthma bronchiale in der Gewerbemedizin (Gentner, Stuttgart).

12) Fuchs, E. (1979) Allergische Atemwegsobstruktion. In: Handbuch der Inneren Medizin, Bd. IV, Teil 2, ed. by W.T. Ulmer, pp. 543-673 (Springer, Berlin).

13) Gronemeyer, W. (1958) Das gewerbliche Asthma. Dtsch. med. Wschr. 83, 30-39.

14) Gronemeyer, W. and Fuchs, E. (1967) Krankheiten durch inhalative Allergeninvasion. In K. Hansen and M. Werner (Hrsg.): Lehrbuch der klinischen Allergie, pp. 122-167 (Thieme, Stuttgart).

15) Gronemeyer, W. and Fuchs, E. (1958) Coffee an occupational vapour allergen. In: Occupational Allergy, pp. 302-305 (Stenfert Kroese, Leiden).

16) Häcki, M., Wüthrich, B. and Hanser, M. (1982) Wildseide: ein aggressives Inhalationsallergen. Dtsch. med. Wschr. 107, 166-169.

17) Karol, M.H., Ioset, H.H. and Alarie, Y.C. (1978) Tolyl-specific IgE antibodies in workers with hypersensitivity to toluene diisocyanate. Amer. industr. Hyg. Ass. J. 39, 454-458.
See also correspondence: Karol, M.H., et al., and Salvaggio, J., et al. (1980) IgE antibodies in TDI workers. J. Allergy Clin. Immunol. 65, 162.

18) Maccia, C.A., Bernstein, I.L., Emmet, E.A. and Brooks, S.M. (1976) In vitro demonstration of specific IgE in phthalic anhydride hypersensitivity. Amer. Rev. resp. Dis. 113, 701-704.

19) Meister, W. (1978) Berufsasthma infolge Daphnienallergie. Allergie u. Immunol. 24, 191-193.

20) O'Brien, I.M., Harries, M.G., Burge, P.S. and Pepys, J. (1979) Toluene diisocyanate-induced asthma. I. Reaction to TDI, MDI, HDI and histamine. Clin. Allergy 9, 1-6.

21) O'Brien, I.M., Newman-Taylor, A.J., Burge, P.S., Harries, M.G., Fawcett, I.W. and Pepys, J. (1979) Toluene diisocyanate-induced asthma. II. Inhalation challenge tests and bronchial reactivity studies. Clin. Allergy 9, 7-15.

22) Oehling, A. (1963) Berufsallergie im Holzgewerbe. Allergie u. Asthma 9, 312-322.

23) Pepys, J., Parish, w.E., Cromwell, O. and Hughes, E.G. (1979) Passive transfer in man and monkey of type I allergy due to heat labile and heat stable antibody to complex salts of platinum. Clin. Allergy 9, 99-108.

24) Prelicz, H., Baur, X., Dewair, M., Tichy, H., Kai, A.B., Tee, R. and Cranston, P.S. (1986) Persistence of hemoglobine allergenicity and antigenicity during metamorphosis of chironomidae (insecta: diptera). Int. Archs. Allergy appl. Immun. 79, 72-76.

25) Rijckaert, C., Thiel, Cl. and Fuchs, E. (1981) Silberfischchen und Staubläuse als Allergene. Allergologie 4, 80-86.

26) Rudolph, R., Kunkel, G., Diller, G. nd Baumgarten, C. (1980) Über einen Fall von Respirationsallergie durch Sensibilisierungen gegen Elefantenschupffen. Allergologie 3, 34-36.

27) Rudolph, R., Muckelmann, R., Kunkel, G., Fahrig, W. and Sladek, M. (1980) Über einen Fall von Löwenallergie. Allergologie 3, 135-137.

28) Schultze-Werninghaus, G., Roesch, A., Wilhelms, O.H., Gonsior, E. and Meier-Sydow, J. (1978) Asthma bronchiale durch berufsbedingte Allergie vom Soforttyp gegen Platinsalze. Dtsch. med. Wschr. 103, 972-975.

29) Thiel, Cl., Fuchs, E., Maasch, H.-J. and Wahl, R. (1986) Allergy to spices: cross-reactivity to other allergens. In: New Trends in Allergy II, pp. 154-164 (Springer, Heidelberg).

30) Thiel, H. and Ulmer, W.T. (1982) Respirationsallergien bei Bäckern. Bücherei des Pneumologen 8 (Thieme, Stuttgart, New York).

31) Wüthrich, B. (1976) Zum Allergenkatalog beruflicher Inhalationsallergien. Berufsdermatosen 24, 123-131.

32) Wüthrich, B., Dietschi, R., Keter, A., Zortea-Calflisch, C. (1985) Das sogenannte "Wildseiden"-Asthma - Eine noch immer aktuelle Inhalationsallergie auf Seidenabfälle. Schweiz. Med. Wschr. 115, 1387-1393.

Fig. 1

Dusts and Aerosols as Common Occupational Allergens

Mixed Allergens	Solitary Allergens
Hair Dust	Raw coffee bean dust
Flour and bran dust	Proteases
Feather dust	Insect dust
Upholstery dust	Castor-oil bean dust
Drug and medicament dust	Raw silk dust (Sericin)
Pollen	Gum arabic dust
Cosmetic dust and aerosols	Lycopodium
Feed dust	
Various fungus spores	
Wood dusts	
Aromatic essence aerosols	

(After Gronemeyer, W., compiled from various publications)

Fig. 2

Occupational Inhalant Allergens Detectable by RAST

Animal hairs (cat, dog, horse, bovine)

Rye and what flour

Cricket (Acheta domestica)

Proteases (bromelain, papain)

Dyestuffs (with azo and anthraquinone groups)

Platinum salts

Trimellitic acid anhydride

Phthalic anhydride

Tolylene diisocyanate (TDI) and other isocyanates

Chloramine-T

EXS 51:
Advances in Aerobiology
©1987 Birkhäuser Verlag Basel

RELATIONSHIP OF AIRBORNE POLLENS AND SPORES TO SYMPTOMS ON THE SKIN AND MU-
COUS MEMBRANES OF PATIENTS IN THE HIGH ALTITUDE CLIMATE IN DAVOS

W. Kneist, H. Düngemann, H. Gehrken, S. Borelli et al
Klinik für Dermatologie und Allergie Davos
CH-7270 Davos/Switzerland

Davos has the world's longest history of continuous measuring of radiation
and metereological data. (3) In 1934 LÜDI and coworkers began with the registration of airborne pollens and spores in Davos. (5) The results of
their registrations have often been used for medical-scientific research.
Comparative measurements of airborne pollens at different locations and
especially in different altitudes have been performed for a long time. In
the high mountain regions there is an essentially smaller amount of airborne pollens which also differ in their distribution according to the vegetation. It is the contribution of Dr. LEUSCHNER to have published these data
for Switzerland. (4)

In our Clinic for Dermatology and Allergy in Davos we register the
airborne pollens and spores since 1973.

In 1963 - 65 we performed an investigation (1): 900 patients had
to registrate frequency and intensity of pruritus, psychic restlessness or
insomnia and dyspnoea every two hours. It was the aim to establish a periodicity of symptoms during the day's and the year's course and put in relation to metereological data. This recording revealed new and highly interesting insights into the periodicity of symptoms in the day's and year's
course. In addition it revealed surprising relationship when compared with
metereological data.

Our recent study has the following background: In our clinic we
treat patients with atopic diseases like neurodermatitis, asthma and rhino-
conjunctivitis who require prolonged treatment for several weeks in the
high mountain climate of Davos due to the severity and chronicity of their
diseases. Most of them require topic or systemic steroids. 30-40% of our
patients suffer from mucous membrane troubles. It is our aim to stabilize
their cutaneous situation after discontinuation of steroids and other antiallergic drugs far enough to enable exact allergy testing. This is often
the first possibility to have these patients tested. (2)

It is the aim of our study to evaluate whether a discontinuation of corticosteroids and other antiallergic drugs will reveal a dependence of symptoms to the pollen- and spore-count which is much lower in Davos than in the lowlands. Secondly we wanted to find out which other factors influenced the course of symptoms.

During the last years we daily registered skin and mucosal symptoms in patients with atopic diseases. Every patient had to record the localisation and the time of occurence of symptoms as well as details of weather and current medication. The symptoms are expressed in per cent and refer to the daily registration in a symptom calendar, which at least 50 up to 100 patients kept simultanuously. These data of 1.032 patients are compared with the results of the pollen trap and several metereological data, mainly temperature, humidity and rain.

When the diagrams of airborne pollens and spores of 1983 and '84 in Davos are compared, unusually large amounts of Cladosporia spores are found in 1983. We have compared the diagrams of airborne pollens and spores with the records of symptoms:

At the end of April 1983 (fig. 1) the skin symptoms start to increase until they reach a peak in the second half of May. At the same time the nasal and bronchial symptoms decrease markedly. The mucous membrane symptoms increase from the end of May and culminate in July, then slowly decrease until September. Looking at the bronchial symptoms there is a relatively uniform course between June and September. The skin symptoms remain quite stable from the end of May on.

The excessively high amounts of Cladosporium from the end of July till the end of August with the maximum of 10.000 spores per cubicmeter on August, 13th did not cause any increase in skin or airway symptoms. The relatively high peak of pollens and spores in the beginning and the second half of June shows a small increase only in the nasal and eye symptoms.

In <u>1984</u> (fig. 2) the total pollen count stays more or less below 100 per cubicmeter. Only in the first half of June and in July the count exceeds a hundred markedly. The Cladosporium counts rise continuously from May on to a maximum of 300 spores per cubicmeter in the middle of July. Following a decrease they culminate again in August. The Cladosporium counts of 300 per cubicmeter in July '84 are clearly lower than those in

eyes
%
30
20
10
0
bronchi
%
30
20
10
0
nose
%
30
20
10
0
skin
%
100
90
80
70
60
50
spores
10000
1000
800
600
400
200
0
pollens
300
200
100
0
May
June
July
Aug.
Sept.

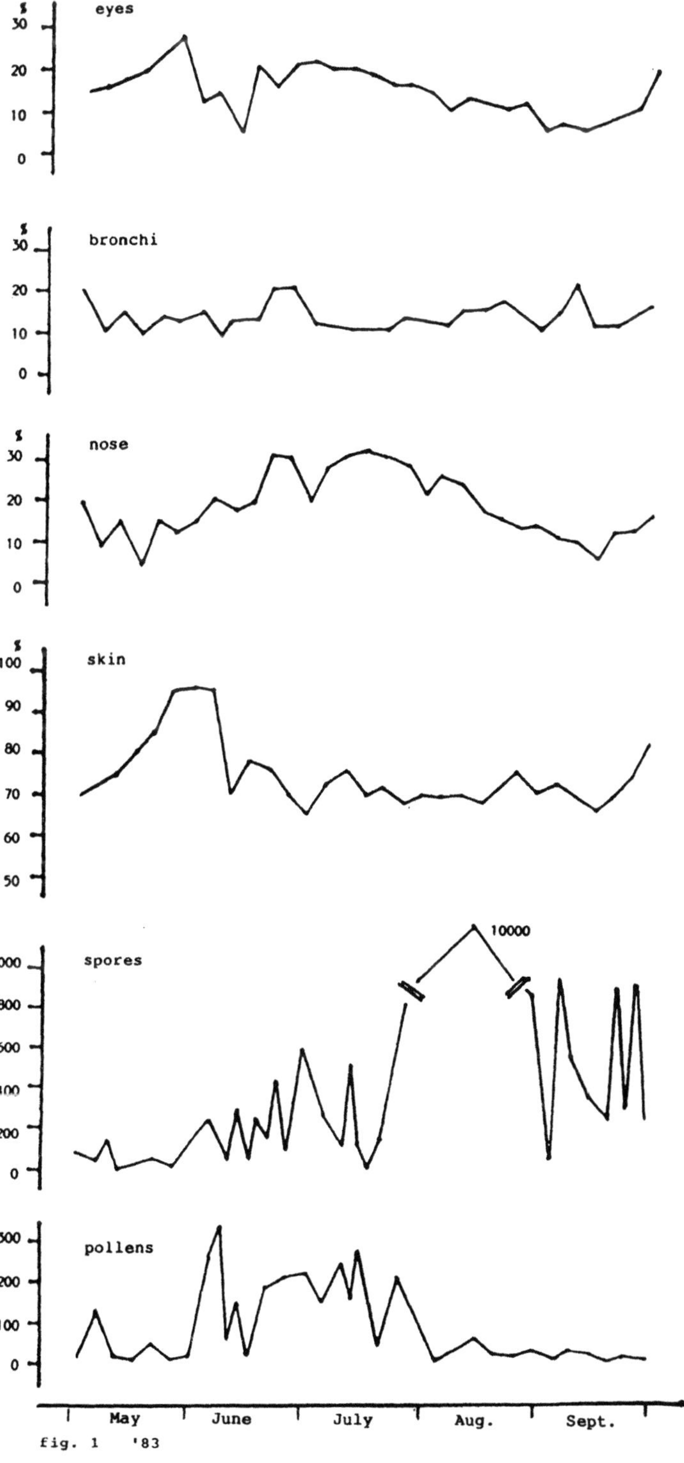

fig. 1   '83

%
30
20
10
0
eyes
%
40
30
20
10
0
bronchi
%
50
40
30
20
10
nose
%
100
90
80
70
60
50
skin
300
200
100
0
spores
300
200
100
0
pollens
May
June
July
Aug.
Sept.

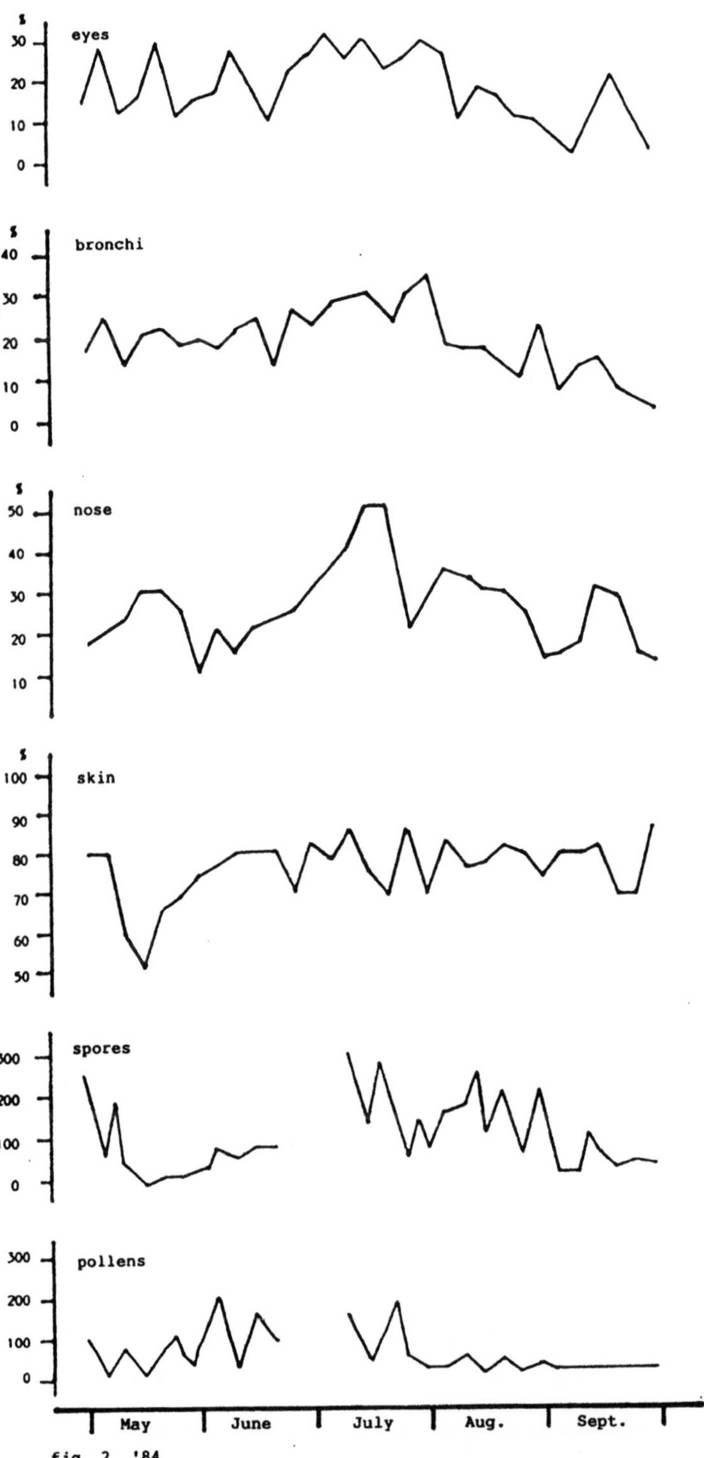

fig. 2 '84

June and July '83, which measured 800 per cubicmeter but went up to the maximum of 10.000 per cubicmeter in August '83. In June and July '84 our pollen trap broke down. We therefore have no registration.

The skin symptoms reach a minimum of 51% in the middle of May, then they rise continuously to a peak in July. The nasal symptoms have a peak in the middle of July. The high pollen count in May and June shows no direct correlation to mucous membrane symptoms. The symptoms of skin and mucous membranes except the nose show no direct correlation to the amounts of pollens and spores.

References:
1. Borelli S, Chlebarov S, Flach E, (1966) Atopische Neurodermitis. Zur Frage ihres 24-Stunden-Rhythmus, ihrer Wetter- und Klima-Abhängigkeit. MüMed Woschr 108:474-480
2. Borelli S, Schnyder UW (1962) Neurodermitis constitutionalis sive atopica. Abhängigkeit von Klima, Jahreszeit, Saisonveränderungen. In: Jadassohn J (Hrsg) Handbuch der Haut- und Geschlechtskrankheiten. Marchionini A (Hrsg) Erg.-Werk II/1,9. Springer, Berlin Göttingen Heidelberg, S 272 ff.
3. Düngemann H (1981) Zur Historie der therapeutischen Bedeutung von Davos. In: Borelli S, Düngemann H (Hrsg) Fortschritte der Allergologie und Dermatologie. IMP-Verlag Basel Neu Isenburg Wien S 451-471
4. Leuschner RM, Boehm (1981) Pollen and inorganic particles in the air of climatically very different places in Switzerland. Grana 20: 161-167
5. Lüdi W (1937) Die Pollensedimentation im Davoser Hochtale. Ber. Geobot. Forschungsinst. Rübel Zürich 1936: 107-127

Dr. med. Werner Kneist, Klinik für Dermatologie und Allergie Davos, Tobelmühlestr. 2, CH-7270 Davos-Platz/Switzerland

EXS 51:
Advances in Aerobiology
©1987 Birkhäuser Verlag Basel

EXPERIENCES WITH THE 'INDIVIDUAL POLLEN COLLECTOR' DEVELOPED BY G.BOEHM

G.Boehm and Ruth M.Leuschner, Basel, Switzerland

Question:

What can we do with problem patients with allergies, in whom it is impossible to make a diagnosis with the ordinary skin test preparations?

Solution:

One solution of this problem can be the 'Individual Pollen Collector' developed by G.Boehm. The device collects the pollen very well, especially, when the patient is walking or cycling. The air comes into the Vaseline-layer on the microscope slide in the device (Fig.1) and the particles are deposited there(1., 2.). The air can escape from the device through four slits at the sides. So there is a continous movement of air.

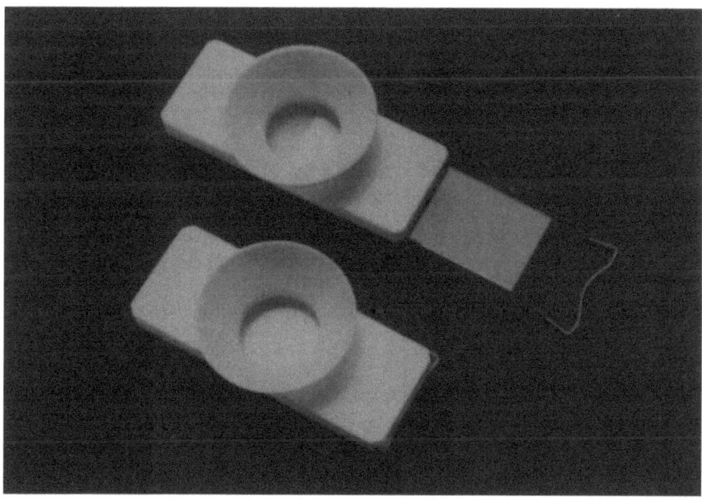

Two pollen collectors. In one of them, a microscope slide is inserted half-way. The spring that holds the slide in place is lying beside the device. The other collector is ready for use.

<u>Results:</u>

When counting the pollen on the slides there may be unexpected results.

Some examples:

Pollen of Dianthus (Carnation)	a insect pollinated plant. Its pollen irritated someone who was working in the garden near the plant of carnation.
Pollen of Carpinus (Hornbeam)	may be the reason for hay-fever This type of pollen was rarely found in the town, but it irritated a patient who was a longer time in the surroundings of a wood.
Pollen of Tilia (Lime tree)	The grains are - as insectpollinated·heavy and therefore not in greater amount caught in the Burkard sampler which was fixed at a height of 15 meters.
Fungus spores of Puccinia striiformis	Caused irritations. The patient lived in the surroundings of a village.
Algae	Caused hay-fever in a person staying on a boat at the weekend.

<u>References:</u>

1. G.BOEHM and Ruth M.LEUSCHNER:"Observations with an 'Individual Pollen Collector'." Berichte 5/79 Umweltbundesamt, Erich Schmidt Verlag Berlin 1979, 411 - 419. (The 1st International Conference on Aerobiology, 13.- 15. August 1978.)

2. Ruth M.LEUSCHNER und G.BOEHM: ' Pollen and inorganic particles in the air of climatically very different places in Switzerland'. Grana 20, 161 - 167, 1981.

<u>Author's address:</u>
Prof.Dr.med.G.Boehm, Herrengrabenweg 51, CH-4054 Basel, Switzerland

AIRBORNE POLLENS AND SYMPTOMS SCORE IN ALLERGIC PATIENTS UNDERGOING IMMUNOTHERAPY

Fasani F., Gorini M.

Clinica Lavoro Foundation, Allergology and Clinical Immunology, Pavia, Italy

Introduction

In the last years standardized monitoring techniques offered means to obtain pollen calendars (1,2).
They are useful in the allergological practice for the diagnosis and the modulation of the therapy.
A good correlation between symptoms of allergic patients and airborne pollen has been found in a number of studies (3,4). The present study reports the aerobiological data referred to pollen of Gramineae and Urticaceae during the year 1985, with reference to the related sensitization in subjects with rhinitis and bronchial asthma and to the symptoms score of patients undergoing immunotherapy.

Materials and Methods

This study was carried out during the year 1985, collecting airborne pollens by a VPPS 2000 pollen trap with a weekly head, placed in the urban area of Pavia town on a roof at about 15 m from the ground.

Microscope slides were prepared according to Kumer and Mandrioli (5) and pollen grains were identified microscopycally
250 X.

Data were processed as ten day averages and the number of
pollen grains were expressed per cubic metre of air.

Temperature max. and min., humidity and rainfall were registered by a meteorological station near the pollen trap.

Symptoms score was obtained, according to a suitable questio-
nar, by rilevation of subjective symptoms of 27 allergic patients, undergoing specific immunotherapy.

Moreover it was considered 540 patients studied in our Section
during 1985with suspected respiratory allergy, undergoing
skin tests and RAST.

Results and discussion

In Fig. 1 we report meteorological data from urban atmosphere
in Pavia during 1985 and the distribution of airborne pollens.
In Fig. 2 we report respectively the concentration of Gramineae
and Urticaceae during the same year and symptom score from 12
patients undergoing immunotherapy to Gramineae and 15 to Pa-
rietaria.

In our region we had an estimation of sensitization of 39%
for the Gramineae and for Urticaceae of 23,7%.

Gramineae become to increase in the second week af April, with
its major peak during the second week of May.

The Urticaceae are the most representative family; they are
consistently present all over the year with the major concentration from June to September.

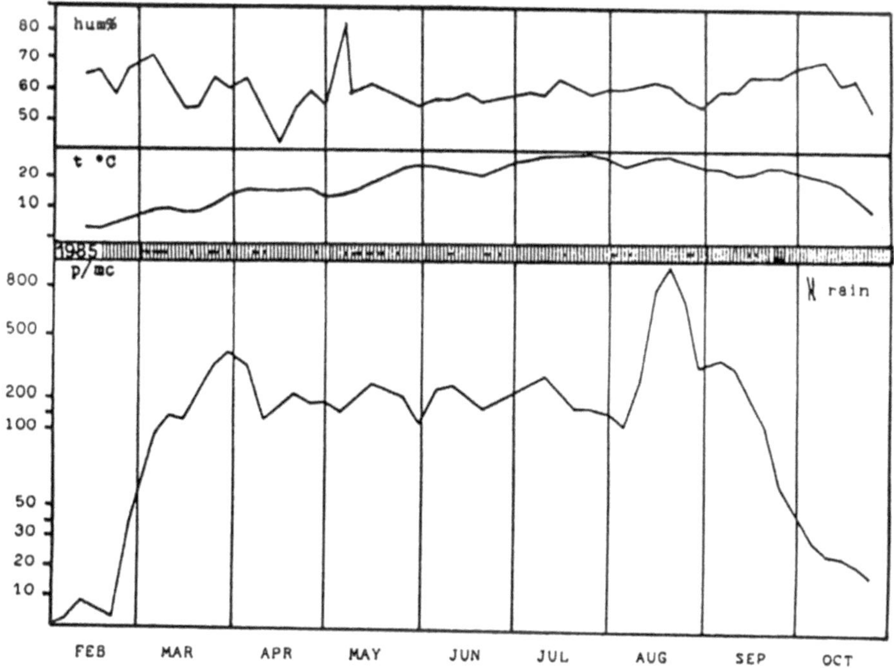

Fig. 1. Distribution of airborne pollens and meteorological
data in Pavia's urban atmosphere during 1985.

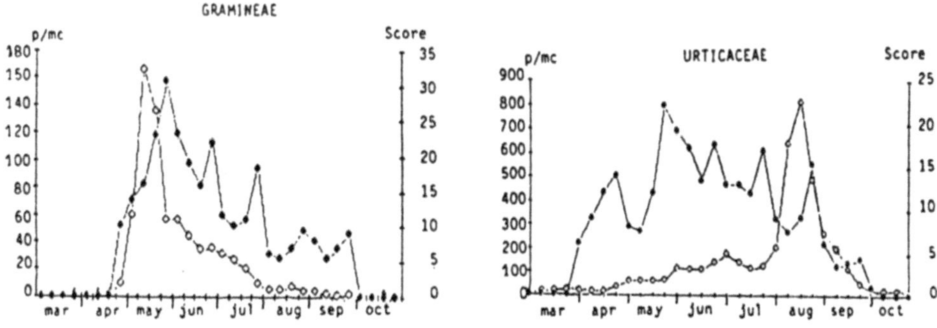

Fig. 2. Diffusion of Gramineae and Urticaceae in Pavia town
during 1985 (black) and symptoms score (white) of 27
patients undergoing immunotherapy.

The symptom score is rather low in both the cases.

The symptoms about Gramineae are shown some days later in comparison with the peaks of concentration, probably due to the necessity of an induction's period; on the contrary the symptoms about Urticaceae are not influenced by pollens concentration but rather by atmospheric precipitation.

Particularly the symptoms are not influenced by the peak that we observed in August.

From the analysis of the graphics we cannot see a clear threshold level and the symptoms appear for concentration major to $40 \div 50$ grains/m^3.

Outstanding studies are done to compare the symptomatology of some patients undergoing immunotherapy with other trated with anti allergic drugs.

References

1. Leuschner R.M., Boehm, G. (1979) Investigation with the "individual pollen collector"and the "Burkard trap" with reference to hay fever patients. Clin. Allergy 9, 175.

2. D'Amato G., Cocco G., Liccardi G., Melillo G. (1983) A study on airborne allergic pollen content of the atmosphere of Neaples. Clin. Allergy 13, 537.

3. Spieksma F.T.M. (1980) Daily hay fever forecast in the Netherlands. Clin. Allergy 35, 593.

4. Hyde H.A. (1973) Atmospheric pollen grains and spores in relation to allergy II. Clin. Allergy 3, 109.

5. Kumer E., Mandrioli P., Tampieri F. (1978) A comparison between the concentration of airborne pollens and the clinical symptomatology in allergic subjects under treatment with hypersensitizing agents. Proc. Intern. Conf. Aerobiol., Munich, 13-15.

2. Special allergens: Pollen

MONITORING OF ATMOSPHERIC CONDITIONS AND FORECAST OF OLIVE POLLEN SEASON

L. Macchia, M. Aliani, M.F. Caiaffa, A.M. Carbonara, E. Gatti, A. Iacobelli, S. Strada, G. Casella*, A. Tursi
Cattedra di Allergologia ed Immunologia Clinica, Medical School, and *Istituto di Coltivazioni Arboree, University of Bari, Italy

Introduction

Olive has just been recognized as one of the most important allergenic pollens in some regions of North America (11, 17, 19, 20) and throughout the Mediterranean basin (1, 3, 7, 16, 18), where large areas are covered with Olive cultivations, mainly in Southern Italy, Southern Spain, Greece, Turkey and Northern Africa. In the district of Bari (Apulia, Southern Italy) (12) the Olive groves occupy the 54.25% of the whole territory, in the countrysides as well as near the principal cities. Therefore, in this area large quantities of Olea Europaea pollen are released in the atmosphere during the flowering season, in May and June. From the aerobiological point of view, the Olive pollen season in Bari appears short enough but very severe. It starts on 10th-15th May, reaches the peak a fortnight later with concentrations usually around many hundreds of grains/cm and decreases more slowly on the last week of June.

According to these findings and in consequence of its high allergenicity, Olea Europaea represents in Bari a serious problem for many thousands of sensitized patients: in this area the frequence of Olive pollinosis is quite high, about 27-30% of all the patients with allergic troubles due to airborne pollens and often manifests itself with quite severe clinical symptoms.

For some kinds of pollen, such as Grass or Ragweed, it has been tried to forecast the seasonal pollen trend on the ground of the meteorological events occurring before the onset of the pollination (4, 5). These studies, interesting both from the aerobiological and allergologic point of view, allow to check environmental and phenological hypotheses (9, 10, 14) as well as to establish etiological and pathogenetical relations (6, 13) and to plan preventive measures and treatments.

This paper reports the results obtained during a long-term investigation on the relationship between weather conditions and pollen counts, in order to forecast onset and severity of the Olive pollen season.

Materials and methods

Meteorological and aerobiological data were collected for 5 years, from 1982 to 1986, in Bari.

Weather conditions

The parameters studied were: temperature, recorded as monthly averages of the daily °C values, and rainfall, recorded as monthly total sums. The data regarding the months of March and April were compared with the averages of the past quarter century, from 1962 to 1986, collected in Bari by the "Stazione Idrografica" of the Ministry of Works and obtained from the local Bureau (years 1982-1986) and from ISTAT - Istituto Centrale di Statistica, Rome (years 1962-1981).

Aerobiological data

Aerobiological surveys were carried out, from 1982 to 1986, using a volumetric spore-trap, Hirst-type, according to the suggestions of the International Biological Program (2). The Olive concentration of the first 100% symptom-day (SY-DAY) was taken as indicating the beginning of the season. This evenience, defined as the day in which all the patients who suffer from a certain pollinosis and live in a certain area show symptoms (5), was observed in Bari when the daily mean concentration of Olive pollen reached 40-50 grains/cm. The peak concentration was taken as indicating the severity degree of the season.

Results

The results are summarized in table I.

Table I. Comparison between meteorological survey and Olive pollen season during 1982-1986

YEAR	METEOROLOGICAL DATA				OLIVE SEASON		
	TEMPERATURE		RAINS (mm)		FIRST 100 %	PEAK CONCEN.	FEATURES
	MARCH	APRIL	MARCH	APRIL	SY-DAY	Grn./cm	
1982	10.7°	13.6°	150.6	4.8	24 MAY	1047	LATE & MODERATE
1983	12.0°	16.0°	81.8	31.0	15 "	1815	EARLY & SEVERE
1984	11.0°	13.7°	27.6	56.6	24 "	667	LATE & MODERATE
1985	11.8°	15.6°	35.1	57.8	17 "	2213	EARLY & SEVERE
1986	11.7°	15.6°	43.0	11.4	13 "	660	EARLY & MODERATE
1962 to 1986	11.8°	14.6°	55.2	39.0	Averages of the past quarter century from 1962 to 1986		
S.D.	1.6	1.2	35.3	32.5	Standard deviations		

The March and April monthly mean temperatures in 1983, 1985 and 1986 were above-average (for April, 1.4 °C, 1 °C and 1 °C respectively) while in 1982 and 1984 they were under-average (1 °C and 0.9 °C respectively). The rains widely varied from year to year, mostly in April; these data are in accord with the records of the past quarter century (see standard deviations). With regard to the seasonal chronology of the aerobiological features, 1983, 1985, and 1986 started early, while 1982 and 1984 seemed more late. The pollen counts were very high in 1983 and 1985 and moderate in 1982, 1984 and 1986.

Discussion and Conclusions

It should be outlined that the data given above refer to a five years survey, a period not long enough for this kind of studies; anyway they could allow at present some preliminary conclusions. It seems that the two aerobiological variables studied (onset of the season and severity degree) are practically independent, as shown by 1986 season. The earliness of the pollen season is directely dependent from the temperature observed in the immediate preceding weeks; high temperatures in March and particularly in April (see 1983, 1985 and 1986) gave rise to an earlier onset of the pollination. Rainfalls in the same period, on the contrary, seem devoid of any relevance (see 1984 and 1985).

From the quantitative point of view, both temperatures and rainfalls (in March and April) seem to play a non determinant role. In fact, it must be noted that Olive presents the peculiar phenomenon of Alternance of Production due to the competition of the fruits, ripening very late, from November to January, versus gemmae, differentiating in February and March. Therefore, if one year the production of fruits is plentifull, the competition will be very strong and in the next year the output of flowers will be very poor as well as the output of fruits; that will lead to a moderate competition and consequently to a good output of flowers and fruits in the successive year. This phenomenon is well studied (8, 15) and seems to be in connection with the availability of nutritional metabolized substances when the differentiation of the gemmae occur; if there is a large availability of these substances the indifferentiated gemmae will evolve in flowers, if not, they will evolve in leaves and wood. Once the imprinting is given, nothing is able to modify the gemmae evolution. Since that happens in February and in March, and since the imprinting depends on events occuring in the previous months (synthesis of nutritional substances, quantity of ripening fruits, time of harvesting etc.), the weather conditions occurring immediately before the start of the pollination will have a small influence on the quantity of pollen.

Therefore, aerobiological data seem to point out that the observed Pollinic Alternance (Table I) corresponds to the Alternance of Production, which seems to be the more effective mechanism of control of the severity of the Olive season.

Finally, it must be emphasized that Olive, unlike other allergenic plants (see wild Grasses, Ragweed, Birch), is a cultivated tree, with a vegetative cycle strongly affected by the human and agronomical activities which are able to modify and correct the influences due to the climate.

References

1. Blanca, M., Boulton, P., Brostoff, J., Gonzales-Reguera, I. (1983) Studies of the allergens of Olea Europaea pollen. Clin. Allergy, 13, 473-478

2. Benninghoff, W.S., Edmonds, R.L. (1972) Ecological systems approaches to aerobiology. I°. Identification of component elements and theyr functional relationships. US/IBP Aerobiol. Program Handb., Publ. Univ. Michigan, Ann Arbor, 2, 158-163

3. Bousquet, J., Guérin, B., Hewitt, B., Lim, S., Michel, F.B. (1985) Allergy in the Mediterranean area. III°: cross-reactivity among Oleaceae pollens. Clin. Allergy, 15, 439-448

4. Buck, P., Leventin, E. (1982) Weather patterns and Ragweed pollen production in Tulsa, Oklahoma. Ann. Allerg. 49, 272-274

5. Davies, R.R., Smith, L.P. (1973) Forecasting the start and severity of the hay-fever season. Clin. Allergy, 3, 263-267

6. Davies, R.R. Smith, L.P. (1973) Weather and grass pollen content of the air. Clin. Allergy, 3, 95-108

7. De Leonardis, W., Longhitano, N., Meli, R., Piccione, V., Zizza, A., Crimi, N., Palermo, F., Mistretta, A. (1985) Flora dei pollini allergizzanti in Italia (Centro S.Bracco, Milano)

8. Gàlan, C., Ruiz de Clavijo, E., Infante, F., Gallego, G. (1986) Annual and daily variation of pollen grains from Olea europaea L. in the atmosphere of Corboba (Spain) along two years of sampling. 3rd Internat. Conference on Aerobiology, August 6-9, 1986, Basel, Abstract Book, pag. 57

9. Jackson, M.T. (1966) Effects of microclimate on spring flowering phenology. Ecology, 47, 407-415

10. Leuschner, R.M., Boehm, G., Brombacher, Chr. (1986) Influence of inversion layers on the daily pollen count in 1 cm of air and on the hay-fever attacks of patients. 3rd Internat. Conference on Aerobiology, August 6-9, 1986, Basel, Abstract Book, pag. 55

11. Lewis, W.L., Vinay, P. (1979) North american pollinosis due to insect-pollinated plants. Ann. Allerg., 42, 309-318

12. Macchia, L., Strada, S., Caiaffa, M.F., Aliani, M., Tursi, A. (1984) The unique high season of Olive pollen in the district of Bari: causes and effects. Internat. Symp. on "Prevention of allergic diseases", Florence, June, 24-27, 1984, Abstract Book, pag. 149

13. McDonald, M.S., O'Driscoll, B.J. (1980) Aerobiological studies based in Galway. A comparison of pollen and spore counts over two seasons of widely differing weather conditions. Clin. Allergy, 10, 211-215

14. Mandrioli, P., Negrini, M.G., Zanotti, A.L. (1982) Airborne pollen from the Yugoslavian coast to the Po Valley (Italy). Grana, 21, 121-128

15. Morettini, A. (1972) Olivicoltura, 1st edn (R.E.D.A., Roma)
16. Negrini, A.C., Arrobba, D., D'Aste, S.M. (1983) Indagine
palinologica nell'atmosfera urbana di Genova e correlazioni
meteorologiche. Folia Allergol. Immunol. Clin., 30, 375-384
17. O'Rourke, M.K., Buchmann, S.L. (1986) Pollen yield of two
cultivars of Olea europaea L.. ("Manzanillo" and "Swan Hill").
3rd Internat. Conference on Aerobiology, August 6-9, 1986,
Basel, Abstract Book, pag. 5
18. Serafini, U. (1974) Pollen calendar for Italy. In "Atlas of
European allergenic pollens", Charpin, J., Surinyach, R., Frankland, A.W. Eds., (Sandoz, Paris), 148-154
19. Unger, D.L. (1977) Allergy practice in the desert. Ann.
Allerg., 39, 300-301
20. Vinay, P., Lewis, W.H. (1986) Pollen morphology, allergenicity and aerobiology of the Oleaceae. 3rd Internat. Conference
on Aerobiology, August 6-9, 1986, Basel, Abstract Book, pag. 56

Dott. Luigi Macchia, Cattedra di Allergologia ed Immunologia
Clinica, Policlinico, 70124 BARI, Italy

This work has been supported by a grant from Consiglio Nazionale delle Ricerche - Rome: Progetto Finalizzato Medicina Pre-
ventiva e Riabilitativa, Obiettivo 4.53

POLLEN YIELD OF TWO CVS. OF OLEA EUROPAEA L. ('MANZANILLO' AND 'SWAN HILL')

Mary Kay O'Rourke* and Stephen L. Buchmann+

*Dept. of Geosciences, Univ. of Arizona, Tucson Az. 85721 USA.
+USDA-ARS Carl Hayden Bee Research Center, 2000 East Allen Road, Tucson, Az.
85719 USA.

ABSTRACT

Airborne pollen concentrations (grains/m^3) under and near trees of two
cultivars of Olea europaea were studied during the 30 day pollination period
in Tucson, Az, USA. Airborne pollen concentration from Manzanillo, the
dominant horticultural cultivar, was compared with that of the fruitless
cultivar, Swan Hill using Burkard samplers. We conclude that airborne
pollen is not produced by the Swan Hill cultivar of Olea. The values
reported from the Swan Hill site reflect background levels of Olea pollen
for the Tucson area. Peak atmospheric Olea pollen concentrations occurred
at both sites on April 14, 1985. Pollen concentrations around the
Manzanillo site ranged from 7 to 6,196 grains/m3/day. At the Swan Hill site
daily totals were an order of magnitude less. Diurnal Olea pollen
concentrations for the Manzanillo site on the peak day varied from 1,000 to
18,133 grains/m3/hour. Both sites exhibited rapidly increasing pollen
concentrations at sunrise with a sharp increase between 1100 to 1300 hours
MST for the Manzanillo site. Seasonal pollen capture was also measured
using Tauber traps placed at various distances from each tree and within the
tree canopy. Both cultivars produced the same number of pollen
grains/anther (85,000). An unknown anatomical or physiological factor in
'Swan Hill' inhibits stomial rupture and thus pollen-shedding. Generally
15% of the anthers partially dehisce in the Swan Hill cultivar. The Swan
Hill cultivar can be useful in reducing pollen concentrations in urban
environments caused by planting of ornamental olive trees.

Introduction

        Pima County Arizona (USA) enacted a law banning the sale or

planting of Olea europaea to protect the health of "allergic" individuals.

Olea is a frequently reported allergen (5,12,15). Most published Olea studies discuss nutrient requirements and growth/fruit production patterns (2,3). Little is known about the physiological/environmental control of anther dehiscence and pollen-shedding, the aerodynamics of pollen dispersal, the efficiency of pollen capture by female flowers, or the minimum number of pollen grains needed to "set" fruit in Olea.

We compared pollen production and dispersal in the common Olea europaea L. cv. Manzanillo with the newly derived fruitless olive tree, Olea europaea L. cv. Swan Hill (4,13). Research strategy, methods of investigation, and pollen trap location have been described previously (11).

Methods

Our studies were conducted at 2 sites in urban Tucson, Pima County, Arizona (USA) during 1985. The study tree at the Swan Hill site was a 4 year old Olea europaea L. cv. Swan Hill. The study tree at the Manzanillo site was a 7 year old Olea europaea L. cv. Manzanillo. Trees were of similar in size and setting. Fig 1. illustrates tree and sampler locations (see 16 for additional information). Burkard and Tauber samplers were employed to capture airborne pollen at both sites from 2 Apr through 1 May, 1985. Twelve vertical transects were counted on the Burkard slides corresponding with odd hours. Pollen from Tauber traps was calibrated using Lycopodium spores .

Equivalent Olea branches were collected from a several individuals of both olive cultivars in urban Tucson. Undehisced anthers from both varieties were placed in FAA until analyzed for pollen content. Anthers were acetolyzed and pollen concentration values were obtained based on addition of Lycopodium spores .

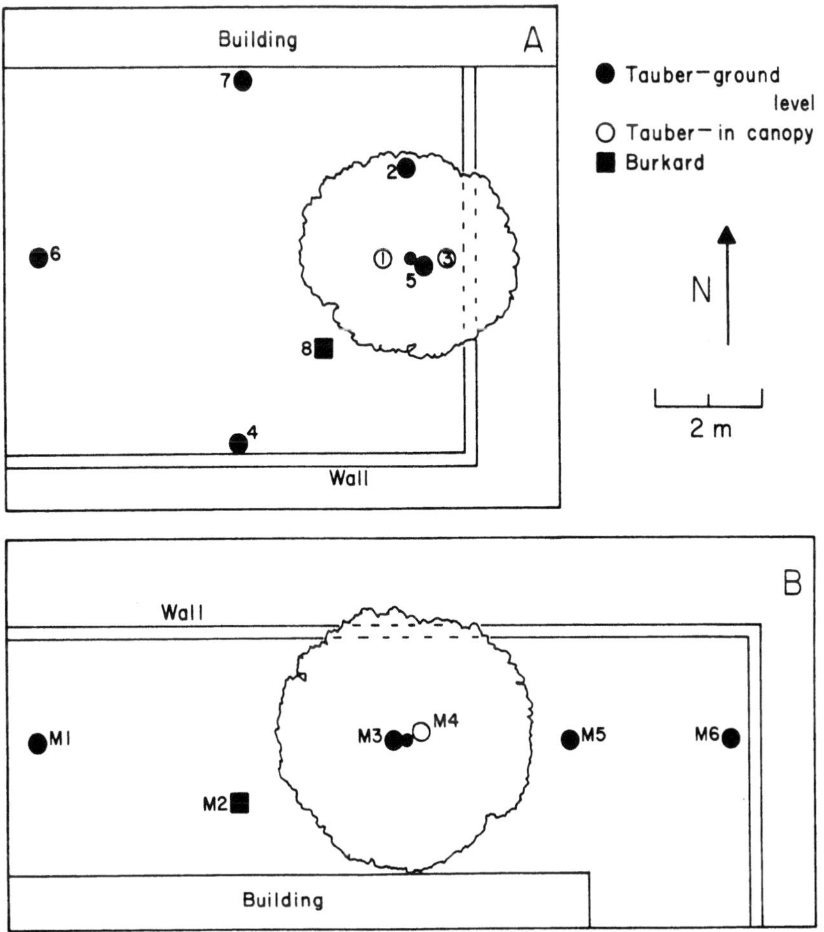

Fig. 1. Placement of Tauber and Burkard traps within the 2 urban courtyards (A=Swan Hill site B= Manzanillo site).

Results

The Manzanillo cultivar produced 83,743 $\pm$ 4416 grains per anther, whereas the 'Swan Hill' produced 85,492 $\pm$ 10,524 grains per anther. _Olea_ pollen concentrations are compared at the Manzanillo and Swan Hill sites in Fig. 2. _Olea_ pollen from the Manzanillo site illustrates the local input of

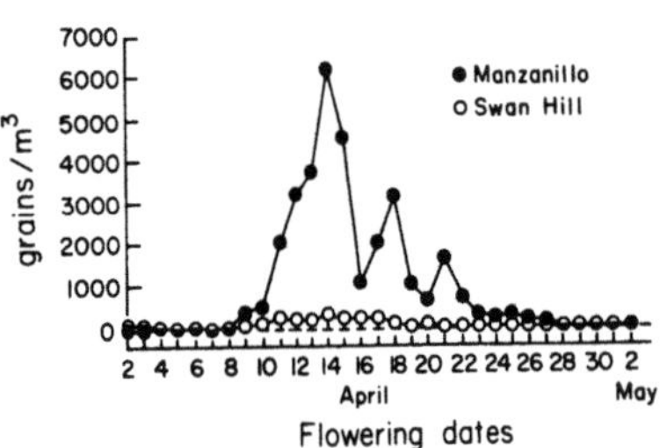

Fig. 2. Daily airborne pollen concentrations for _Olea_ _europaea_ cultivars.

Fig. 3. Hourly pollen concentrations for _Olea_ on 14 April 1985, the day of peak abundance.

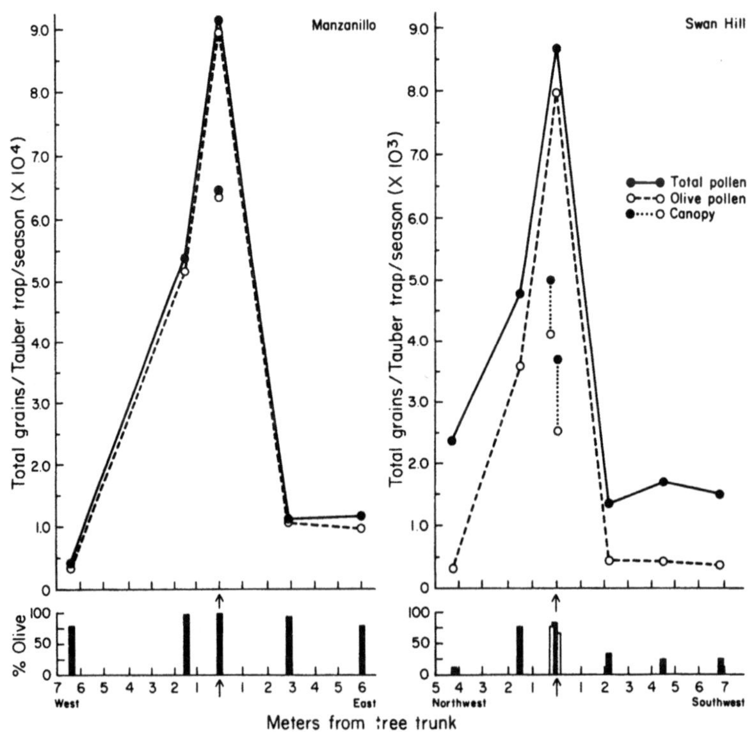

Fig. 4. Seasonal pollen concentrations (above) and percentages (below) for the Manzanillo and Swan Hill sites based on Tauber trap collections. The solid line represents the total pollen collected at ground level. The dashed line represents the total _Olea_ pollen collected. Dots within the curve represent pollen collected in the canopy.

pollen from a flowering tree. On the same scale the Olea pollen from the Swan Hill site is roughly equivalent to background pollen levels. It is readily apparent that the 'Swan Hill' tree yields little if any airborne pollen when compared with the 'Manzanillo' tree.

Field observations corroborate these measurements. Nine 'Swan Hill' trees were visited in urban Tucson. Only about 15% of the anthers on the trees appeared to partially dehisce and violent shaking of limbs elicited no "yellow pollen cloud." All 'Manzanillo' anthers examined had stomial ruptures and shaking of limbs produced large visible yellow clouds of pollen.

Fig. 3 illustrates the diurnal variability of Olea pollen concentrations at the 2 sites. Pollen concentration at the Manzanillo site is high with a midday peak; pollen concentrations at the Swan Hill site are uniformly low and represent background levels of Olea pollen.

Fig. 4 represents the pollen capture by the Tauber traps for both sites. The pollen capture is an order of magnitude higher at the Manzanillo site. The similarity of the total pollen curve and the Olea pollen curve is indicative of the dominant role local pollen production plays at the Manzanillo site. Olea percentage values at the Manzanillo site are similar under the tree and several meters away.

At the Swan Hill site the pollen totals are much lower and greater distance is found between the total curve at the Swan Hill site and the Olea curve. This emphasizes the lesser role played by olive. Lower levels of Olea generated by a less local source are also evident from the Olea percentages collected a short distance from the tree. High percentage values of Olea pollen in and directly under the canopy are due to whole anthers falling into the open mouth of the Tauber traps. The pollen input from the anthers is responsible for the pollen peak at both sites. Even with direct input of pollen from anthers, the concentrations at the 'Swan Hill' tree never achieve the concentrations reached at the 'Manzanillo' tree.

As might be predicted, pollen concentrations collected in the canopy are lower than those collected directly under the tree for both tree types. Reasons for this include; lower air movement within the canopy, flowers growing toward the apex of twigs, and less direct anther input into canopy Tauber traps than those located directly under the tree.

<u>Discussion</u>

Morettini (6,7,9) measured pollen concentrations at various distances between two orchards 701 m apart using gravity slides. Morettini's values ranged from 0.2 to 0.66 grains/cm^2/day. Later, Morettini and Pulselli (8) found that under ideal conditions olive pollen was sometimes carried up to 16 km. They stated that such long distance transport cannot be relied upon for fruit production. They emphasized the need for interplanting cultivars for cross-pollination. Morettini (6) recommended planting 1 "pollinizer" tree for every 10-15 fruit bearing trees.

Griggs <u>et al</u>. (3) performed further pollen dispersal assessments in orchards from 1951-54. They found a strong leptokurtic distribution of <u>Olea</u> pollen (i.e., pollen concentrations varied inversely with the square distance from the source). They recommended interplanting "pollinizer" varieties within orchards to enhance airborne pollen concentrations.

Recently, Cour and Villemur (2) have found a direct relationship between airborne pollen concentrations and fruit production in three crops including <u>Olea</u>. They report regional atmospheric pollen concentrations using a Cour trap of 10 to 80 grains/m^3 for the Montpellier region and estimate 2 to 8 million pollen grains produced per inflorescence.

Regional _Olea_ concentrations have been measured at 100 to 500 grains/m^3 in Tucson (10). These regional values are similar to the _Olea_ values from the Swan Hill site (Fig. 1).

Concentrations of atmospheric pollen determined using Tauber traps for forest of uniform species composition (i.e., all _Quercus_, all _Fagus_, etc.) have been determined by Andersen (1). These assume pollen collection 30 m from the forest edge, and range from 3300 to 48,400 grains/cm^2/yr. More appropriate numbers for comparison were generated by Tauber (14) above the canopy and in the trunk space of deciduous forests. Trunk space values of _Quercus_ were 6022 grains/cm^2/yr; for _Fagus_ they were 3105 grains/cm^2/yr. These are very similar to the canopy and trunk space values for 'Manzanillo' 3316 and 3571 grains/cm^2/yr. 'Swan Hill' values are 1 to 2 orders of magnitude lower.

Conclusions

We found equivalent numbers of flowers and quantities of pollen per anther in _Olea europaea_ L. cvs. Manzanillo and Swan Hill. We observed about 15% partial dehiscence in the 'Swan Hill' anthers. When 'Swan Hill' anthers split, individual grains or clumps of grains were not released. Our pollen collections indicate a 10 fold difference in pollen values at the 2 sites, with greater pollen concentrations at the Manzanillo site. Evidence is overwhelming that _Olea europaea_ L. cv. Swan Hill yields little if any airborne pollen. We believe that planting the 'Swan Hill' olive tree will not affect atmospheric pollen levels. Further, we find only background levels of _Olea_ pollen in closed patios where 'Swan Hill' is planted. Laws limiting sale of plant material should be carefully researched and written.

References

1. Andersen, S.Th. (1974) Wind conditions and pollen deposition in a mixed deciduous forest. II. Seasonal and annual pollen deposition 1967-72. Grana 14, 64-77

2. Cour, P. and P. Villemur (1985) Fluctions des emissions polliniques atmosphériques et prévisions de récoltes de fruits. Proceedings: 5 th colloque sur les recherches fruits. Bordeaux, 5-20.

3. Griggs, W.H., H.T. Hartmann, M.V. Bradley, B.T. Iwakiri and J.E. Whisler (1975) Olive pollination in California. Calif. Agr. Exp. Sta. Bulletin 859.

4. Hartmann, H.T. (1967) 'Swan Hill' -- a new ornamental fruitless olive for California. Calif. Agr. 21, 4-5

5. Kessler, A. (1958) Sensitivity to olive pollen (Olea europaea) as the cause of allergic disease. Dapim Refuiim 17, 3

6. Morettini A. (1941) L'incremento produttivo negli olivi Moraiolo e Frantoio con l'impiego di adatte varietà impollinatrici. L'Italic Agricola 9, 631-639.

7. Morettini, A. (1950) Olivicoltura, (Ramo Editoriale Degli Agricolotri, Rome).

8. Morettini, A. and A. Pulselli (1953) L'Azione del vento nel trasporto del polline dell'olivo, (Estratto dagli Annali della Sperimentazione Agraria, Rome).

9. Morettini, A. and A. Valleggi (1940). Ricerche sull'autofertilità e sull'autosteritita delle varieta di olivo nel Pesciatino. L'Olivicoltore 17, 12-17.

10. O'Rourke, M.K. (1986) The implications of atmospheric pollen rain on fossil pollen profiles in the arid Southwest. (Univ. of Arizona, Ph.D. Dissertation, Tucson).

11. O'Rourke, M.K. and S.L. Buchmann (1986) Pollen yield from olive trees cvs. Manzanillo and Swan Hill in closed urban environments. J. Amer. Soc. Hort. Sci. 111, (2).

12. Phillips, E.W. (1932) Pollen incidence in central Arizona. J. Allergy 3, 489-494

13. Rallo, L., G.C. Martin, and S. Lavee (1981) Relationship between abnormal embryo sac development and fruitfulness in olive. J. Amer. Soc. Hort. Sci. 106, 813-817

14. Tauber H. (1977) Investigations of aerial pollen transport in a forested area, Dansk Bot. Arkiv. Bd. 32, 1-121.

15. Unger, D.L. (1977) Allergy practice in the desert. Ann. Allergy 39, 300-301.

EXS 51:
Advances in Aerobiology
©1987 Birkhäuser Verlag Basel

ANNUAL AND DAILY VARIATION OF POLLEN FROM OLEA EUROPAEA L. IN THE ATMOSPHERE
OF CORDOBA (SPAIN) ALONG TWO YEAR OF SAMPLING

C.Galán (1), E.Ruiz de Clavijo (1), F.Infante (1) and G.Gallego (2)

(1) Faculty of Sciences, Department of Botany, University of Córdoba,
(2) Clinical Hospital, University of Córdoba, Spain.

Abstract

We carried out a study on the annual variation of pollen from Olea europaea L.
var. europaea in the atmosphere of the city of Córdoba along 1.982 and 1.983.
The samples were collected with a Burkard spore-trap. Pollen from this
species, one of the major agents causing pollinosis, occurs in very high
concentrations in the atmosphere of Córdoba, reaches its highest incidence in
may and the first half of june. We have also studied the variation along a day
of "Olive tree" pollen. The graphs obtained for both years are virtually
coincident and reveal the greatest occurrence of the grains at about two or
three in the afternoon.

Introduction

The city of Córdoba lies in a chiefly olie-growing region, pollen from Olea
europaea L. var.europaea being the most abundant in the atmosphere, together
with that of Platanus hybrida and Gramineae.
Several authors have quoted its allergenic nature: TAS & FEINBRUNN (1962)
consider it the second major cause of hey fever in Jerusalem, after Gramineae;
IZCO & al. (1972), STANLEY & LINSKENS (1974), SAENZ (1978), LEWIS & VINAY
(1979), SUBIZA (1980) and CANDAU & al. (1981) also regard it as a
allergy-inducer. We thus believe that the pollen of Olea europaea can be one
of the chief agents responsible for pollinosis in Córdoba.
Olea europaea has been included in the list of aerowandering species in various cities of the mediterranean area such as Barcelona, Spain (MONTSERRAT,
1953 and SUAREZ-CERVERA & SEOANE-CAMBA 1983); Lisboa, Portugal (PINTO DA
SILVA, 1954); Jerusalem (TAS & FEINBRUNN 1962); Montpellier, France (COUR &
al., 1973); Córdoba, Spain (DOMINGUEZ & al., 1984) and in the catalonian
towns, Spain (BELMONTE-SOLER & ROURE-NOLLA, 1985).
The aim of this work was to study the seasonal and daily variation of these
pollen grains in order to contribute data which may help allergenists of this
city in their investigations.

Material and methods

Córdoba is a semirural city of 265.000 inhabitants, located in the southwest of Spain, at 120 m above sea level, and with mediterranean climate featuring continental characteristics.
The sampling operation was carried out along two consecutive years (1982 and 1983) at the Faculty of Veterinary building by using a Burkard recording volu- metric spore-trap placed 15 m above the ground.
The daily pollen concentration was expressed as the number of grains found per cubic metre of air.
In determining daily variation, counts were performed hourly throughout the day, and the mean concentration corresponding to each hour provided a means of establishing an "ideal" day.

Results

The data obtained in the daily aeropallynological sampling of pollen from Olea europaea are plotted in Fig.1. Althouhg the study was carried out along two whole years, the graph only includes those months in which "olive" pollens are detected in the atmosphere.
Pollen grains from Olea europaea begin to be detected in the air of Córdoba around mid-april and attain their highest concentrations in may and the first half of june, starting to decrease gradually in the second half of this month until becoming virtually unappreciable.
We also recorded the daily variation of the concentration of "olive" pollen grains along the two above-mentioned years (Fig.2). Such variation was virtually the same along both years.

Discussion

As can be seen in Fig.1, pollen grains from Olea europaea only occur in the atmosphere of Córdoba in spring. Although the climatic conditions during this period were very similar in both years, there was a significant difference in the concentrations detected in the two years, which were higher in 1982 (maximum concentration 1387 grains/m^3 on may 24th) than in 1983 (maximum concentration 408 grains/m^3 on june 4th). This difference can be attributed to the fact that, as a rule, the "olive trees" of our region yield large and small masses of flowers alternately. Nineteen eighty-two was a year of excellent flowering (the "olive" harvest amounted to 668.073 tm), while in 1983 the harvest was somewhat poorer (184.327 tm).
The pattern of the daily variation of the concentration of pollen from Olea europaea was very similar in the two years. The occurrence of grains in the air reaches its peak at about two or three p.m. and is minimum early in the morning.

REFERENCES

BELMONTE-SOLER, J. & J.M. ROURE-NOLLA (1985) Contenido polínico de la at- mósfera de Cataluña. Resultados año 1.983. An. Asoc. Palinol. Leng. Esp. 2, 319-328.

CANDAU, P., J. CONDE & A. CHAPARRO (1981) Palinología en Oleaceae, incidencia de su polen en el aire de Sevilla, clínica de la polinosis. Bot. Macaronés. 8-9, 89-102.

COUR, P., D. DUZER & N. PLANCHAIS (1973) Analyses polliniques de l'atmophére de Montpellier. Document correspondant a la phénologie de la floraison de la vigne, en 1.972. Nat. Monspel. 23-24, 225-229.

DOMINGUEZ, E., J.L. UBERA & C. GALAN (1984) Polen alergógeno de Córdoba, (Publ. Monte de Piedad, Córdoba).

IZCO, J., M. LADERO & C. SAEZ DE RIVAS (1972) Flora alergógena de España. Distribución, descripción e interés médico-alergológico de las especies responsables de síndromes alérgicos. An. Real Acad. Farm. 38, 521-570.

LEWIS, W.H. & P. VINAY (1979) North American Pollinosis due to insect-pollinated plants. Ann. Allerg. 42, 309-318.

MONSERRAT, P. (1953) El polen atmosférico de Barcelona en 1.951. Inst. Biol. Apl. 13, 121-128.

PINTO DA SILVA, Q.G. (1955) Le conteu pollinique de l'air á Lisbonne. Agron. Lusit. 5-16.

SAENZ, C. (1978) Polen y Esporas. (Ed. H.Blume, Madrid).

STANLEY, J. & C. LINSKENS (1974) Pollen biology, biochemistry, management. (Springer, New York).

SUAREZ-CERVERA, M. & J.A. SEOANE-CAMBA (1983) Estudio del contenido polínico de la atmósfera de Barcelona según un nuevo método de filtración. Collec. Bot. 14, 587-615.

SUBIZA, E. (1980) Incidencia de granos de pólenes en la atmósfera de Madrid. Método volumétrico. Allergol. et Immunopathol. (supp. 7).

TAS, I. & N. FEINBRUNN (1962) A survey of airborne pollen in Jerusalen. Act. Allergol. 17, 246-267.

Prof. Carmen Galán-Soldevilla, Departamento de Botánica, Facultad de Ciencias, Universidad de Córdoba, 14004 Córdoba, ESPAÑA.

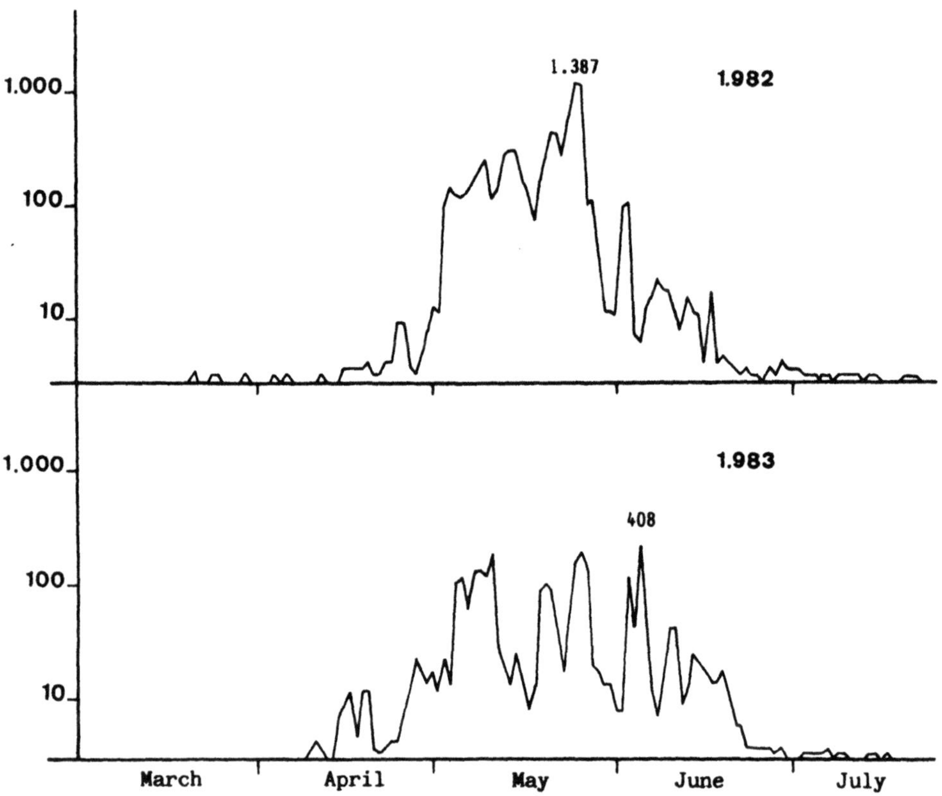

Figure 1. Seasonal variation of pollen from Olea europaea L.

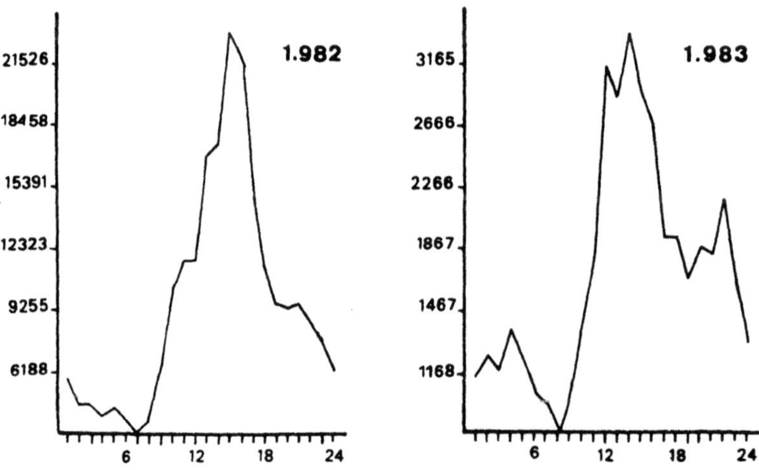

Figure 2. Daily variation of pollen from Olea europaea L.

EXS 51:
Advances in Aerobiology

CHESTNUT POLLEN COUNTS RELATED TO PATIENTS POLLINOSIS IN
PARIS

J-P. Sutra, M-R. Ickovic, H. De Luca, G. Peltre, B. David

Institut Pasteur, Paris, France.

Introduction

From the pollen counts recorded in Paris in 1984 and 1985,
it has been observed that the chestnut tree pollination is
very important. It follows the birch and the grass pollen
seasons. In this preliminary work, we studied the
incidency of the chestnut tree pollen on the patients
allergic symptomatology in patients sensitive to grass
pollens and still sick until mid July.

Material and Methods

Pollen grains were counted after their capture by a Hirst-
Burkard trap placed in the Pasteur Institute inside Paris.
Temperatures, rainfalls and winds were recorded in Paris.
Pollen sensitivities were checked on 67 patients by skin
prick tests, confirmed by RAST values to chestnut pollen
extracts.
The heterogeneity of the allergens in chestnut and 8 other
tree pollens were compared by immunoprints (Peltre et al.,
1982).

Results and Discussion

The chestnut pollen season lasts 20 days in July with peak

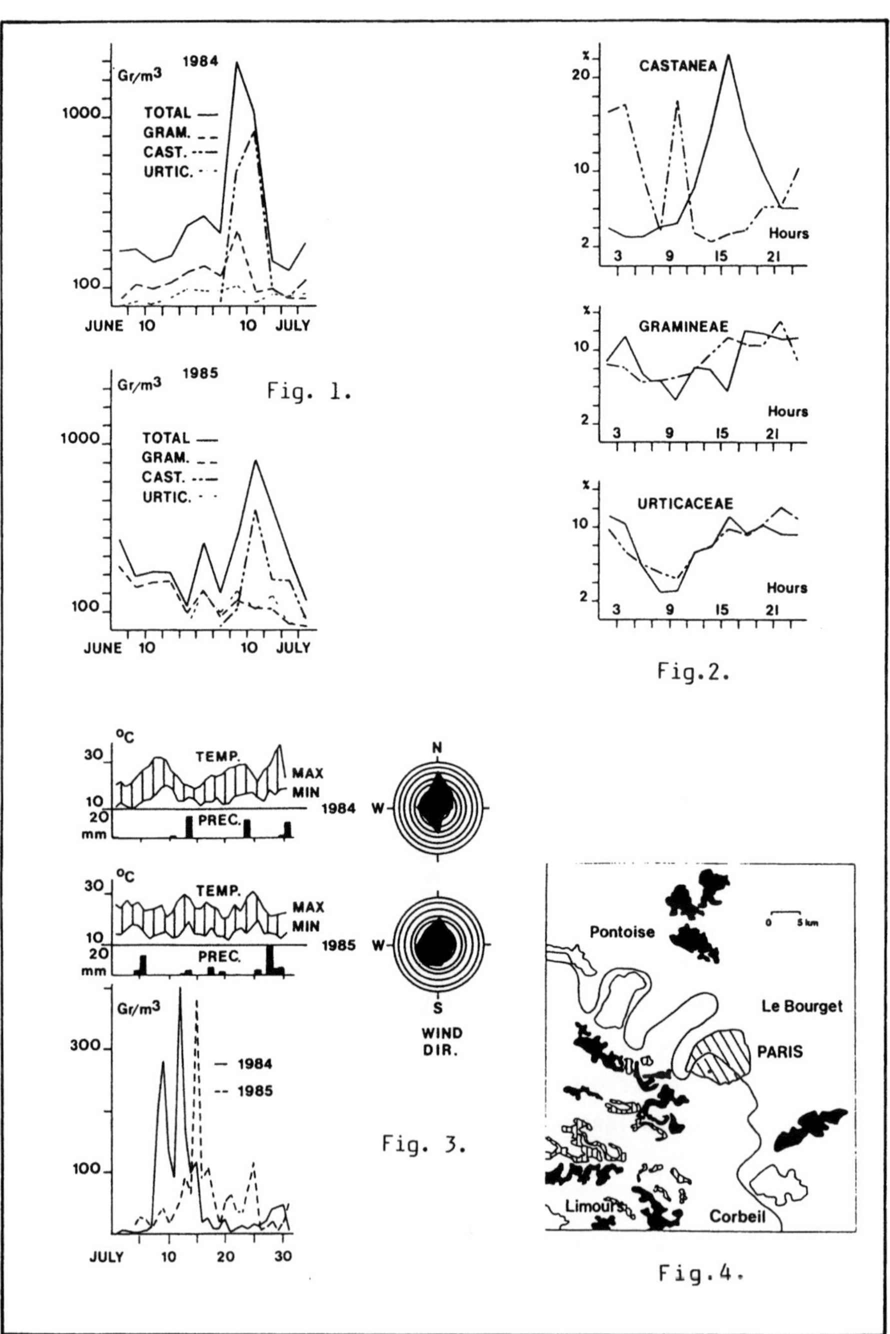

1984
Gr/m3
1000
TOTAL
GRAM.
CAST.
URTIC.
100
JUNE 10 JULY
Fig. 1.
1985
Gr/m3
1000
TOTAL
GRAM.
CAST.
URTIC.
100
JUNE 10 JULY
CASTANEA
20
10
2
Hours
3 9 15 21
GRAMINEAE
10
2
Hours
3 9 15 21
URTICACEAE
10
2
Hours
3 9 15 21
Fig.2.
°C
30
TEMP.
MAX
10
MIN
20
1984
PREC.
mm
°C
30
TEMP.
MAX
10
MIN
20
1985
PREC.
mm
Gr/m3
300
1984
1985
100
JULY 10 20 30
N
W
S
WIND
DIR.
Fig. 3.
Pontoise
0 5 km
Le Bourget
PARIS
Limours
Corbeil
Fig.4.

values up to 600 grains/m^3 on 5 day periods. In 1984, these pollen counts were higher than in 1985 (Fig. 1). The influence of the weather conditions on the pollination was analyzed. It was obvious that rainfalls dropped the pollen counts (Fig. 2), that marked increases in temperature induced peaks in pollen numbers as we showed for the grass pollen grains (Donini et al., 1985). Means of pollen captured every two hours periods over July for grasses and nettles showed only minor differences between 1984 (solid line) and 1985 (doted line) (Fig. 3) reflecting the variation of temperature and hygrometry due to the day and night alternance. The totally different pattern observed for chestnut is mainly explained by the dominant winds at the peak of the pollination : blowing from south and south-west and bringing to Paris the pollens from forests located in south west of the parisian area and where chestnut trees are present (Fig. 4) (high chestnut tree frequency in black, medium in hachures, low in small dots). This effect of wind on the small chestnut pollen grains was also showed by Mandrioli et al. (1980) and Frenguelli et al. (1983).
We studied 67 patients. 63 of them were sensitive to gras-ses. Figure 5 represents the patient numbers proportional to the circle areas bearing the pollen names. 44 patients were sensitive both to grasses and to birch pollen. 24 pa-tients had positive prick skin tests to chestnut pollen (CH), 14 of them having positive RAST values to this pollen. Figure 6 gives more precisions on this last sensitivity. 15 patients are sensitive to oak pollen, 11 of them being also sensitive to birch pollen. 5 patients are sensitive to plane tree as well as to oak, birch and chestnut. 3 patients are sensitive to lime tree pollen,

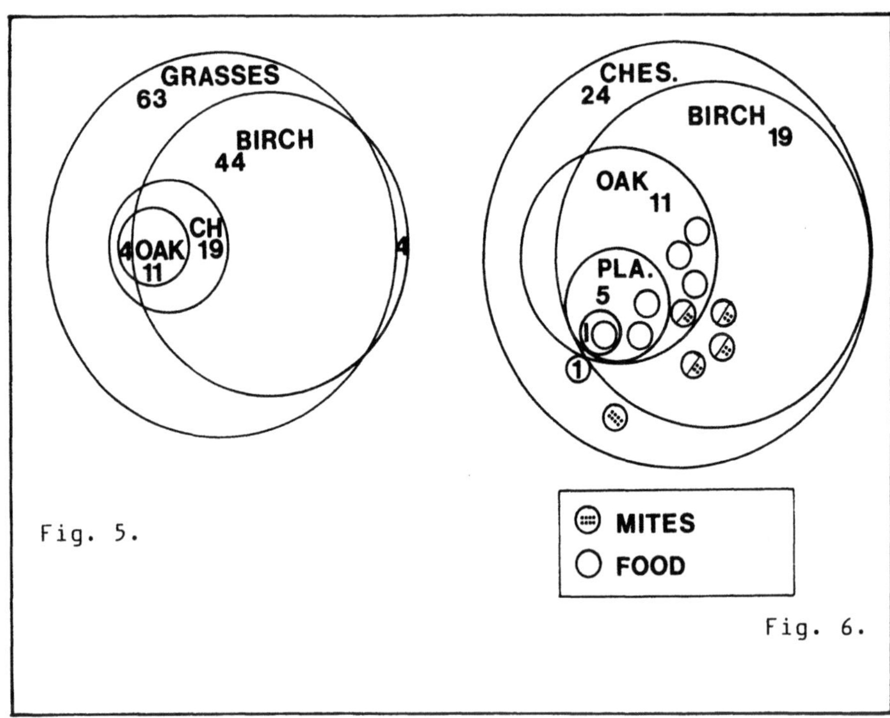

Fig. 5.

Fig. 6.

one of them being not sensitive to plane tree, oak and birch pollens. Other allergies to food or mites are noted, each small circle representing one patient.

Are these different sensitivities observed in individuals due to the existence of a common allergen to the different pollens or to unrelated sensitivities added in the polysensitive patients ? We studied by immunoprints the allergens from chestnut, birch, oak, plane-tree, lime-tree, hornbeam, poplar and hazel recognized by patients. We can find 3 groups. Group A recognized a common allergen in birch,oak and chestnut. Group B patients recognized in addition to the previous allergen some others which were species specific. Finally, group C patients recognized even more heterogeneous allergens unrelated to each other.

Conclusion

Pollen counts analysis led us to study the clinical incidency of chestnut tree pollen, abundant but not usually considered as a major sensitizing allergen source. This work induced a careful survey of the co-occuring sensitivities to other tree pollens and grass pollens. A preliminary immunochemical research showed the existence of crossreacting allergens in several pollen sources and this type of research subject has to be developped in the future.

References

- Donini,D.; Galiez,N.; Herman,D.; Sutra,J.P. (1985). comptes polliniques parisiens : saison 1984 - Corrélation avec la clinique pour les pollens de graminées. Rev. fr. Allergol., 25 (n°3), 145-149.
- Frenguelli,G.; Mincigrucci,G.; Romano,B.; Bricchi,E. (1983) Census of airbone pollen grains in the atmosphere of Ascoli Piceno (Central Italy). New Phytol, 95, 147-151.
- Mandrioli,P.; Scarani,C.; Tampieri,F.; Trombetti,F. (1980) Airbone pollen source modelling. In : the 1st Intern. Conf. on Aerobiology, 393-406. Munich 13-15 August 1978, Proceedings; E.Schmidt Verlag, Berlin.
- Peltre,G.; Lapeyre,J.; David,B. (1982) Heterogeneity of grass pollen allergens recognized by IgE antibodies in human patients sera by a new nitrocellulose immunoprint technique. Immunol. Letters, 5, 127-131.

J-P. Sutra, Institut Pasteur, 75724 Paris Cedex 15, France.

POLLEN COUNTS OF RAGWEED AND MUGWORT (COUR COLLECTOR) IN 1984 MEASURED AT 12 METEOROLOGICAL CENTERS IN THE RHÔNE BASSIN AND SURROUNDING REGIONS.

C.Déchamp* and P.Cour**

*Association Française d'Etudes des Ambroisies (AFEDA) F 69780 St Priest
** Laboratoire de Palynologie CNRS F 34060 Montpellier

Introduction

The analysis of pollen (p) contained in the atmosphere, recorded at the same place over several annual cycles, has enabled the authors to measure the spread, the stability or the regression of p emissions produced by the various plants present in Lyon-Bron. Thanks to the Public Health Authorities in the Rhône District, an uninterrupted study of the p in the Lyon-Bron Meteorological Center has been conducted for the past 5 years. The analysis have shown that the atmosphere in Lyon is indeed polluted by Ragweed (R) p and that concentrations reach or supersede the levels measured in the Bethesda and Baltimore areas situated on the East Coast of the USA (2, 5, 6).
R first appeared in Europe around 1870 : it arrived in the Lyon area in the 1930s, along with potato plants and grain seeds imported from North and South America (I have spoken to the farmers who bought these supplies and saw the weed growing in their fields). It appears that American vehicles were brought by air to the Lyon-Bron Airport in 1944 : this hypothesis is unproven, but is a widespread legend among the local population. Now, the entire area where these potatoes and grains were planted and where the American tanks were said to have landed (damaged and unusable tanks were subsequently stored in the same area), became the site of a vast building development project in 1964. It is known that R grows on recently cleared

ground and made soil, so the building sites offered favourable conditions.
An immunological survey conducted by D.Marsh (7) has confirmed our own
conclusions, drawn from observation, that is , that Ambrosia
artemisiaefolia is the only variety to be found in very large quantities.
Virtually no other varieties of R have been identified.

1. Aims of the survey

**1.1 To detect the presence of R p at the outer limits of the zones
traditionally contaminated by this weed,**that is the area mentioned by a
survey conducted among allergists (French Society of Allergology) in 1981
(3).
**1.2 To continue taking p counts in future years in any regions where R p
has been detected in significant quantity.**

2. Study methods

A continuous sample of p content in the air is obtained by a Cour
collector. (1) Installation of p collectors was dependent on many factors of
an economic, administrative or practical nature. Ten collectors were set up
in July 1984 at the weather stations chosen for this survey (Fig.1). The
network of AFEDA was reinforced by two additional stations, one at Gréoux
les Bains, the other at Clermont-Ferrand. These stations run in cooperation
with our association. All these stations, except Le Puy, Lus la Croix
Haute, Tarare are situated at an altitude of less than 500 meters.
Since summer pollinosis are due not only to R but also to Mugwort (M), the
p counts of both these Compositae species were recorded at the same time.
We have shown in 1982 that patients suffering from R pollinosis present an
associated pollinosis to M in 65% of cases. (4) Pollination of these two
weeds almost takes place at the same time .In Lyon, Artemisia vulgaris
before R.(the end of July), Artemisia annua at the same time et after R (6)
It should be remembered that M species are present throughout France,
unlike R.

<u>3. Results</u> (Fig.2)

We classify the towns studied by order of intensity of atmospheric
pollution by R p. The results cover analysis of filters exposed to p during
weeks 32 to 40 : 30 July to 1st October. As compared to the calendar year
for 1984, this numerotation is "out" by a week, due to fact that week-by-
week, comparisons with the data obtained at LYON-BRON during previous years
made it impossible to add on a 53rd w to "even things up". Results are
expressed in average number of grains in one cubic meter of air during one
week. The order of intensity of pollution by M p in the same regions is
shown in brackets : LYON-BRON (1), MONTELIMAR (5), AMBERIEU (3), GRENOBLE-
SAINT GEOIRS (8), NEVERS (4), CLERMONT-FERRAND (2), CHAMBERY (6), SAINT
ETIENNE-BOUTHEON (9), MACON (7), GREOUX LES BAINS (11), TARARE (10), LUS LA
CROIX HAUTE (12).
LYON came first of all the stations covered by the survey for both R and M
p, with a maximum of 175 grains (g) of R p detected in week (w) 37 (3 to 10
September) and a maximum of 69 g of M p observed in w 40 (end Sept.).
MONTELIMAR came second with a peak of 94 R p g recorded in w 36 and came
5th for M pollution with a peak of 20 p g in week 33 (mid-August). AMBERIEU
came third with a maximum of 40 R p g recorded in w 37, same position for M
with a peak of 25 p g in w 33. GRENOBLE came fourth with a peak of 23 R p g
in mid September, 9 M p g were recorded in week 34 (8th position). NEVERS
came fifth, with 21 R p g recorded in week 37 and 20 M p g during the w 13-
20 August (4th position). CLERMONT-FERRAND came only in 6th position, this
region showed the earliest appearance of R p, with a peak of 30 g in mid
August and came 2nd for the emission of M p, with 25 g recorded in the w of
6 to 13 August. CHAMBERY scored 10 R p g in weeks 35 and 37; 12 M p in w
34. SAINT ETIENNE reached a level of 9 R p from 3 to 10 September, 7 M p in
week 34. MACON scored 8 R p in week 35 and virtually nothing after, 11 M p
in the same w. GREOUX LES BAINS recorded 4 R p in early September virtually
no M p. TARARE noted 3,7 R p from 3 to 10 Sept., 3 M p ,the preceding w.
LUS LA CROIX HAUTE had the least polluted atmosphere as regard both R and M
: the p count was less than 1g at all times.

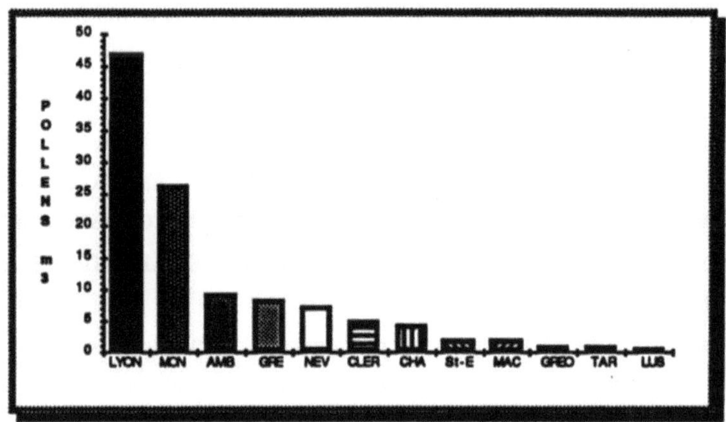

Fig. 1 Distribution by location of average density of ragweed pollen emission into the atmosphere during the period 30/07-1/10/1984.

Conclusion

Ragweed has caused pollinosis in France especially in the Lyon area only since about 1964. Works for town extension are the main factor. Nevertheless, in spite of research, French Association for the Study of Ragweed, has not established the reason why this weed grows in such large quantities only in this region of France. A fact is sure, this plague is increasing and spreading to the South.

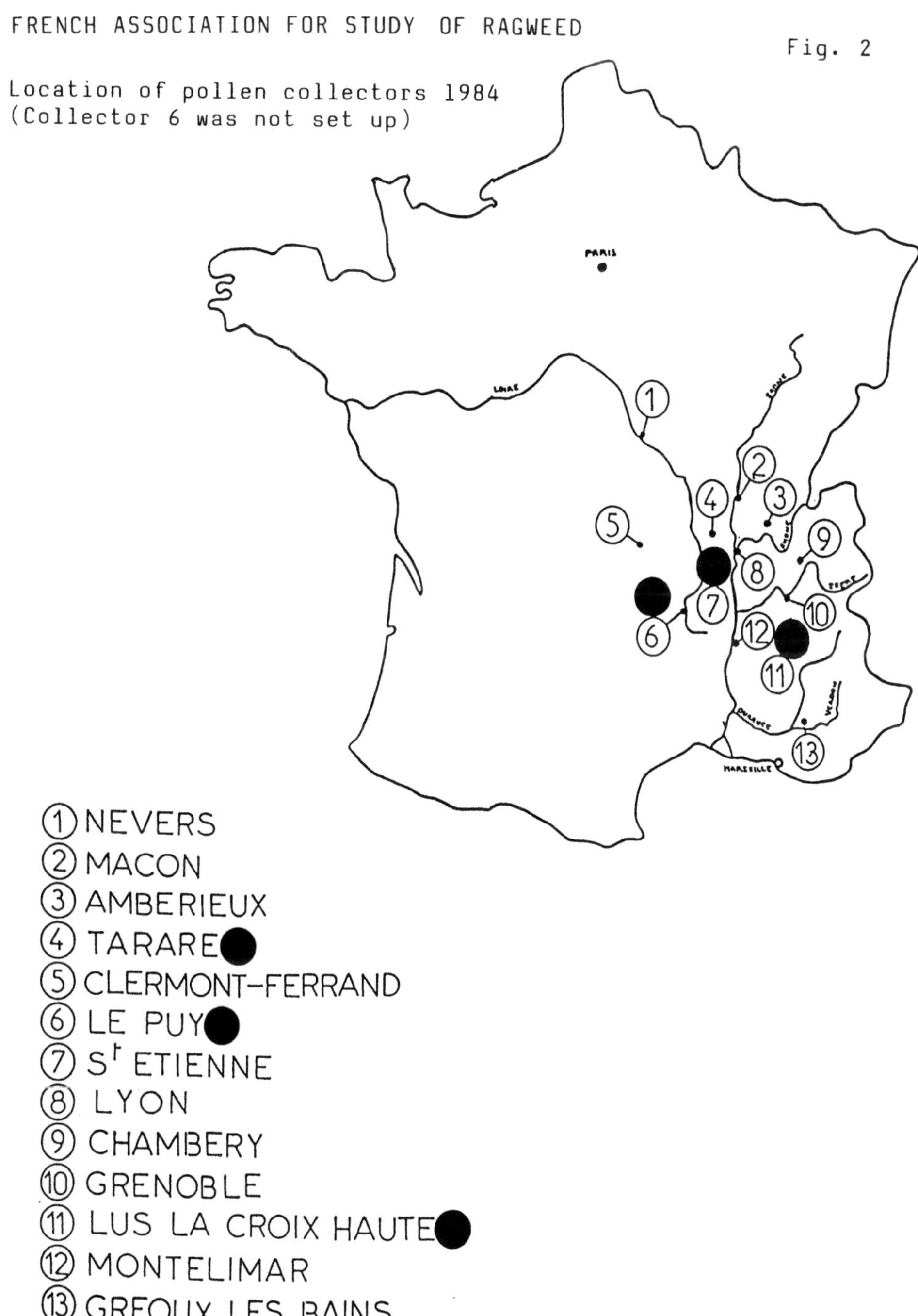

① NEVERS
② MACON
③ AMBERIEUX
④ TARARE ●
⑤ CLERMONT-FERRAND
⑥ LE PUY ●
⑦ St ETIENNE
⑧ LYON
⑨ CHAMBERY
⑩ GRENOBLE
⑪ LUS LA CROIX HAUTE ●
⑫ MONTELIMAR
⑬ GREOUX LES BAINS

● Collectors sited at over 500 m altitude

4. Acknowledgments

-Financial support: French Ministery of Research and Technology (1984)
Mr F. JUILLET and Mrs SCHLEICHT Ex Rhône Alpes Regional Direction for
Research and Technology
-Mrs J.CAMPOS, Mr P.RICHARD (treatment and pollen analysis), Laboratory of
Palynology, USTL, F 34 060 MONTPELLIER CEDEX
-Mr J.DROUILLET, Director of the Rhône Alpes Regional Meteorological
Center.

5. References

1. Cour, P. (1974) Nouvelles techniques de détection des flux et retombées
polliniques : étude de la sédimentation des pollens et des spores à la
surface du sol. Pollens et Spores 16, 1, 103-141.

2. Déchamp, C., Hoch,D., Perrin, L.F., Loublier, Y., Cour, P. (1982). Place
des pollens d'ambroisies dans la Ière étude informatisée du calendrier
pollinique de Lyon-Bron. (1982) Lyon Médical 253, 173-178.

3. Déchamp C., Boyer J., Perrin L.F., Touraine R. (1983) Résultats d'une
enquête auprés des allergologues français sur la pratique des tests aux
pollens d'ambroisies en France. Rev.Franc. d'Allergol. 23, 189-192.4.

4. Déchamp,C., Cohen A., (1985), Etude multicentrique informatisée 1982 de
la pollinose due aux ambroisies. Second meeting of the AFEDA, Domaine de
Rajat, F 69780 Saint Pierre de Chandieu, OCT.27, 1984. Allergie et
Immunologie 17, 8, 464-480.

5. Déchamp C., Chouraqui M., (1985) Pollens d'ambroisies et calendrier
pollinique de Lyon-Bron en 1983. Etude informatisée. Lyon Médical 253, 6,
173-176.

6. Déchamp C., Cour P. (1985) Analyse quantitative des émissions
polliniques atmosphériques des ambroisies et des armoises en 1984 dans 12
centres météorologiques du bassin rhôdanien et des régions limitrophes.
Second meeting of the AFEDA, Domaine de Rajat, F 69780 Saint Pierre de
Chandieu, OCT.27, 1984. Allergie et Immunologie 17, 8, 438-461.

7. Déchamp C., Cour P., Bousquet J., Marsh D.G. (1986) Relationship between
atmospheric levels of Amb a I AgE and short ragweed pollen grains in
central FRANCE. Forty second annual meeting of the AAAI, NEW ORLEANS, LA.,
USA, MAR. 21-26. J ALLERGY IMMUNOL 77 (1 Part 2). 198.

6. Author's adress

Docteur Chantal DECHAMP ASSOCIATION FRANCAISE D 'ETUDES DES AMBROISIES B.P.
622 69800 SAINT PRIEST CEDEX FRANCE.

EXS 51:
Advances in Aerobiology
©1987 Birkhäuser Verlag Basel

ALLERGIC REACTIVITY TO <u>PARTHENIUM HYSTEROPHORUS</u> POLLEN:
AN UNRECOGNIZED TYPE I ALLERGEN

H. James Wedner, Vincent E. Zenger, and Walter H. Lewis
Department of Medicine, Washington University School of Medicine, and
Department of Biology, Washington University, St. Louis, MO., USA.

Introduction

<u>Parthenium</u> (feverfew) is a small genus of Western Hemisphere plants of the
Asteraceae included in the subtribe Ambrosiinae along with such well known
genera as <u>Ambrosia</u> (ragweed) and <u>Iva</u> (marsh elder). One of the most common
and biologically isolated species is <u>P. hysterophorus</u> L. (Santa Maria
feverfew), a ubiquitous annual weed native to the Gulf of Mexico region and
West Indies but now widely distributed from Texas to southern Florida, and
inland about 300 miles, especially in areas disturbed by humans (PICMAN and
TOWERS 1982). <u>P. hysterophorus</u> (Ph) has been introduced elsewhere; for
example, to the Poona region of India, and its subsequent spread in that
subcontinent has given rise to a serious medical hazard as a major source of
allergic contact dermatitis (ARLETTE and MITCHELL 1981).

   The biology of Ph is poorly understood and what little has been
reported is conflicting. Pollination has been described as entomophilous
with insects attracted to the pollen-containing disk flowers by glandular
hair secretions (ROLLINS 1950). Yet others found no insect visitors to the
plants (M.C. Johnston, Univ. of Texas, pers. comm.; W.H. Lewis, pers.
observ.), and still others reported that the species "produces an enormous
quantity of pollen (an average 624 million/plant) which is carried away at
least for short distances in clusters of 600-800 grains..." (TOWERS and
MITCHELL 1983) and that its pollen becomes airborne to great heights in

significant amounts either as individual grains or in clumps (SEETHARAMIAH ET AL. 1981).

Methods

Pollen from $\underline{P.}$ $\underline{hysterophorus}$ and $\underline{Ambrosia}$ $\underline{psilostachya}$ DC. (Ap), both common in the Texas Gulf coast, was suspended in 10 mM phosphate buffer and gently stirred at a ratio of 1 g of pollen to 10 ml of buffer for 48 hr at 4°C. Pollen extracts were clarified by a combination of filtration and centrifugation.

The 1:10 pollen extracts were coupled to cyanogen bromide derivatized paper disks at a ratio of 2.0 ml of extract to 2.0 g of disks using the method of CESKA (1983). The derivatized disks were blocked with 0.5 M ethanolamine in 0.1 N $NaHCO_3$ for 3 hr, washed, and stored in 0.05 M phosphate buffer pH 7.4 containing 0.9% NaCl, 0.3% human serum albumin, 0.05% NaN_3, and 0.5% Tween 20 (assay buffer). The RAST was performed using 50 µl of patient sera diluted 2:3 with assay buffer. Bound IgE was detected with [125] anti-human IgE using polyclonal epsilon chain specific anti-IgE (Bio-Rad).

Results

Eight patients from Corpus Christi, TX and 10 patients from Austin, TX were studied using the RAST as outlined above (table I). When the actual RAST counts bound are compared to the Pharmacia standard disks, 9 out of the 10 patients from the Austin area had RAST scores of 2+ or greater to Ap and the same number of patients was also sensitive to Ph. However, patient 10 had a positive RAST score to Ap but not to Ph, while patient 6 was RAST positive to Ph and not to Ap. All patients in the Corpus Christi group were sensitive to Ap at a level of 2+ or greater and all were also sensitive to Ph. Patients 2, 4, and 6 had higher counts bound to Ph than to Ap, while patient 3 had identical counts bound to both disks.

The marked similarity between the number of patients who were sensitive to Ap and Ph in both the Austin and Corpus Christi areas led us to

Table I. RAST to <u>Ambrosia</u> <u>psilostachya</u> and <u>Parthenium</u> <u>hysterophorus</u> in patients from Austin and Corpus Christi, Texas

		_A_. psilostachya		_P_. hysterophorus	
		CPM Bound	% Counts Bound	CPM Bound	% Counts Bound
I.	Austin				
	1	23105	27.4	10737	12.7
	2	25885	30.7	19268	22.9
	3	8379	9.9	5018	5.9
	4	29744	35.3	15137	17.9
	5	8577	10.2	5698	6.8
	6	3739	4.4	9513	11.3
	7	35351	41.9	20379	24.2
	8	42338	50.2	28896	34.3
	9	43879	52.1	36793	43.7
	10	5936	7.0	1924	2.3
II.	Corpus Christi				
	1	15799	18.7	6228	7.4
	2	34141	40.5	39694	47.1
	3	19442	23.1	19680	23.3
	4	14902	17.7	31756	37.7
	5	21347	25.3	10943	13.0
	6	8687	10.3	12304	14.6
	7	43204	51.3	33240	39.4
	8	37755	44.8	18663	22.1
III.	Pharmacea Reference Disks			Counts Bound	% Counts Bound
	A			33095	39.3
	B			16384	19.4
	C			5476	6.5
	D			3583	4.2

analyze the data further in order to be assured that the data did not simply indicate cross-reactivity between Ph and Ap pollen proteins. We, therefore, compared the RAST scores for the patients from the Austin and Corpus Christi areas. Statistical analysis from scattergrams for absolute counts showed that for the Austin group, where there are as many or more Ap than Ph plants, the slope of the regression line is significantly slanted towards ragweed sensitivity (S=0.33) and there is a remarkable concordance between Ap and Ph RAST values (r=0.921, t=6.71, p=0.00015). In contrast, data from Corpus Christi, where there is a far greater preponderance of populations of Ph, demonstrated no correlation between Ap and Ph RAST values. The slope of the regression line is intermediate (S=0.46), and there is no statistical relationship of the RAST values (r=0.55, t=1.63, p=0.152).

To further explore cross-reactivity between Ph and Ap pollen proteins, sera from 3 patients from Corpus Christi were incubated with paper disks coupled to Ph or Ap pollen extracts in the presence of graded concentrations of the same or the alternative pollen extract. Three patterns of <u>RAST inhibition</u> emerged. The first pattern (patient 8, table I) demonstrated a serum that was competed off both the Ap and Ph disks more efficiently by Ap than Ph. The second pattern (patient 3, table I) was a serum in which Ph pollen extract was significantly better at preventing binding to the Ph disk, while Ap was significantly more potent in inhibiting binding to the Ap disk. The third pattern (patient 4, table I) showed antiserum in which Ph pollen extract was more potent in preventing binding to both the Ph and Ap coupled disks.

Discussion

We analyzed sera from 18 patients, conceivably exposed to high levels of Ph and Ap pollen, for sensitivity by RAST and RAST inhibition testing. Results demonstrated that 17 of the 18 patients were RAST positive to Ph and a like number were also sensitive to Ap. When the data were analyzed by scattergram analysis there was a significant correlation in the 10 patients from Austin, but there was no correlation in the group from Corpus Christi. This suggests that there was significant cross-reactivity between the two pollen extracts, but that in areas where the frequency of Ph, and by analogy the amount of airborne Ph pollen, is very large, sensitivity to Ph might predominate. Nevertheless, simple RAST did not allow us to differentiate individuals sensitized to Ph from those sensitized to Ap.

The RAST inhibition analysis did demonstrate that there were significant differences among the sera tested. The sera were selected based upon differences in the absolute number of counts bound in the initial RAST study. When this was used as the criterion we could identify individuals whose sera showed greater avidity for the Ph antigens, those whose sera showed greater avidity for Ap antigens, and those who appeared to have antibodies directed against antigens found in both the Ph and Ap extracts.

A number of conclusions can be drawn from this analysis. First, it is clear that there is significant cross-reactivity between Ph and Ap pollen. Secondly, there appear to be patients in whom the immune response is largely directed against Ap pollen with cross-reactivity to proteins contained within the Ph extract. Another group of patients appear to have high levels of antibody directed against both Ap and Ph pollen, while a third group of patients appear to have made an immune response which is largely directed against Ph pollen proteins.

Based on the RAST inhibition studies, performed on this relatively small group of patients, approximately 30% of the entire group were allergic to Ph while another 10% showed sensitivity to both Ph and Ap. The remainder were individuals whose major reactivity was directed against Ap with cross-reactivity against the Ph proteins. In addition to demonstrating that a significant portion of the population is indeed sensitive to Ph, either solely or in combination with Ap sensitivity, these data illustrate one of the difficulties in attempting to assess overall allergic sensitivity using RAST when patients are present in an environment with pollen containing cross-reacting proteins. Indeed simple RAST to Ambrosia in the group from Austin and Corpus Christi suggested that all of these patients were in fact sensitive to ragweed. It was the failure of a group of these patients to respond to appropriate immunotherapy and the finding of the prevalence of Ph within the area that suggested that some of these patients may in fact be sensitive to Ph rather than Ambrosia. In summary, pollen of Ph is an important environmental allergen with a potency approximating that of Ap. Patients from areas where Ph and Ap are common may be sensitive to one or both of these weeds and our data demonstrate that sensitivity to Ph is not simply the results of cross-reactivity between pollen proteins from these two species. Thus, Ph represents a unique allergen whose importance in type I hypersensitivity has been largely overlooked.

References

Arlette J. and Mitchell, J.C. (1981) Compositae dermatitis. Current aspects. Contact Dermat. 7, 129-136.

Ceska, M. (1983) Radioimmunoassay of IgE using paper disks. Meth. Enzymol. 73, 646-656.

Picman, A.K. and Towers, H.N. (1982) Sesquiterpene lactones in various populations of Parthenium hysterophorus. Biochem. Syst. Ecol. 10, 145-153.

Rollins, R.C. (1950) The guayule rubber plant and its relatives. Contr. Gray Herbarium (Harvard Univ.) 172, 1-73.

Seetharamiah, A.M., Viswanath, B. and Rao, P.V.S. (1981) Atmospheric survey of pollen of Parthenium hysterophorus. Ann. Allergy 47, 192-196.

Towers, G.H.N. and Mitchell, J.C. (1983) The current status of the weed Parthenium hysterophorus as a cause of allergic contact dermatitis. Contact Dermat. 9, 465-469.

Dr. H. James Wedner, Division of Allergy and Immunology, Department of Internal Medicine, Washington University School of Medicine, St. Louis, Missouri 63110, USA.

3. Different aeroallergens

4. Special allergens:
    Fungus spores

AEROALLERGENS IN NEW YORK INNER-CITY APARTMENTS OF ASTHMATICS

I.F.Goldstein*   C.E.Reed**   M.C.Swanson**   J.S.Jacobson***

  * Columbia University School of Public Health,
     Epidemiology, New York, NY USA
 ** Mayo Medical School and Allergic Diseases Clinic,
    Rochester, MN USA
***Dept. of Neurology, Cornell University Medical College
   New York, NY USA

For presentation at the International Aerobiology Conference
         Basel, Switzerland, August 6, 1986

Introduction

Asthma is a heterogeneous condition of multifactorial etiology.
Worldwide, the prevalence of asthma varies from population to
population, and with such factors as inbreeding, low altitude,
damp climate, urbanization, and industrialization. In New York
City, asthma is the second major cause after accidents of
pediatric hospitalizations[1] and is the most frequent reason
for school absenteeism.[2]
        Some data suggest that asthma is more prevalent,[3,4,5]
more severe,[6] and more of a drain on health care resources[8]
in poor families and in the inner city than elsewhere. In New
York City for example, the lower socio-economic districts of
Washington Heights and Harlem whose population is mostly black
and Hispanic has been shown to have a 4-fold rate of asthma
compared to other areas of the northeastern United States.
        We at the Environmental Epidemiology Research unit at
Columbia University have been studying the epidemiology of New
York inner-city asthma for the past 12 years. We have searched
for environmental factors that might be responsible for the high
prevalence of asthma in that population and/or might constitute
the immediate triggers of asthmatic symptoms or attacks in
individuals who already have asthma. Our epidemiologic studies
have provided strong indications that environmental factors play
important roles in the temporal and spacial patterns of
inner-city asthma and pointed to the indoor environment as the
source of these patterns.[7,8,9] As a result, for the past
three and one-half years we have intensively investigated the
indoor environment of our study population. We have monitored
nitrogen dioxide in the kitchens, living rooms, bedrooms, of
120 inner-city apartments, in which we have also studied the

vertical and horizontal stratification of nitrogen dioxide whose
indoor source is the unvented gas cooking stove and
oven.[10,11,12,13]

Most asthma attacks are triggered by an identifiable
allergen; many attacks however, especially those in children,
occur during or after a respiratory infection.[14] The results
of research with animals suggest an association between exposure
to nitrogen dioxide and impaired resistance to bacterial and
viral infections.[15]

In addition to nitrogen dioxide we have also monitored
respirable particulates, carbon monoxide and a variety of
aeroallergens. Such indoor air pollutants may be particularly
common in apartments occupied by families that recently have
moved to New York from countries with warmer climates, such as
the Caribbean and Latin American countries. These recent
immigrants suffer from the cold and keep their windows tightly
closed in cold weather. To compensate for the dryness of
artificial heat, they keep pans filled with water on their
radiators and pots with boiling water on their stoves and
vaporizers or humidifiers in several locations in their
apartments. The combination of heat and humidity are known to
provide good breeding grounds for the growth of moulds and
fungi. In this report we will briefly summarize our findings on
nitrogen dioxide and focus on our findings on the indoor
aeroallergens that are likely to play a role in asthma in our
study population. [16,17,18,19]

The causal link between rises in specific aeroallergen
levels and specific asthma attacks is not always easy to
demonstrate. However, in several studies asthma patients have
shown improved lung function and reduced severity of symptoms in
settings chosen for their clean air and lack of allergens and
pollutants. [16,17] These findings are the basis for a pilot
study to evaluate the efficacy of Control Resource Systems, Inc.
(CRSI) home air filtration systems in reducing aeroallergens and
nitrogen dioxide levels in four inner-city homes of asthma
patients. Preliminary findings of this study will also be
reported in this paper.

Methodology

The following aeroallergens--cockroach, mouse and rat
urine, house dust mite, alternaria and cat antigen--were
monitored using the Air Sentinel, which is an aerollergen
sampler developed at the Mayo Clinic for home use . It captures
airborne particles greater than 0.3 microns in diameter on
fiberglass sheets and has a flow rate of about 3 liters per
second. The filter media are eluted with buffered saline by
descending chromatography. The elutes are dialyzed,
lyophilized, and tested for allergenic activity by
radioallergosorbent test (RAST) inhibition assays. One or two
of these instruments were placed in each study home; sampling
was conducted for a week at a time. The collection medium was
replaced and sent to the Allergic Diseases Research Laboratory

of the Mayo Clinic for analysis. An automatic timer and clock both regulated the length of time sampling was conducted each day and recorded the total number of hours of sampling that took place during the week.

In four of our study apartments in which asthmatic patients resided, an air filtration system was installed to evaluate the efficacy of a home air filtration system in reducing aeroallergen levels as measured with the Air Sentinel and nitrogen dioxide levels measured with Palmes tubes.

The CRSI systems that were used in this study provide three-stage air filtration. The prefilter is one-half inch thick and is designed to trap large particles and to protect the other filters. A three-inch thick activated carbon filter absorbs odors and vapors. A High Efficiency Particulate Absolute (HEPA) component consisting of an eleven and one-half inch thick aluminum separator fiberglass media filter is designed to remove 99.97 percent of all particles 0.3 microns or larger. The whole unit is tightly sealed to prevent filter bypass.

Results

The results of our nitrogen dioxide monitoring study have been presented elsewhere[10,11,12,13] and will only be summarized here.

48-hour average levels of nitrogen dioxide sampled with Palmes tubes[20] in most of our kitchens exceeded the United States Environmental Protection Agency's Air Quality Standard for nitrogen dioxide of 100 μg/m^3 and in some kitchens they exceeded 300 μg/m^3. Continuous monitoring procedure for nitrogen dioxide showed that women preparing a meal on an unvented gas stove can be exposed to levels of nitrogen dioxide exceeding 3000 μg/m^3.[11]

All twenty of the apartments where aeroallergens have been monitored had very high levels of cockroach allergen, even when no cockroaches were visible and the apartment residents were not aware of an infestation. Many of these homes also had high levels of mouse or cat allergen (but not both), and most had high levels of house-dust mite at certain seasons of the year.

Table I summarizes the data collected for cockroach antigen in 7 of our New York City study homes by season in 1985. Similar monitoring was conducted in homes in New Orleans, LA, and Rochester, MN; cockroach antigen was _not_ detected in Rochester and in the few homes that it was detected in New Orleans, levels were usually orders of magnitude lower. Table II summarizes the mean levels of mouse urine antigen in the same study homes in New York City; again the levels obtained were high compared to the other two cities where this antigen was generally not detected. Table III summarizes the levels of alternaria antigen for the same New York City apartments; an increase in the fall season is apparent for this antigen. The fall season is also the season in which asthmatic attacks and symptoms are more prevalent in our study population. A similar

trend is noticeable in the seasonal distribution of the levels
of the house-dust mite antigen which is shown on Table IV.
The relatively high levels of indoor aeroallergens
obtained in the homes of our study population provided a good
testing ground for the CRSI air filtration system which was
tested in homes in the fall and winter season of 1985-1986.
Levels of NO_2 in 4 study homes measured in periods when the
CRSI filtration systems were operating were 25% lower than when
they were not operating. There was also a small reduction in
the levels of some of the aeroallergens in some of the study
homes when the filtration system was turned on. The efficacy of
the air filtration was far below expectation, which might be
explained by the following: (1) Installation of the filtration
system may not have provided for thorough air filtration
throughout our study apartments. (2) Aeroallergens were
measured on floor level which might differ from the levels
obtained at breathing levels of the apartments' residents. We
plan to do further work to answer some of these questions.

Conclusion

Our findings of unusually high levels of aeroallergens and
nitrogen dioxide inside inner-city apartments might help explain
the high rate of asthma in inner-city areas (approximately 4
times that of the rest of the population of the northeastern
United States). The seasonal distribution of the house-dust
mite and alternaria antigens in these homes might provide an
explanation for the seasonal distribution of asthma in this
population.

I. F. Goldstein, Dr.P.H., Department of Epidemiology, School of
Public Health, Columbia University, New York, NY 10032, USA.

Acknowledgments

This study was supported by funds from Control Resource
Systems, Inc., Michigan City, Indiana, USA, and from the
Electric Power Research Institute, Palo Alto, California, USA.

References

1. Karetsky, M.S. (1977) Asthma in the South Bronx: clinical
   and epidemiological characteristics. _J. Allergy Clin._
   Immunol. **60**, 383.
2. Freudenberg, N., Feldman, C.H., Clark, N.M., _et al_. (1980)
   The impact of bronchial asthma on school attendance and
   performance. _J. School Health_, 522.
3. Brunswick, A.F. (1980) Health stability and change: a study
   of urban black youth. _Am. J. Pub. Health_, 70, 504.
4. Brunswick, A.F. and Josephson, E. (1972) Adolescent health
   in Harlem. _Am. J. Pub. Health Supplement_, Part Two, 62
   pages (monograph).

5. Ng. S. and Gergen, P. personal communication, based on unpublished results of the 2nd National Health and Nutrition Examination Survey, 1976-1980.
6. McCarthy, P., Byrne, D., Harrison, S., and Keithly, J. (1985) Respiratory conditions: effect of housing and other factors. J. Epidemi. Comm. Health 39, 15.
7. Goldstein, I.F. and Dulberg, E. (1981) Air pollution and asthma: search for a relationship. J. of Air Pollution Control Assoc. 31(4), 370-376.
8. Goldstein, I.F. and Currie, B. (1984) Seasonal patterns in asthma, clues to etiology. Environmental Research 33, 201-215.
9. Goldstein, I.F. and Cuzick, J. (1983) Daily Patterns of Asthma in New York City and New Orleans: an epidemiologic investigation. Environmental Research 30, 211-223.
10. Goldstein, I.F., Hartel, D., Andrews, L.R. (1985) Monitoring personal exposure to nitrogen dioxide. Proceedings of the Air Pollution Control Assoc.
11. Goldstein, I.F. and Andrews, L.R. (1986) Peak exposures to nitrogen dioxide. Presentation Air Pollution Control Assoc.
12. Goldstein, I.F. (1986) Design strategies for epidemiologic studies of environmental impacts on health. Presentation Air Pollution Control Assoc.
13. Goldstein, I.F., Hartel, D., and Andrews, L.R. (1984) Indoor exposure of asthmatics to nitrogen dioxide. Proceedings of the 3rd International Conference on Indoor Air Quality and Climate, Stockholm, Sweden.
14. Busse, W.W. (1985) The precipitation of asthma by upper respiratory infections. Chest 87:44S
15. Henry, M.C., Ehrlich, R., and Blair, W.H. (1969) Effect of nitrogen dioxide on resistance of squirrel monkeys to Klebsiella Pneumoniae infection. Arch. Environ. Health 18: 580.
16. Platts-Mills, T.A.E., Mitchell, E.B., Nock, P. et al. (1982) Reduction of bronchial hyperreactivity during prolonged allergen avoidance. Lancet Sept. 25, p. 675.
17. Murray, A.B., and Ferguson, A.C. (1983) Dust-free bedrooms in the treatment of asthmatic children with housedust or housedust mite allergy: a controlled trial. Pediatrics 71:418.
18. Ohman, J.L., Jr., Lowell, F.C. and Bloch, K.J. (1974) Allergens of mammalian origin III. properties of a major feline allergen. J. Immunol. 113: 1668.
19. Maunsell, K., Wraith, D.G. and Cunnington, A.M. (1968) Mites and house-dust allergy in bronchial asthma. Lancet 1:1267.
20. Palmes, E.D., Gunnison, A.E., Dimattio, J. and Tomcyk, D. (1976) Personal sampler for nitrogen dioxide. Am. Ind. Hyg. Assoc. J. 36:570.

Table I Cockroach

	Spring	Summer	Fall	Winter
Fam. 2				
Mean	---	2888.9	11012.5	5883.3
±S.D.	---	954.2	6309.3	5569.0
N	---	9	8	6
Fam. 3				
Mean	19250.0	18836.4	12271.4	21166.7
± S.D	4062.0	13541.4	8709.9	10144.9
N	8	11	7	6
Fam. 3				
Mean	---	12500.0	20188.9	---
± S.D.	---	7801.1	31694.5	---
N	---	8	9	---
Fam. 7				
Mean	14500.0	8250.0	18444.4	---
± S.D.	707.1	4136.9	7124.8	---
N	2	12	9	---
Fam. 8				
Mean	---	17250.0	12100.0	3700.0
± S.D.	---	9795.8	9161.9	6408.6
N	---	10	7	5
Fam. 9				
Mean	38000.0	14272.7	16409.1	---
± S.D.	45254.8	34586.4	6416.6	---
N	2	11	11	---
Fam.10				
Mean	26400.0	86863.6	65900.0	17433.3
± S.D.	29988.3	82393.8	47476.3	15797.6
N	5	11	5	3

Table II Mouse

	Spring	Summer	Fall	Winter
Fam. 2				
Mean	---	84.0	5.8	98.4
±S.D.	---	96.6	10.4	135.4
N	---	9	8	6
Fam. 3				
Mean	0.0	37.4	5.6	9.2
± S.D	0.0	123.9	11.2	22.5
N	8	11	7	6
Fam. 3				
Mean	---	22.4	4.0	---
± S.D.	---	44.1	10.2	---
N	---	8	9	---
Fam. 7				
Mean	44.0	68.7	1.5	---
± S.D.	62.2	123.6	2.2	---
N	2	12	9	---
Fam. 8				
Mean	---	87.5	4.8	1158.3
± S.D.	---	135.0	9.9	2316.5
N	---	10	7	4
Fam. 9				
Mean	0.0	26.6	1.3	---
± S.D.	0.0	46.7	1.8	---
N	2	11	11	---
Fam.10				
Mean	467.8	1168.6	1099.8	645.0
± S.D.	295.4	1296.0	1567.6	559.2
N	5	11	5	3

Table IV House Dust Mite

	Spring	Summer	Fall	Winter
Fam. 2				
Mean	---	880.6	4695.4	3197.0
±S.D.	---	1045.0	2211.2	787.1
N	---	9	8	6
Fam. 3				
Mean	1960	1657.9	6049.7	5844.0
± S.D	1695.6	1727.7	2177.5	3325.5
N	8	11	7	6
Fam. 3				
Mean	---	1207.6	7368.0	---
± S.D.	---	1150.5	3308.5	---
N	---	8	9	---
Fam. 7				
Mean	1417.5	492.5	4616.4	---
± S.D.	557.9	613.7	3559.9	---
N	2	12	9	---
Fam. 8				
Mean	---	1235.2	2406.7	3939.4
± S.D.	---	1195.4	1307.4	3273.4
N	---	10	7	5
Fam. 9				
Mean	0.0	978.2	4542.2	---
± S.D.	0.0	2519.4	2037.4	---
N	2	11	11	---
Fam.10				
Mean	684.0	3107.0	14988.0	22980.3
± S.D.	944.3	8389.6	4164.2	7511.4
N	5	11	5	3

EXS 51:
Advances in Aerobiology
©1987 Birkhäuser Verlag Basel

INDOOR MOLD SPORE EXPOSURE: CHARACTERISTICS OF 127 HOMES IN
SOUTHERN CALIFORNIA WITH ENDOGENOUS MOLD PROBLEMS

J. Gallup, P. Kozak, L. Cummins, S. Gillman
Specialty Labs of Orange County, Orange, CA   USA

Introduction

Physicians are becoming increasingly aware of the importance of
fungi as etiologic agents of a variety of diseases. In the past,
fungi were considered to be causes of inhalant respiratory allergies and rare, frequently obscure, infections. With recent advances in the use of chemotherapy and radiation treatment, oppor-
tunistic fungal infections are increasing. Recent clinical observations also suggest that excessive environmental exposure
to fungi may cause a variety of symptoms compatible with a toxic
reaction.

We are normally and continuously exposed to a variety of viable
and nonviable microorganisms. The consequences of this exposure
will vary from one individual to another. Studies have been
conducted to quantitate exposure to mold spores, but these studies need to be greatly expanded before comprehensive knowledge
of normal and abnormal exposure can be defined. Our group has
conducted over 450 volumetric mold surveys in homes, schools and
professional buildings in southern California. The following
information attempts to characterize problem interiors in this
coastal and relatively dry region.

Materials and Methods

We are presently evaluating the home or work environment using
two different volumetric sampling techniques in addition to
Scotch-tape imprints of suspected mold-contaminated items. The
Anderson air sampler (a volumetric six stage cascade sampler) is
used to determine the viable mold population indoors and outdoors. The rotorod (a volumetric impaction sampler) is used to
identify viable, nonviable and parasitic fungal flora. We have
found the Scotch-tape imprint technique to be especially helpful,
and we use it for rapid identification of mold colonies growing
on organic materials. Air samples are taken at each problem location as well as a non-problem location indoors. A single sample is taken outdoors. Interpretations of the indoor air flora
are made generally on the basis of spore levels indoors in comparison to spore levels outdoors at the time of sampling.
Sampling is conducted only during periods of stable weather.

<u>RESULTS</u>

As a frame of reference we have made some determinations of what constitutes normal mold spore exposure in southern California. We consider an interior to be normal if there is no history of chronic or acute water damage, good dust control and no unusual structural problems. Spore levels in these interiors average 30% to 70% of the outdoor spore count at the time of sampling. The distribution of genera is also very similar to the outdoor distribution.

Indoor mold problems are generally associated with either chronic water spills or a disastrous structural problem in the home. Jute-backed carpeting can be damaged by broken plumbing, malfunctioning appliances, chronic water spills on bathroom carpet, overflow from overwatered plants, or rain entering the home through open doors or windows. Water seepage through damaged roofs leads to superficial mold growth on walls and ceilings. Any organic material exposed to chronic dampness can also be the source of significant fungal contamination. These could include paper, leather, cork, wood, and wicker baskets.

A water problem existing for longer than three days was invariably associated with increased spore levels. Indoor spore levels in these problem homes were significantly higher than the outdoor spore level at the time of sampling with counts averaging more than 300% of the outdoor count.

When indoor organic materials were damp longer than two weeks the mean spore count was 4,654 spores/M3, with a range of 83 to 64,312 spores/M3 (P<0.00005). When organic materials were damp 3 to 14 days the mean spore count was 4,893/M3 with a range of 436 to 22,682 spores/M3 (P=0.0325). When all materials were dried within three days no spore problem developed. Spore counts in these homes ranged from 118 to 3,659 spores/M3, with a mean count of 914 spores/M3. These counts fell into a range determined to be normal for southern California (P=0.77).

The material most likely to be damaged was the jute-backed carpeting (51%) followed by wicker baskets (18%) and walls and ceilings (14%). The most likely area to have a problem was the bathroom (35%) followed by the living room (17%), family room/den (16%) and bedrooms (14%).

The cause of the problems were chronic water spillage such as bathing or plant watering (39%), recurrent leaks in plumbing (20%), structural defects or landscaping problems (16%), and acute problems such as an overflow or flood from an appliance (9%).

The following table attempts to summarize in absolute numbers the spore levels found in problem homes vs. normal homes using two different sampling techniques.

GENERA	ANDERSON STUDY		ROTOROD STUDY	
	Non-Problem Homes	Problem Homes	Non-Problem Homes	Problem Homes
	MEAN SPORE COUNT/M3		MEAN SPORE COUNT/M3	
Alternaria	51	56	165	133
Basidiospores			53	53
Aspergillus sps.	14	893		
Penicillium sps.	164	3004		
Hyaline spores (Asp/Pen)			78	2418
Chaetomium			70	97
Cladosporium	375	1080	448	1016
Epicoccum	35	17	110	79
Rusts			60	71
Smuts			98	99
Stachybotrys			19	309
Torula herbarum			19	2487
Sterile mycelia	44	115		
TOTAL SPORES/M3	716	5950	1139	6831

DISCUSSION

Summary analyses of problem interiors are very complex and multi-faceted. We have found that although approximately 50% of the homes have a single problem area, 35% have two problems and 15% have three or more problem areas. Some of these spores are being distributed and some are not. In addition, sampling methods detect certain genera in preference to others. Circumstances and conditions vary widely.

No one sampling technique can uncover all the many types of environmental problems. Advantages and disadvantages are associated with each. Problems encountered with Anderson sampling include non-viability of spores, non-sporulation of mold colonies, differences in growth rates of mold genera and overcrowding of colonies on sampler plates of media. Problems with the rotorod include imperfect collections due to the aerodynamic characteristics of mold spores and difficulties in identification of small, non-distinctive spore types.

Some problems are solved by comparisons of simultaneous Anderson and rotorod studies. Scotch-tape sampling provides direct information regarding actual populations present without laboratory skewing. For problems of overcrowding we have used incubation at 37° C. as well as 26° C. to favor isolation of certain genera as well as using all six stages in the Anderson sampler. In one such study we isolated 74,345 spores/M3 of _Penicillium_ and no colonies of _Aspergillus fumigatus_ incubating at 26° C. In the same location we isolated 8,486 spores/M3 of _Penicillium_ and 8,204 spores/M3 of Aspergillus fumigatus incubating at 37° C.

The benefits of home environmental studies can best be illustra-by an example. The home of A.M. was initially surveyed because of clinical deterioration in the control of this child's asthma. No initial historical environmental information was obtained that would have indicated an endogenous mold problem. No evidence of a problem was found during the sample taking period by either the aerobiologist or the mother. However, after the results of the Anderson study were available the home was visited a second time. The family had converted the garage into a bedroom. When the wall-to-wall carpet was removed, a crack in the concrete slab was evident. Insidious water damage had occurred to the jute-backing in the bedroom with subsequent growth of predominantly hyaline spores. Removal of the damaged carpet, repair of the concrete slab, and the correction of a second minor water problem in the bathroom was associated with clinical improvement in the patient. The following is the result of the Anderson study.

MOLD GENERA	SPORE ISOLATES BY ANDERSON SAMPLING/M3		
	Bedroom	Bathroom	Outdoors
Aspergillus	35,275	2,519	35
Cladosporium	318	306	3,354
Penicillium	1,754	1,177	0
All others	330	366	353
TOTAL SPORES	37,677	4,368	3,742

Awareness of the need for microbiological air sampling is increasing. When such testing is carried out the results must be interpreted intelligently and wisely with an understanding of what is normally all around us. It has not been unusual for wildly worded reports to be given to employees working in a building under investigation when in fact, spore levels are quite normal or even low. Such reports cause panic and misunderstanding in the workplace. It is sometimes very difficult to correct those first impressions given by those using non-volumetric equipment and by those with no real knowledge of normal air flora. It is hoped that this study will add to the growing volume of data regarding spore levels in normal and problem environments.

SUMMARY

We are constantly being exposed to molds in our environment. Indoor mold problems occur after prolonged or chronic water damage to a variety of organic materials such as unfinished wood, jute-backed carpeting, wallpaper, books, cardboard, leather, cork, paper, wallboard, and wicker baskets. Mechanisms for spore dispersal (such as air currents or foot traffic on carpets) must also be present. The presence of these organic materials and dispersal mechanisms leads to significant increases in indoor spore levels.

Janet M. Gallup, Aerobiologist, Specialty Labs of Orange County, 725 W. LaVeta, Suite 90, Orange, California 92668, USA.

EXS 51:
Advances in Aerobiology
©1987 Birkhäuser Verlag Basel

INTRINSIC VARIABILITY IN AIRBORNE FUNGI: IMPLICATIONS FOR ALLERGEN STANDARDIZATION.

Harriet A. Burge, Emory G. Simmons, Michael Muilenberg, Marion Hoyer, Janet Gallup, William Solomon.

University of Michigan, Division of Allergy, Ann Arbor, MI; University of Massachusetts, Department of Botany, Amherst, MA; Garden City, CA.

Fungus spores are among the most abundant and least well-known of airborne allergens. While ragweed pollen reaches a peak of $5000/m^3$ of air, fungus spores often exceed $100,000/m^3$. A majority of pollen types can be identified from air samples and many common and important pollen allergens have been characterized. On the other hand, only a small fraction of different kinds of fungus spores are recognizable and allergen characterization is only beginning. In addition, it has become apparent that fungus allergen preparations from commercial sources vary widely.[1,2,3] Among the possible causes for this variability are: 1)intrinsic variation between spores and mycelium with respect to antigen content; 2) somatic mutations causing allergen content changes within single cultures; 3) aging and culture variable changes; 4) strain to strain variability; 5) interspecific variability.

The airborne exposure unit for most fungi is the spore. In spite of this, most commercial allergen preparations utilize mycelium, often metabolic products, and occasionally, variable numbers of spores. Differences in protein and antigen content between spores, mycelium and metabolic products has been demonstrated by several investigators [4,5] and is demonstrated in Figure 1 for <u>Alternaria</u>.

Somatic mutations and cultural and aging variables combine to produce batch to batch variability [6] which is demonstrated in Figure 2. These isoelectric focusing patterns were produced using three strains of <u>A. alternata</u> grown in two separate batches each using identical culture methods. Immunoblotting where proteins from IEF gels are transferred to nitrocellulose and stained with specific human antibodies has produced parallel results and indicates that significant changes in antigen content occur over time even within one pure culture.

Each spore from an Alternaria colony can theoretically grow to produce a new strain. Furthermore, <u>Alternaria</u> <u>alternata</u> for example, is a cosmopolitan fungus of world-wide distribution. Spores from colonies growing on natural substrates in

Spokane Washington may produce strains that are significantly different from those produced in artificial culture in Sweden. We have also demonstrated this strain to strain variability using IEF and immunoblotting, comparing 5 environmental isolates of A. alternata. Major antigen bands were present in some strains and not in others.

Finally, interspecific variability can be a serious problem where isolates used for allergen preparations are misidentified. The group of isolates known to the medical community as <u>Alternaria alternata/A. tenuissima</u> actually comprises a dozen or more morphologically distinguishable species. The identification errors that have caused this variable have arisen due to the widely variable expertise of those identifying the isolates. The fact that an isolate has a name in a large commercial culture collection does not insure that the name is accurate.

Source, batch, strain and identification variability factors are additive and create a situation where documentation of real differences (between species, or between spores and mycelium, or different preparative methods) is extremely difficult if not impossible, and calls into serious question most of the work on antigen differences in the fungi, which in general does not document these sources of variability.

Steps toward solving these problems for all fungal types include: 1) Use of isolates identified by a recognized expert: This can be achieved by checking the mycology literature and finding who is most currently publishing taxonomic treatments of the taxon. 2) Use of morphologically and biochemically stable strain: Each strain should be monitored for morphological and biochemical changes over a year of subculturing under conditions prescribed by the expret in that taxon. 3) Industry wide standardization of methods for preparing allergens including control of spore/mycelium ratios. 4) Use of multiple stable strains in each extract to ensure that all relevant allergens are included. 5) Standardization of each extract based on allergen content as measured by an allergen-specific in vitro assay, and either IEF or SDS-PAGE followed by immunoblotting, and in vivo skin testing.

References

1. Kim SJ, Chaparas SD, Buckley H: Characterization of antigens from <u>Aspergillus fumigatus.</u> IV. Evaluation of commercial and experimental preparations and fractions in the detection of antibody in aspergillosis. Am Rev Respir Dis 120:1305-1311, 1979.

2. Yunginger JW, Jones RT, Gleich GJ: Studies on Alternaria allergens II. Measurement of the relative potency of commercial Alternaria extracts by the direct RAST and by RAST inhibition. J Allergy Clin Immunol 58:405-413, 1976.

3. Aas K, Leegaard, Aukrust L, Grummer O: Immediate hypersensitivity to common molds. Comparison of different diagnostic materials. Allergy 35:443-451, 1980.

4. Solomon WR, Burge HA, Muilenberg, ML: Allergenic properties of <u>Alternaria</u> spore, mycelial and "metabolic" extracts. J Allergy Clin Immunol 65:229, 1980.

5. Hoffman DR, Kozak PP, Gillman SA, Cummins LH, Gallup J: Isolation of spore specific allergens from <u>Alternaria</u>. Ann Allergy 32:245-270, 1961.

6. Aukrust L, Almeland T, Aas K: Batch to batch variations in allergens in Mucor racemosus. Allergol Immunopathol 8:306, 1980.

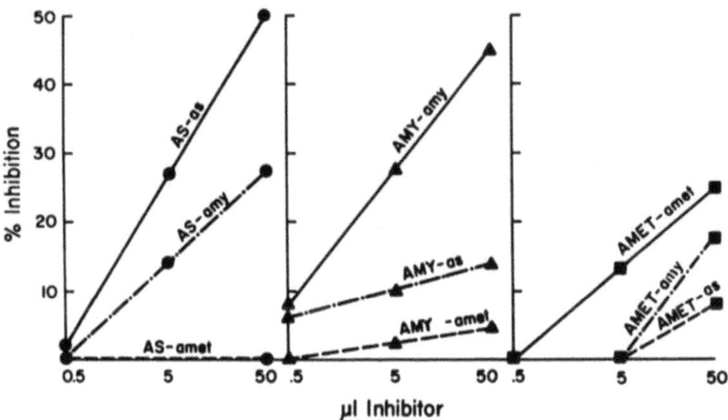

Fig. 1. _Alternaria_ RAST Inhibition of Spore (AS), Mycelial (AMY), and Metabolic (AMET) Solid Phases by Purified Spore (as), Mycelial (amy) and Metabolic (amet) Extracts.

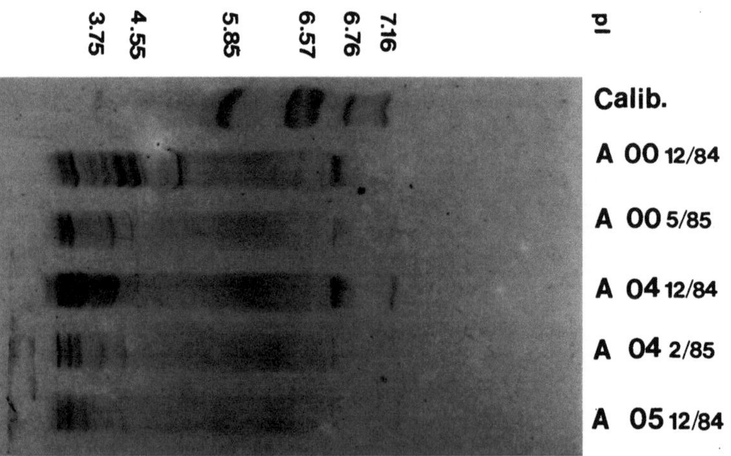

Fig. 2. Isoelectric focusing patterns for three strains of Alternaria alternata (two batches each for strains A00 and A04.

EXS 51:
Advances in Aerobiology
©1987 Birkhäuser Verlag Basel

CLINICAL CROSS-REACTIONS BETWEEN VARIOUS MOULDS WITH SPECIAL REFERENCE
TO FUSARIUM

A. W. Frankland, 139, Harley Street, London W1N 1DJ, UK

When investigating the possibility that Fusarium could be an
ingestant as well as an inhalant allergen in man, it seemed reasonable
to investigate those patients who were already known to have fungal spore
inhalant problems. Cladosporium is the most common fungus found in the air
followed by Alternaria, Penicillium and Aspergillus. It may be of some
surprise that Fusarium is the fifth commonest. There is often a diurnal
rhythm in the number of spores in the air. This depends on the environment rather than the fungus, but Fusarium as well as yeasts and basido-
spores show a night time peak. Although the four commonest moulds have
been extensively studied as allergens, amazingly little work has been done
on Fusarium as a possible inhalant allergen. The mycoprotein of Fusarium
is now available in the UK as an ingestant. Some moulds, such as yeasts,
can act both as inhalant and ingestant allergens. The present work was
undertaken to see whether the Fusarium mycoprotein might have allergenic
properties, particularly in those patients who had inhalant mould spore
sensitivities. Apparent cross-reactivity between related and unrelated
moulds is a commonly observed phenomenon [1]. Because of the cross-reactivity of related and unrelated moulds, it was an obvious step to investigate the potential allergenicity of the Fusarium mycoprotein in those
patients who already were clinically allergic to Alternaria.

Common allergens have been shown between Alternaria and Stemphylium [2] and between Didymella exitialis and Alternaria alternata [3].
As will be described, Fusarium also crossreacts with Alternaria and Stemphylium, as do Cladosporium and Alternaria [4] [5]. There seems to be a
cross-reaction between Aspergilli (Aspergillus fumigatus Aspergillus
niger and Aspergillus flavus and Alternaria tenuis) [6]. More surprising
is the fact that Alternaria alternata cross-reacts with ragweed pollen [7].

Fusarium

Fusarium occurs widely in the soil both as mycelium and chla-
mydospores. Over fifty species and varieties have been recognised, alt-

hough over 1000 published names have appeared [8]. Interest in <u>Fusarium</u> <u>graminareum</u> has arisen because it can very easily and cheaply be grown to produce a mycoprotein on continuous culture that provides a high quality protein food for man.

It seemed likely that if patients did react clinically to <u>Fusarium,</u> it was possible that they would react immunologically and, possibly, clinically to related moulds. A preliminary study [9] has already shown that <u>Fusarium</u> does indeed cross-react with <u>Alternaria.</u> Further studies were therefore carried out by skin and immunological tests to see whether the mycoprotein of <u>Fusarium</u> might produce allergic symptoms.

<u>Material and methods</u>

As in the previous study [9], skin prick tests were carried out, using a commercially available <u>Fusarium</u> oxysporum extract (Miles Laboratories) as well as the usual battery of routine skin tests. Patients were also tested with a 10mg/ml strength of the mycoprotein grown from <u>Fusarium</u> graminareum (extracts made and standardised by Miles Laboratories). In all 2500 patients were routinely screened for a possible <u>Fusarium</u> sensitivity. Selected patients who gave positive skin tests to <u>Fusarium</u> and to the mycoprotein were bled for immunological studies, as well as a controlled series of patients who had worked in the factory producing the mycoprotein and who had for some months eaten the mycoprotein.

<u>Results</u>

Skin tests showed, as in previous studies, that when a patient gave a positive skin test to one mould, usually - but not always - they reacted to other fungal spores. All patients who reacted to <u>Alternaria</u> also did to <u>Stemphylium.</u> Indeed the two <u>Stemphylium</u> moulds used for skin testing, <u>Stemphylium</u> solani and <u>Stemphylium</u> botryosum always gave much larger skin test responses to these moulds than they did to <u>Alternaria</u> which clinically caused them symptoms. When carrying out spore counting, using a Burkard trap, only very occasionally is <u>Stemphylium</u> seen in the UK, so unlike <u>Alternaria,</u> it has no clinical relevance, in spite of large positive skin prick test results in patients who are clinically allergic to <u>Alternaria.</u>

Although all patients, whatever their allergic complaint, were routinely skin tested with fungal extracts, all but one who were found to react to <u>Fusarium</u> and/or the mycoprotein of <u>Fusarium</u> also reacted to <u>Alternaria.</u> On the other hand, although there were often multiple mould spore positive skin tests, particularly in patients with clinical mould spore inhalant allergy, there was in some patients no cross-reactivity with closely related spores.

Immunologically it will be seen (Table I) that patients who already form IgG antibodies to inhalant fungal spores, do so to the <u>Fusarium</u> mycoprotein if they eat it. Those patients who form IgE antibodies to <u>Alternaria</u> (Table II) also do so to <u>Stemphylium</u> and also to <u>Fusarium</u> species and to Fusarium mycoprotein. Clinically and immunologically there is a cross-reaction between different mould spores. This, however, is not absolute because the one patient who had a gastro-intestinal reaction from eating the <u>Fusarium</u> mycoprotein, although giving large positive skin responses to many fungal skin prick tests and also to <u>Fusarium,</u> did not react to <u>Alternaria.</u>

Discussion

Although skin and other provocation tests have been carried out with commercially available mould spore extracts, these extracts are sometimes wrongly named and are unstandardised and often of very weak strength and possibly unstable. Schumacher and colleagues [10] noted that A. tenuis cross-reacted or shared antigens with three other species of moulds: <u>Stemphylium</u> sp.. <u>Curvularis</u> sp. and <u>Aspergillus fumigatus</u> but not with many other fungi. Hoffmann and Kozak [11] noted 51% of mould allergic patients reacted to <u>Fusarium</u> and 100% to <u>Alternaria</u> tenuis. Agarwal and colleagues [12] have characterised the major allergen of <u>Alternaria</u> tenuis but it remains to be seen how many antigens are common between the common moulds. <u>Fusarium</u> I have called a neglected mould [9]. It not only causes disease but it may be of great benefit to man as a cheap food.

Acknowledgments

Thanks are due to

Dr. Ursula Allitt for Discussion

Dr. B.J.F. Goodwin of Unilever Research for the immunological work

REFERENCES

1.   Jones HE et al. Apparent cross-reactivity of airborne
      molds and dermatophytic fungi. J Allergy Clin Immunol
      1973; 52 :346

2.   Agarwal MK,Jones RT, Yunginger JW. Shared allergenic
      and antigenic determinants in Alternaria and Stemphylium
      extracts. J Allergy Clin Immunol 1982; 20: 437-44

3.   Harries MG,Lacy J,Tee RD,Cayley GR,Newman Taylor AJ.
      Didymella Exitialis and late summer asthma. Lancet
      1985;i :1063-66

4.   Salvaggio J,Aukrust L. Mould induced asthma. J Allergy
      Clin Immunol 1981; 88 :327

5.   Karr RM,Anicetti VR,Lehrer SB,Butcher BT,Wilson MR,
      Salveggio JE. An approach to fungal antigenic
      relationships by radioallergosorbent test inhibition.
      J Allergy Clin Immunol 1981; 67 :194

6.   Vijay HM,Tsang P,Young NM,Copeland DF,Bernstein IL.
      Isolation of allergens from Cladosporium aladosporiodes
      J Allergy Clin Immunol 1985; 75 (Part 2) :117

7.   Santilli J,Rockwell WJ,Collins RP, Marsh DG. Coprinus
      Micaceus, Fuligo septica and Alternaria alternata
      spore allergy in highly ragweed-sensitive subjects.
      J Allergy Clin Immunol 1985; 74 (Part 2) :118

8.   Booth C. The genus Fusarium. Commonwealth Mycological
      Institute, Kew, England 1971

9.   Frankland AW. Fusarium: A neglected mould. Grana 1983;
      22 : 167-170

10.   Schumacher ML,Clatchy JK, Farr RS,Minden P. Primary
      interaction between antibody and components of Alternaria.
      J Allergy Clin Immunol 1975; 56 :39-53

11.   Hoffman DR and Kozak RP. Shared and specific allergens
      in mould extracts. Abs.No.267 J Allergy Clin Immunol
      1979; 63 :213

12.   Agarwal MK,Jones RT and Yunginger JW. Immunological
      and physiochemical characterization of commercial
      Alternaria extracts: a model for standardization of
      mould allergen extracts. J Allergy Clin Immunol
      1982; 70 : 432-436

Table I   Incidence of IgG antibodies to Fusarium Mycoprotein and fungal extracts

	Mycoprotein of Fusarium	Fusarium sp (Bencard)	Fusarium oxysporum	Stemphylium solani	Alternaria tenuis
Controls (N=10)	0	1	1	0	0
Mould allergic (N=11)	0	1	1	2	3
Mycoprotein eaters (N=20)	4	4	5	3	6

Table II   Uptake of anti-IgE (cpm) Mean and standard deviation

	Mycoprotein of Fusarium	Fusarium sp. (Bencard)	Fusarium oxysporum	Stemphylium solani	Alternaria tenuis
Controls (N=10)	206+52	337+ 51	61+ 33	32+ 20	89+ 37
Mycoprotein eaters (N=20)	147+31	309+ 17	50+ 28	37+ 18	82+ 62
Mould allergic (N=11)	314+93	418+112	192+150	352+118	381+494

EXS 51:
Advances in Aerobiology
©1987 Birkhäuser Verlag Basel

CLINICAL SIGNIFICANCE OF AIRBORNE ALTERNARIA TENUIS-SPORES: SEASONAL SYMPTOMS, POSITIVE SKIN AND BRONCHIAL CHALLENGE TESTS WITH ALTERNARIA IN SUBJECTS WITH ASTHMA AND RHINITIS

G. Schultze-Werninghaus, J. Lévy, E.-M. Bergmann, A.D. Kappos, J. Meier-Sydow

Hospital of the J. W. Goethe-University, Department of Internal Medicine, Division of Pneumology, Frankfurt/Main, FRG

To date the clinical feature of respiratory allergy to spores of Alternaria tenuis has remained ill-defined. Therefore, we analyzed data from two series of clinical investigations together with Alternaria spore-counts.

Methods

<u>Spore counts</u> were performed with a Burkard-trap on the roof of our hospital in 1985 and 1986.

<u>Allergic history</u> was explored by a) a standardized questionnaire and b) specialized colleagues of the out-patient department.

<u>Skin tests</u> (modified prick method) were performed with common allergens (standard concentration, 5000 PNU/ml; various pollens including mugwort and plantain, mites, and animal danders) and mould spores, evaluated acc. to (2), where the wheal size of the histamine control test is used as reference (= 1.0).

<u>Bronchial provocation tests</u> were performed acc. to (1).

<u>Extracts</u>: Allergopharma, Reinbek, FRG.

Subjects

<u>Study 1 (Skin test study)</u>: Skin prick tests with Alternaria were performed in 611 out of 1745 out-patients (1/1983 - 6/1986) because of rhinitis and/or asthma. In 102 patients (5.9 % of all patients) a positive (> 2 mm wheal-diameter) test with Alternaria

was found. These patients were further analyzed. Two groups
of patients were compared: group A (71 patients): "strongly
positive skin test" = diameter of Alternaria wheal >= 1.0 (histamine
control); group B (31 patients): "weakly positive skin test"
= wheal size < 1.0.
Study 2 (Bronchial provocation test study): 110 patients were
challenged with Alternaria (1/1976 - 6/1986). Two groups were
compared: group A (72 patients) with a positive and group B
(38 patients) with a negative provocation test.

Results and Discussion

Study 1 (Skin test study): The overall frequency of respiratory
symptoms (rhinitis, asthma) was identical in both groups. Skin
test results with common allergens were equally frequent in
both groups. Positive skin tests with further moulds were only
seen in a proportion of cases (maximum: group A, Cladosporium
in 36.8 %). "Isolated" skin test reactions to Alternaria without
pollen allergy were infrequent (7 out of 102 positive tests). The
most impressive difference between both groups was the frequency
of seasonal asthmatic symptoms (group A: 66.2 %, group B: 41.9%),
alone or in combination with occasional perennial symptoms. In
contrast, the seasonal difference of nasal symptoms was less
marked (A 63.4 %, B: 54.8%). The maximum of seasonal respiratory
symptoms was seen between may and august, significantly more
marked in group A (peak in july, 70.4%) than in in group B (peak
in june and july, 48.4%). In "isolated" Alternaria allergy without
pollen allergy these findings were confirmed (figure 1).
Study 2 (provocation test study): In spite of the different
origin of the data, the results resembled closely those of study
1. Overall frequency of respiratory symptoms and sensitizations
to common allergens were not significantly different in both
groups. Seasonal asthmatic (but not nasal) symptoms, were more
frequent in group A (69.4 and 51.4 %, resp.) than in group B
(47.4 and 47.4 %, resp.). Seasonal symptoms were most prominent
between may and august (group A: peak July, 65.3 %; group B: peak
June, 50.0 %). - There was a relation between skin and bronchial

155

65.3 %; group B: <u>peak</u> June, 50.0 %). - There was a <u>relation</u>
<u>between skin and bronchial provocation tests</u>: the larger the
skin test reaction was, the more frequent were positive bronchial
provocation tests; at a reaction to Alternaria >= 1.0, most
provocation tests were positive.

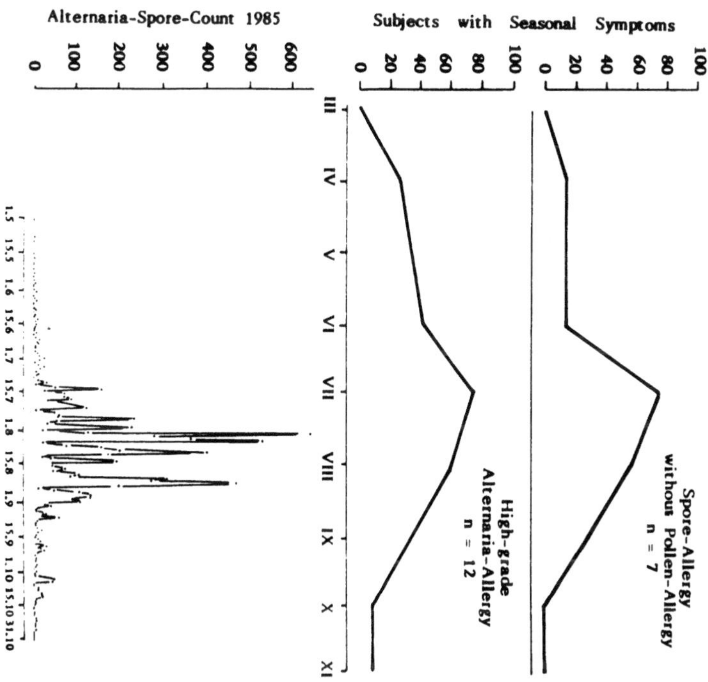

<u>Figure 1</u>: Seasonal asthmatic complaints in subjects with "isolated"
positive skin test to Alternaria (see text) (top) or with large
wheal-diameter (> 2.0) (mid), and daily Alternaria spore counts.

<u>In conclusion</u>, in two different approaches, excluding a significant
interference with allergies to pollens or other moulds, we
could demonstrate that a) patients with respiratory allergies
to Alternaria spores are usually atopic, b) Alternaria allergy

is in two thirds of the patients characterized by a seasonal
increase of bronchial (and less nasal) symptoms, with a peak
in july, in accordance with Alternaria spore counts. Therefore,
in Central Europe an immediate type allergy to spores of Alternaria
tenuis should be suspected in each case of "july-asthma".

References
1.Gonsior, E., Krüger, M., Meier-Sydow, J. (1976) Die Durch-
führung inhalativer Antigen-Provokationsproben mit Hilfe
der Ganzkörperplethysmographie. Acta allergol. **31**, 283-296

2.Schultze-Werninghaus, G., Gonsior, E. (1976): Antigenspektrum
und Reaktionsbeurteilung im Routine-Hauttest. Mschr. Kinder-
heilk. **124**, 231-234

Address: PD Dr. G. Schultze-Werninghaus, Klinikum der Universität, ZIM, Theodor-Stern-Kai 7, 6000 Frankfurt am Main, FRG

EXS 51:
Advances in Aerobiology
©1987 Birkhäuser Verlag Basel

OCCURRENCE OF ALTERNARIA Nees ex Fr. IN INDOOR AND OUTDOOR HABITATS IN CORDOBA (SPAIN)

F.Infante(1), E.Ruiz de Clavijo(1), C.Galán(1) and G.Gallego(2).

(1) Faculty of Sciences, Department of Botany, University of Córdoba,
(2) Clinical Hospital, University of Córdoba, Spain.

Abstract

We have studied the occurrence of a genus of allergic interest, Alternaria, in the outdoor and indoor atmosphere of 14 homes in Córdoba (Spain). The sampling was carried out by the method of sedimentation on a mycological medium twice a month for a period of one year. We detected an overall 3955 colonies of Alternaria belonging to seven different species of this genus (A.consortiale, A.crassa, A.dendritica, A.harzii, A.japonica, A.tenuis and A.tenuissima), some of which are proven allergenic agents. A.tenuis was by far the most frequent of the above mentioned species, both indoors and outdoors. Three of the species identified (A.consortiale, A.crassa and A.harzii) occurred in greater concentrations indoors than outdoors; the remaining four were found to occur preferentially outdoors.

Introduction

The knowledge of the aeromycoflora of urban areas is currently of great interest as it is related directly to fields such as medicine, meteorology or environmental sciences.
As early as 1926, Floyer quoted fungi as the causing agents of pathological processes, and in 1860, Pasteur noted the presence of bacteria, yeasts and moulds in the air of different regions of France. Later, in 1873, Blackley demonstrated the involvement of spores from Chaetomium and Penicillium expansum in respiratory diseases. Already in this century (1935),Feinberg revealed the significance of fungi in allergic diseases. From then on, many authors have shown the occurrence of fungal spores responsible for different pathological processes affecting both animals and plants.
Alternaria Nees ex Fr. (Or. Moniliales) lives as a saprophyte or parasite on various vegetable substrates such as fabrics, clothes or foods, so commonly four in home environments. Hence the frequent presence of its spores in the air. This genus has been quoted by a number of authors (HOPKINS & al.,1930; BONILLA-SOTO & al.,1969; CAPLIN & HAYNES, 1970; GREGORY, 1973; PETERSEN & al.,1976; GRAVENSEN, 1979, 1980; PETERSEN & SANDBERG, 1981; MUÑOZ & al.,1983) as a human allergenic, sporosis-inducer, responsible for diseases similar to hay fever and even related to Baker asthma or one of the cause of non-seasonal conjunctivitis and other pathological processes.

Many authors have studied the evolution of fungal spores such as those of
Alternaria in the air of different places all over the world. To name a few,
KRAMER & al.(1959) studied volumetrically the variation of this genus along
two years and quoted three species (A.tenuis, A.consortiale and A.cucumerina)
as the most common in Kansas. CALVO & al.(1979), by gravimetry on a culture
medium, isolated six species (A.alternata, A.chartarum, A.saponariae,
A.oleracea, A.tenuissima and A.consortialis) from the atmosphere of Barcelona
and reflected their variation along the two sampling years. In Arizona,
SNELLER & al.(1981) studied the annual variation of Alternaria from data
corresponding to 20 years and reported that the spores were found in greatest
numbers from late summer to early winter. In a recent study on the atmosphere
of Córdoba, NOGALES & al.(1986) identified six species (A.alternata,
A.dianthi, A.radicina, A.raphani and A.tenuissima) and compared two different
sampling methods, finding the gravimetric technique on a culture medium to be
more efficient in the sampling of Alternaria spores in the air.
The aim of this work was to study the Alternaria species present in the
atmosphere of Córdoba and to determine their variation throughout 1984, both
indoors and outdoors.

Material and methods

The sampling was carried out in Córdoba, a semirural city of 265.000 inhabitants lying in the southwest of Spain at about 120 m above sea level and
featuring a mediterranean climate tending to continental.
Samples were collected in sterile Petri dishes (Ø 90mm), which were exposed
simultaneously for 20 minutes inside and outside the homes (3 each).
The sampling was carried out fortnightly throughout 1984 at 14 sampling
spots in different areas of the city, covering the different physical and
social conditions as far as possible.
The culture medium used consisted of 2% agar Malt Extract with oxytetracy-
cline, the pH being adjusted to 5. The exposed plates were incubated at 27ºC
for seven days and then the colonies were counted, isolated and identified by
the usual procedures.
The different species were identified according to the criteria set in the
monograph by JOLY (1964) and the papers by ELLIS (1971, 1976) and SIMMONS
(1967).
The gravimetric methods is commonplace in the quantitation of the fungal
populations of the air and has been used by authors all over the world: DRAN-
SFIELD (1966), GRAVESEN (1972), COLLINS-WILLIAMS & al.(1973), LUMPKINS &
al.(1973), CALVO & al.(1978), AL-DOORY & al.(1980), AL-TIKRITI & al.(1980),
etc.

Results

We detected a total of 3955 Alternaria colonies (727 indoors and 3228 outdoors), corresponding to seven different species. These figures account for
5'2% and 13'6% of all fungal colonies isolated indoors and outdoors, respectively (these data and the relative occurrence of each species are listed in
Table I, which also shows the number of homes in which each taxon was found
and compares it with the results found outdoors).
The Figure 1 reflect the variation of the occurrence of each species and the
genus as a whole, both indoors and outdoors along the period investigated (we

have supressed the graphs corresponding to <u>Alternaria harzii</u> because of its scant occurrence).
In Table II are listed the climatological data corresponding to 1984 as kindly provided by the "Observatorio Especial y Medio Ambiente" of Córdoba. Such data are representative of the climate in the sampling area.
The species identified were: <u>Alternaria consortiale</u> (Thüm.)Hughes, <u>A.crassa</u> (Sacc.)Rands, <u>A.dendritica</u> (S.Cam.)Joly, <u>A.harzii</u> Joly, <u>A.japonica</u> Yoshii, <u>A.tenuis</u> Nees and <u>A.tenuissima</u> (Fr.)Wiltsh.

T A B L E I

TAXA	Nº colonies		%		Nº homes		%	
	Int	Ext	Int	Ext	Int	Ext	Int	Ext
A.consortiale	95	72	13'1	2'2	14	5	100	36
A.crassa	25	9	3'4	0'3	9	2	64	14
A.dendritica	58	297	0'0	9'2	12	13	86	93
A.harzii	2	0	0'3	0'0	1	0	7	0
A.japonica	59	162	8'1	5'0	12	12	86	86
A.tenuis	323	1263	44'4	39'1	14	14	100	100
A.tenuissima	164	489	22'6	15'2	14	14	100	100
Alternaria ssp.	1	936	0'1	29'0	1	14	7	100
Gen. Alternaria	727	3228	100'0	100'0	14	14	100	100

T A B L E II

	Jan.	Feb.	Mar.	Apr.	May.	Jun.	Jul.	Aug.	Sep.	Oct.	Nov.	Dec.
Total rainfall (mm)	17	43	87	92	90	15	0	0	4	22	201	11
Mean humidity (%)	82	70	70	74	72	59	43	46	51	66	87	83
Maximum mean temperatures (ºC)	13	16	17	23	21	29	37	35	33	25	17	15
Minimum mean temperatures (ºC)	4	2	5	11	10	14	17	18	16	10	9	5
Mean of mean temperatures (ºC)	9	9	11	17	15	22	27	27	24	18	13	10

Discussion

<u>Alternaria</u> is a common constituent of the aeromycoflora, both indoors and outdoors, as it occurs throughout the year at higher or lower levels. It represents 5'2% and 13'6% of the fungi detected indoors and outdoors, respectively. The latter data coincides with that reported by KRAMER & al. (1959) for Kansas (12'6%) and is much higher than that reported by ALLER & al.(1971) for the air of León (1'9%), by LARSEN (1981) for Copenhagen (9'4%) and by NOGALES & al.(1986) for Córdoba (4'3%). However, its occurrence indoors is much lower than that claimed by GRAVESEN (1972) (16%).

We should emphasize the commoned occurrence outdoors than indoors, probably due to the type of habitat required by this fungus. There is a well-defined seasonal variation, both indoors and outdoors, with minima corresponding to later summer and maxima appearing during the spring and early summer. This seasonal variation coincides with that reported by NOGALES & al.(1986) for Córdoba, in which May is quoted as the month during which the highest concentrations are reached, and the winter as the period over which such concentrations are minimum, consistent with the results found by SOLOMON (1976) who, in his work on indoor air in winter pointed out the scant occurrence of <u>Alternaria</u>, both indoors and outdoors. However, this variation does not coincide with the maxima reported by KRAMER & al.(1959) for the air of Kansas, GRAVESEN (1972) for dannish homes, CALVO & al.(1979) for Barcelona and LARSEN (1981) for Copenhagen, which are shitfed to the summer.

SNELLER (1981) found the autumn to be the period of greatest occurrence as the average of the results compiled along twenty years in Tucson.

These apparent discrepancies may be the result of the different climatological conditions prevailing in different parts of the world. It is thus important and advisable to carry out repeated samplings in as many different areas as possible in order to be able to establish parameters describing in a more factual way the evolution of the different species of fungi and increase our knowledge on this topic.

We think it advisable to state the species detected as the taxonomic likeness does not necessarily have to result in similarities in the variation patterns, we also report the overal variation of the genus.

We should note that only two species (<u>A.consortiale</u> and <u>A.crassa</u>) are commoner indoors than outdoors; the remainder are more frequently found outdoors. Thus <u>A.consortiale</u> appeared in 100% of the indoor habitats and in 36% of the outdoor ones, the corresponding figures for <u>A.crassa</u> being 64% and 14%, respectively. The greatest occurrence of <u>A.consortiale</u> outdoors corresponded to February, while it indoor peak concentration was reached from the end of spring to the summer.

<u>A.harzii</u> was detected very sparsely indoors.

<u>A.tenuis</u> is by far the most abundant species, both indoors (44'4%) and outdoors (39'1%), consistent with the results of KRAMER & al.(1959) for the USA, and CALVO & al.(1979) and NOGALES & al.(1986) for Spain. In addition to its abundance, this genus is the most frequently quoted as inducer of pathological processes in humans (BONILLA-SOTO, 1961; MUÑOZ & al.,1983).

<u>A.tenuissima</u> is the second species in quantitative significance as it represents 22'6% and 15'2% of the <u>Alternaria</u> colonies isolated indoors and outdoors, respectively. Both species could be labelled as the most cosmopolite as they appear in 100% of the sampling spots, both indoors and outdoors. Both species have similar seasonal evolutions: they reach a peak indoors during the spring and early summer, while their highest concentrations outdoors are usually detected in the spring and (to a lesser extent) in the autumn and the beginning of winter.

A.dendritica and A.japonica have similar values. Each represents 8% of the total of colonies isolated indoors and 9'2% and 5% respectively, of those collected outdoors. Both have very similar occurrences in the different types of sampling. Thus, they appear in 86% of indoor spots and in 93% and 86%, respectively of outdoor ones. Their distribution curves, more uneven than in the previous cases, show occurence maxima in the spring and early summer.

REFERENCES

AL-DOORY, Y., J.F. DOMSON, W.A. HOWARD & M.R. SLY (1980) Airborne fungy and pollens of the Washington DC., metropolitan area. An. Allerg. 45, 360-367.

ALLER, B., M. REY & A. MARTINEZ (1971) Estudio de la incidencia de los hongos en el aire de León durante un año. An. Fac. Vet. León 17, 13-20.

AL-TIKRITI, S.K., M. AL-SALIHI & G.E. GAILLARD (1980) Pollen and Mold survey of Baghdad, Iraq. An. Alerg. 45, 97-99.

BONILLA-SOTO, O., N.R. ROSE & C.E. ARBESMAN (1961) Allergenic Molds. Antigenic and allergenic properties of Alternaria tenuis. J. Allerg. 246-270.

CALVO, M.A., J. GUARRO, G. SUAREZ & C. RAMIREZ (1979) Air-borne fungi in the air of Barcelona (Spain).II. The genus Alternaria. Mycopath. 69, 137142.

CALVO, M.A., J. GUARRO, E. VICENTE & G. SUAREZ (1978) Estudio comparativo de la micoflora atmosférica de dos ciudades del área mediterránea. Rev. Clin. Esp. 151, 203-206.

CAPLIN, I. & J.T. HAYNES (1970) Mold allergy. An. Aller. 28, 87-91.

COLLINS-WILLIANS, C., H.K. KUO, D.N. GAREY, S. DAVIDSON, D. COLLINS-WILLIANS, M. FITCH & J.B. FISCHER (1973) Atmospheric mold counts in Toronto, Canada. An. Allergy. 31, 69-71.

DRANSFIELD, M. (1966) The fungal air-spora at Samaru Northerm. Nigeria. Trans. Br. Mycol. Soc. 49, 121-132.

ELLIS, M.B. (1971) Dematiaceous Hyphomycetes, (Commonwealth Mycological Institute, Kew).

ELLIS, M.B. (1976) More Dematiaceous Hyphomycetes, (Commonwealth Mycological Institute, Kew).

GRAVESEN, S. (1972) Identification and quantitation of indoor airborne micro-fungi during 12 months from 44 Danish homes. Act. Allergol. 27, 337-354.

GRAVESEN, S. (1979) Fungi as a cause of allergic disease. Allergy, 34, 135-154.

GRAVESEN, S. (1981) On the connection between the ocurrence of airborne microfungi and allergy symptoms. Grana, 20, 225-227.

GREGORY, P.H. (1973) The microbiology on the atmosphere, (Leonard Hill Ltd., London).

HOPKINS, J.G., R.W. BENHAN & B.M. KESTEN (1930) Asthma due to a fungus Alternaria. Jour A. M. A. $\underline{4}$, 6-10.

JOLY, P. (1964) Le genre Alternaria, (P. Lechevalier, Paris).

KRAMER, C.L., S.M. PADY & C.T. ROGERSON (1959) Kansas Aeromycology IV. Alternaria. Trans. Kansas Acad. Sci. $\underline{62}$, 252-256.

LARSEN, L.S. (1981) A three-year-survey of Microfungi in the air of Copenhagen 1977-79. Allergy, $\underline{36}$, 15-22.

LUMPKINS, E.D., S.L. CORBIT & G.M. TIEDEMAN (1973) Airborne fungi survey. 1.Culture-plate survey of the home environment. An. Allerg. $\underline{31}$, 361-370.

MUÑOZ, F. & M.A. MARTIN (1983) Alergia a hongos, (Publ. Sandoz, Barcelona).

NOGALES, M.T., E. DOMINGUEZ, C. GALAN & E. RUIZ DE CLAVIJO (1986) Variación estacional del contenido de esporas del género Alternaria Nees ex Fr. en el aire de la ciudad de Córdoba (España). Allergol. et Inmunopath. $\underline{14}$, 115-119.

PEDERSEN, N., P.A. MARDH, T. HALLBERG & N. JONSSON (1976) Cutaneous alternariosis. Brit. J. Dermatol. $\underline{94}$, 201-209.

PETERSEN, B.N. & J. SANDBERG (1981) Diagnostic in allergic diseases by correlating pollen/fungal spore counts with patient scores of symptoms. Grana, $\underline{20}$, 219-224.

SIMMONS, E.G. (1967) Typification of Alternaria, Stemphylium, and Ulocladium. Mycologia, $\underline{59}$, 67-92.

SNELLER, M.R., H.D. HAYES & J.L. PINNAS (1981) Frequency of airborne Alternaria spores in Tucson, Arizona over a 20-years period. An. Allerg. $\underline{46}$, 30-33.

SOLOMON, W.R. (1976) A volumetric study of winter fungus prevalence in the air of midwestern homes. An. Arbor. $\underline{57}$, 46-55.

Prof. Félix Infante-García-Pantaleón, Departamento de Botánica, Facultad de Ciencias, Universidad de Córdoba, 14004-Córdoba, ESPAÑA.

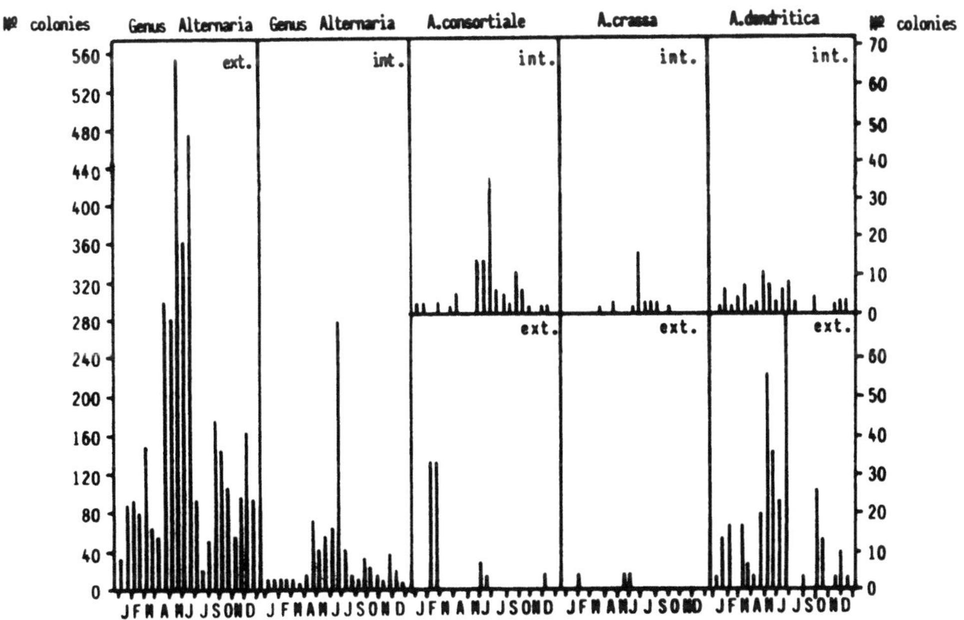

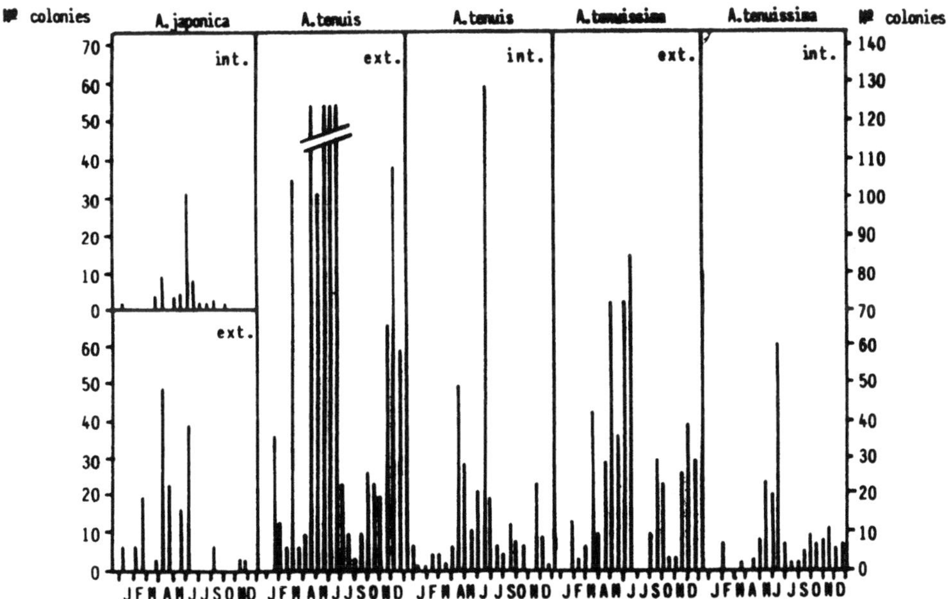

Fig. 1. Incidence of the different _Alternaria_ taxa collected in outdoor (ext.) and indoor (int.) of cordoban atmosphere throughout 1.984.

CONCENTRATIONS OF AIRBORNE *BOTRYTIS* CONIDIA, AND FREQUENCY OF ALLERGIC
SENSITIZATION TO *BOTRYTIS* EXTRACT.

F.Th.M. Spieksma[1], N. Nolard[2], F. Beaumont[3], and P.H. Vooren[1]

[1]Dept of Pulmonology, University Hospital, Leiden, The Netherlands
[2]Lab. of Mycology, Inst. Hygiene and Epidemiology, Brussels, Belgium
[3]Dept of Pulmonology, University Hospital, Groningen, The Netherlands

Introduction

Botrytis cinerea is a mould with a world-wide distribution, living mostly as
a parasitic fungus on leaves and fruits of plants, also well-known cultivated
products like lettuce, grapes, and strawberries. It can develop into a
serious (economic) pest, also in greenhouses and is known as "grey mould".
Its spores are one-celled, hyaline, pale brown, ellipsoid conidia, with a
smooth wall, and measure approximately 8 by 12 μm. These conidia could become
airborne in relatively high numbers to be distributed by air to other target
plants.
Allergic sensitization to extract of *Botrytis* material has been reported
incidentally, but its relative importance is still unclear. This preliminary
report attempts to compare the airborne concentrations and the frequency of
allergic sensitization to *Botrytis* with those to some other taxa of moulds.

Methods

Aerobiological observations on airborne mould spore concentrations were
carried out with the continuous, volumetric Burkard® spore trap, and with the
discontinuous Viable Andersen sampler.
Allergic sensitization was recorded with a direct bio-assay in the form of an
intracutaneous test with allergenic extracts from cultivated mould mycelium.

Results

Airborne occurrence

Table I presents the results of aerobiological observations in the period
1974-1977 at Brussels (NOLARD, 1977). It shows that *Botrytis* conidia are
among the spore types with the highest year totals.
From Table II, in which the Ascospores, the Basidiospores, and *Aspergillus/
Penicillium* are left out of consideration, it can be seen that also at Leiden
high *Botrytis* conidia concentrations are recorded. This occurs especially in
late summer. From a preliminary European survey it has appeared that
particularly in Western Europe, between July and October, high daily
concentrations are observed not unfrequently (SPIEKSMA, 1984).

Table I: Season totals of daily concentration per m³ of mould spores, at Brussels

Taxa	1977	average 1974-1977
Cladosporium	635,109	483,095
Ascospores	72,625	153,875
Basidiospores	106,465	107,519
Ustilago	24,019	35,675
Alternaria	21,855	22,820
Botrytis	12,781	19,440
Asp./Pen.	36,182	17,195
Tilletiopsis	6,856	9,605
Epicoccum	3,035	4,140
Erysiphe	2,639	2,245
Entomophthora	1,370	1,510
Stemphylium	893	1,380

Table II: Season totals of daily concentrations per m³ of mould spores in 1980, at Leiden and Antwerp

Taxa	Leiden	Antwerp
Cladosporium	812,575	440,770
Botrytis	25,665	22,020
Ustilago	12,535	25,550
Alternaria	11,485	13,635
Epicoccum	5,095	4,690
Erysiphe	3,685	1,050
Torula	2,980	1,625
Entomophthora	2,665	3,575
Stemphylium	1,410	215

Allergic sensitization

From two independent studies at Leiden and Groningen (BEAUMONT, 1985)
comparisons between airborne occurrence and allergic sensitization frequency
can be deduced. The result of this comparison is given in Table III, showing
very high percentages of skin reactivity to extracts of *Botrytis*, accounting
4,9% at Groningen, in 692 patients with "suspected allergy", and 24% at
Leiden, in 180 patients with "suspected mould allergy".

Table III: Comparative results of airborne concentrations and allergic skin reactivity of various mould spores

Groningen : * Anderson sampler * 692 patients ("suspected allergy")			Leiden : * Burkard sampler * 180 patients ("suspected mould")		
	airb.spore concen-tration %	allergic sk.react. %		airb.spore concen-tration %	allergic sk.react. %
Cladosporium	42.6	2.3	*Cladosporium*	53.6	14
Botrytis	8.6	4.9	*Sporobolomyces*	5.2	4
Yeasts	7.0	0	*Botrytis*	2.7	24
Penicillium	5.8	3.0	*Ustilago*	2.3	8
Aspergillus	3.7	4.7	*Alternaria*	2.1	13
Alternaria	0.9	1.1	*Asp./Pen.*	1.4	23/27
Basidiospores	5.7	N.D.	Ascospores	8.3	N.D
			Basidiospores	8.1	N.D.

Conclusion

Although mycelium material from which the extracts are prepared might not be allergenically identical to the conidia sampled from the air, it can be concluded that the relative importance of *Botrytis* as a possible cause of respiratory mould allergy is underestimated compared to some other better known mould taxa such as *Cladosporium* or *Alternaria*.

References

Beaumont, F. (1985) Aerobiological and clinical studies in mould allergy. Thesis, Groningen

Nolard-Tintigner, N. (1977) Biologische Luchtverontreiniging te Brussel, Antwerpen en Charleroi. Lab. Parasitologie, U.L.B., Brussel

Spieksma, F.Th.M. (1984) Comparative European spore counts. In: Atlas of moulds in Europe causing respiratory allergy. Ed.: Wilken-Jensen and Gravesen. ASK Publishing, Copenhagen.

Authors address

Dr F.Th.M. Spieksma, Lab. Aerobiology, Dept Pulmonology (C3-P-14),University Hospital, Rijnsburgerweg 10, NL-2333 AA LEIDEN, The Netherlands

5. Special allergens: Algae

MICROALGAE AS AEROPLANKTON AND ALLERGENS

Ebba Tiberg

Institute of Physiological Botany, University of Uppsala, Uppsala, Sweden

Many studies of the airborne algal flora have been performed in this century (see e.g. Schlichting, 1969). The samplings have been carried out with all types of methods, from stationary culture plate exposures to aeroplane sampling and the use of different spore traps. The effect of meteorological conditions on the algal dispersal has also been investigated.

The results of measuring the amount of airborne algae are varying. Some studies show a relatively high amount of algae at certain times. For example, an average of 110 Gloeocapsa cells/m^3 air has been measured in England and 100-200 impactions/m^3 air in Texas. Schlichting states, that a man can inhale 240 algal cells/hour but other authors mean that the algal frequency in the air never exeeds the frequency of pollen and fungi spores. This is in accordance with our own studies in middle Sweden in 1982 and 1983 (Tiberg et al., 1984). We used an Andersen sampler with an inorganic medium for green algae. The plates were incubated in light for three weeds. The results show a maxima of 50 cfu/m^3 air and an average of 0-8 cfu/m^3 air.

The results of registrating the composition of species, however are uniform. The green algae are dominating, followed by the blue-green algae, the diatoms and the Xanthophyceae. In tropical areas, the blue-green algae seem to bee more important. Schlichting has listed 152 species, found in the air, which probably gives a very good picture of the airborne algal flora. The most dominant genera in the temperated area are, among the green algae; the chlorococcean genera Chlorella, Chlorococcum, Neochloris and <u>Nannochloris</u>, among the blue-greens; <u>Anabaena</u>, <u>Nostoc</u>, <u>Oscillatoria</u> and among the diatoms; <u>Navicula</u>.

The reason for sampling airborne algae has often been the suspicion, that they can cause some hazard to the human health, especially inhalant allergy. Therefore even the indoor algal flora has been investigated and we found algae in 90% of all dust samples, collected from homes of allergy patients (Tiberg <u>et</u> <u>al</u>., 1984). The genera are mostly the same as found in outdoor air, with slight higher frequency of blue-green algae.

The first report about algae as an inhalant allergen came in 1962 (McElhenney <u>et</u> <u>al</u>., 1962). The allergenicity has been studied by clinical trials, mostly by intracutaneous testing with algal extracts. The results of the different trials vary from 2-82% positive patients against algae, depending on the use of different algae, extraction procedures and concentrations as well as different testing methods and variable criteria for selecting patients and for positivity. The allergenicity has also been demonstrated by bronchial and nasal provocations (see e.g. Bernstein and Safferman, 1966).

We have chosen a <u>Chlorella</u> extract for a clinical trial on allergic children in the town Linkoping in Sweden. The algae were grown under sterile conditions in batch cultures. The extract was made 1:50 in ammonia-bicarbonate buffer, ultrasonicated, dialyzed and stored freeze-dried. The patients were divided into three groups and tested with the skin prick method in an extract concentration of 1 mg dry weight/ml. I do present the results so far but the study will continue. The first group consists of children, searching for the first time at the hospital allergy clinic and here we have found no positive reaction. In the next group we have children being treated (immunotherapy) for pollen and animal allergies and here is just 1 out of 30 tested positive. In the third group we have children, being treated for mould allergy and interestingly 3 out of 7 tested are positive. The material is very small but it seems insidious. In an earlier study, we have shown by RAST-testing sera from 33 children, that 21% were positive against <u>Chlorella</u> or <u>Anabaena</u>. 57% of the positive children had one or more aquarias at home.

By screening atopic sera with RAST-Chlorella, I found, that positive
patients often are multiallergic and have high total IgE. In order to
reveal the IgE-binding pattern of those Chlorella-positive patients, I
have used the immunoblotting method with 20 sera. This shows, that the IgE-
binding proteins are present in the whole molecular weight range, from
approx. 10-100 kd. and that the patients are reacting very individually to
the different proteins. Some react on just a few proteins, while other are
more broadspectra, reacting on several proteins. It is also possible to
inhibit the IgE-binding by preincubating patient sera with allergen extract.

Conclusions

. Microalgae are present in both out- and indoor environments.
. The amount of airborne algae (max 200-300 impactions/m^3 air) does not
  exeed the amount of pollen and fungi spores.
. The dominating genera belong to the Chlorococcales. Other representatives
  of the Chlorophyceae as well as Cyanophyceae, Xanthophyceae and
  Bacillariophyceae are also present.
. Atopic patients can form IgE against the green alga Chlorella. This is
  especially expressed among multiallergic patients with high total IgE.
. There is probably a correlation between mould and green-algal allergy.

References

Bernstein, I.L., Safferman, R.S. (1966) Sensitivity of skin and bronchial
mucosa to green algae. J. Allergy 38, 166-173.

McElhenney, T.R. et al. (1962) Algae: A cause of inhalant allergy in
children. Ann. Allergy 20, 739-743.

Schlichting, H.E.Jr. (1969) The importance of airborne algae and protozoa.
JAPCA 19, 946-951.

Tiberg, E. et al. (1984) Occurrence of microalgae in indoor and outdoor
environments in Sweden. Nordic Aerobiology (ed. Nilsson&Raj), 24-29.

address: Ebba Tiberg
           Institute of physiol. botany
           Box 540
           S-751 21   UPPSALA
           SWEDEN

6. Special allergens:
   Insects and mites

EXS 51:
Advances in Aerobiology
© 1987 Birkhäuser Verlag Basel

SENSITIZATIONS AGAINST TRIBOLIUM CONFUSUM DU VAL IN PATIENTS
WITH OCCUPATIONAL AND NON - OCCUPATIONAL EXPOSURE

R. Rudolph , E. Stresemann , Beatrice Stresemann and Meike Haupthof

Asthma - and Allergy Clinic , Bad Salzuflen , FRG

Introduction

The flour beetles of the genus Tribolium (Tribolium confusum DU VAL
("confused flour beetle") , Tribolium madens CHARPENTIER ("black flour
beetle") , Tribolium casteneum HERBST ("rust-red flour beetle"), Tri =
bolium destructor UYTTENBOOGAART ("rice flour beetle")) belong to the
family of Tenebrionidae ("black beetles") which includes wellknown
species as Blaps mortisaga , Pedinus femoralis , Opatrum sabulosum ,
Phaleria cadaverina , Scaphidema metallicum , Diaperis boleti , Alphi =
tobius diaperinus , Gnathocerus cornutus , and , above all , the
notorious Tenebrio molitor ("meal worm") (1).
The tribolium beetles play an important role as common pests on stored
products of different origin : all kinds of cereals (grains , brans ,
flours , pastries , dog biscuits) , peanuts, nutmegs, almonds, copra,
plant seeds, spices , expellers. Therefore , permanent or occasional
exposure to tribolium beetles , larvas and/or exuvias can be expected
in several places and professions (3). The aim of this study was to
find out whether any risk in exposed patients do exist or not.

Methods , materials , patients

1. 430 patients (210 males , 220 females ; aged from 10 - 58 years , mean
age 35 $\pm$ 9,4 years) suffering from perennial asthma and/or rhinitis with
proven sensitizations to indoor allergens underwent additional intra =
dermal tests with commercial (Bencard) whole body extracts of Tribolium

confusum 1 : 10 000 w/v. The results were compared to former experiences (Asthma-Policlinic , Free University Berlin (West)) with commercial prick test solutions (Bencard) (2).

2. The intensity of intradermal reactions were analysed for three groups of patients (group A : various occupations ; group B : bakers ; group C : farmers).

3. The incidence of concomitant sensitizations to other groups of immediate - type - allergens was registrated for patients with negative and positive reactions to tribolium.

4. A further differentiation was performed for allergies to mould species.

5. 36 patients with marked positive skin tests undenwent intradermal endpoint - titration (dilution steps from 1 : 10^4 - 1 : 10^8).

6. In 18 patients with strong skin reactions , in 15 with mild skin reaction and 19 controls intranasal provocation tests with 5 % w/v glycerinated tribolium extracts as well as physiological saline /glycerol 50 : 50 % control solutions were performed.

Results

1. The comparison (see table I) demonstrates a much higher (30,7 %) ratio of positive skin reactions in intradermal than in prick tests series (16,5 %).

Table I. Comparison of prick tests (Berlin) and intradermal tests (Bad Salzuflen) with tribolium extracts

	Prick test (n = 1040)	Intradermal test (n = 430)
Tribolium confusum	16,5 % positive	30,7 % positive
	83,5 % negative	69,3 % negative

2. Table II shows a much higher incidence of positive intradermal tests in bakers (54,5 %) and farmers (43,2 %) than in a collection of other professions (23,9 %).

Table II. Intensity of skin reactions with respect to occupation

skin reactions	group A (n = 305)	group B (n = 44)	group C (n = 81)
negative (%)	77,1	45,5	56,8
positive (total) (%)	23,9	54,5	43,2
weak (% pos.)	26,0	4,2	28,6
mild (% pos.)	17,8	45,8	25,7
hist.equiv. (% pos.)	35,6	8,3	28,6
strong (% pos.)	20,5	41,7	17,2

Table III. Sensitization patterns in patients with positive and negative skin tests to tribolium

positive intradermal tests	tribolium positive (n = 132) %	tribolium negative (n = 298) %
"house dust "	59,8	53,0
Dermatophagoides pteronyssinus	65,2	43,0
Dermatophagoides farinae	56,8	33,6
Glyciphagus destructor	23,5	10,4
Tyrophagus putrescentiae	14,4	9,4
Acarus siro	19,7	10,1
animal dander	37,1	28,5
feathers	15,2	11,4
food	13,6	15,1
grass pollen	42,4	24,4
yeasts	16,7	13,4
moulds	44,7	30,5

3. A significant higher percentage of sensitizations to various arthropods, danders , feathers , grass pollen , and moulds could be proven in patients

with positive skin tests to tribolium but not for "unspecific" allergens
like house dust , food and yeasts (see table III).
4. A higher ratio of positive skin reactions to penicillium, fusarium, paeci=
lomyces, phoma, aspergillus, botrytis, aureobasidium, chaetomium , and hel=
minthosporium could be demonstrated in patient with intradermal sensiti =
zations to tribolium.

Table IV. Mould sensitization patterns in patients with positive and
negative intradermal tests with tribolium

mould species	tribolium positive (n = 132) %	tribolium negative (n = 298) %
Curvularia lun.	9,1	12,1
Epicoccum purp.	12,1	12,8
Penicillium not.	18,2	11,7
Serpula lacr.	8,3	14,4
Fusarium sol.	31,8	19,8
Mucor muc.	15,2	10,7
Paecilomyces marqu.	28,8	10,1
Phoma bet.	28,0	14,8
Neurospora sit.	10,6	8,4
Ustilago trit.	6,8	9,7
Alternaria alt.	5,3	10,7
Cladosporium herb.	7,6	11,4
Aspergillus fum.	16,7	8,1
Botrytis cin.	13,6	7,4
Aureobasidium pull.	15,2	6,7
Rhizopus nigr.	5,3	6,4
Chaetomium glob.	11,4	5,7
Helminthosporium hal.	8,3	-

5. The dose—titration (see table V) shows for most of the studied patients
a step − by − step "diminuendo" of intensity of the weal − and − flare −

reaction which may be much more indicative for immunological than for
irritation mechanisms.

Table V. Results of endpoint titration in 36 patients with marked skin
reactions to tribolium

reaction	dil.step 1 : 10^4	: 10^5	: 10^6	: 10^7	: 10^8
strong	11	2	1	0	0
histam.equiv.	25	14	5	2	0
mild	0	14	3	4	1
weak	0	4	8	5	1
negative	0	2	19	25	34

Table VI. Results of nasal provocation tests in patients with strong ,
mild and negative skin reactions to tribolium

| Nasal test | Skin test positive | | | negative |
	total n = 33	strong n = 18	mild n = 15	n = 19
positive	10 (30,3 %)	7	3	0
negative	23 (67,7 %)	11	12	19

6. In 30,3 % of 33 patients with positive skin reactions to tribolium the
clinical actuality of sensitization could be proven by means of nasal
provocation tests. Unspecific irritations could be excluded by provocation
tests in 19 controls using undiluted tribolium provocation extract (see
table VI).

Conclusions

Our results demonstrate the superiority of intradermal test techniques
compared to the common prick tests . They should additionally be performed
in patients with negative prick tests; but persisting suspicion of

tribolium exposure. According to the role of tribolium as a widespread pest
on stored products in some professions a high risk of sensitization is to
be expected (e.g. bakers, farmers, millers, silo workers, animal breeders,
animal holders , plant breeders, pet shop workers, pharmacologists , pharma=
cists, biologists, customs officers , vermin killers, garbage men , and
even hoesewifes (!)). The results of the endpoint – titration indicate
that specific mechanisms are involved and therfore IgE–determination should
be performed. A significant risk of tribolium sensitization must be ex =
pected in patients with polyvalent sensitization to typical immediate –
–type– allergens like grass pollen and house dust mite as well as in
patients with sensitizations to "indoor" moulds. According to our ex =
periences the sensitizing potency of tribolium for the lower airways is
rather low , but on the other hand a marked percentage of positive skin
test reactions corresponds to nasal allergies.

In summary it is our impression that tribolium may not be underestimated
as a cause of nasal allergy in patients with occupational and/or
domestic exposure and therefore should be introduced to the diagnostic
routine programme.

References

1. Harde,K.W., F.Secera (1984) Der Kosmos–Käferführer – Die mitteleuropä=
ischen Käfer , 2nd ed. (Franckh , Stuttgart) 226 ff.

2. Rudolph,R. (1984) Möglichkeiten und Grenzen der Hauttestungen.
Allergie Kolloquien Dome Hollister Stier (Tropon, Köln) 15 – 23

3. Rudolph,R. (1985) Zur ökoallergologischen Bedeutung von Trogoderma
angustum SOLIER in Berlin (West) . Habilitationsschrift (Thesis), Freie
Universität Berlin.

Dr.med.Reimar Rudolph , Asthma– und Allergieklinik, Abteilung für klinische
und experimentelle Forschung , Parkstr. 25 – 27 , D 4902 Bad Salzuflen 1 ,
Federal Republic of Germany

EXS 51:
Advances in Aerobiology
© 1987 Birkhäuser Verlag Basel

GIBBIUM PSYLLOIDES ('MITE BEETLE') AS AN INHALATIVE ALLERGEN

E. Fuchs[1], J. Bossert[1], K. Honomichl[2], H.-J. Maasch[3], H. Risler[2], R. Wahl[3]

[1]Deutsche Klinik für Diagnostik, Fachbereich Allergologie, Wiesbaden (D)

[2]Institut für Zoologie der Universität, Mainz (D)

[3]Allergopharma Joachim Ganzer KG, Abteilung für Forschung und Entwicklung von Allergenextrakten, Reinbek (D)

Case History

9 years ago at the age of 18 the patient came to our clinic the first time. He had a history of relapsing rhinitis and occasional asthmatic reactions. Nocturnal dyspnoea occured first in May 1974. During two vacation periods the patient was symptom free. The patient lived in the house of his parents, an older house in a suburb of the city of Mainz. Following his own statements there were no infestations with moulds in the house.

When the patient came back to our clinic in July 1977 we could diagnose an accentuation of the symptoms during the months May to July. Again during vacations in the mountains the patient was symptom-free. Due to the seasonal accentuation of the symptoms which coincidenced with the grass pollen season we did not look for perennial allergens as causative agents.   Only when the perennial symptoms became dominating we has a closer look into the living conditions of our patient.

Since 8 years the patient lived in an attic under the roof of his parents' house. The whole place was totally infested with beetles. As these beetles were regarded as harmless they were tolerated by the inhabitants of the house. These clumsy beetles lived in large numbers on the walls, in the carpets, in the built-in cupboards and were even found in the toothbrush-glass. The beetles had to be removed from clothes before putting them on.

The beetles were recognized as Gibbium psylloides, the English common name is mite beetle, belonging to the family of the Ptinidae. Their size is about 2-3 mm. Their glassy-brown elytrons are grown-together forming a globe with a smooth surface.

The beetles are cosmopolites and have a preference for warm temperatures. Therefore in our regions they are mostly found in houses. They live on plant particles, hay, wool, cereals, etc. As they do not destroy wood they are regarded as harmless, despite of their mass-development. Clefts of the wood they use only for depositing of their eggs. Their lifespan is approximately 18 month. Typical for the family of Ptinidae is the bent down head which is vaulted over by a chitin collar. The hairless elytrons are grown together, so the only way to move is by use of their long legs. Noteworthy is the marked formation of sensilles on the cuticula of the antennae and the legs (fig. 1, 2, 3) (2, 6).

Diagnostic

1. Prick tests were done with extracts of house dust mites, moulds, feathers, grass- and rye-pollen, weed- and flower-pollen. The patient demonstrated a moderate skin reactivity and had only positive reactions to mugwort- and chrysanthemum-pollen. The suspected sensitization to grass- and rye-pollens could not be confirmed in December 1981.

2. A rub-test with three crashed beetles was strongly positive with confluent wheals and with asthmatic reaction.

3. We prepared an extract from the mite beetle by the Frugoni method, that is extraction of the crashed material with 12 % alcohol at a concentration of 1:99. With this extract we did a careful intracutaneous skin titration at the forearm of the patient. At a dilution of 10^{-11} we had a clear wheal reaction, at a dilution of 10^{-10} the wheal had a mean diameter of 8 mm which is comparable to the size of the histamine control solution. Unspecific reactions were achieved in a number of control persons at dilutions of 10^{-2} to 10^{-3}.

4. A bronchial provocation test with the aerosol of the mite beetle extract was clearly positive.

5. In the Prausnitz-Küstner assay we could provoke a positive reaction in the skin 48 hours after the transfer of the patient's serum to the skin of another persons. This proves that the specific antibodies could be transfered passively to other persons.

6. Specific IgE-determinations were done with two different RAST-disks. In one case we coupled the Frugoni-extract to the disks, in the other case we took a Coca's extract for this purpose. Both disks yielded

positive reactions. Frugoni disks gave a mean of 15.5 % binding which means RAST-class 3. The other disks gave also RAST-class 3, however with approximately 10 % more binding of specific IgE. This is probably due to the higher amount of protein which could be extracted by the Coca's method.

In addition we performed a high performance thin layer chromatography (HPTLC). By the negative results we could exclude the presence of histamine or histamine like substances which could give false positive results.

Discussion

We here demonstrated a case of asthma which developed in the course of eight years with increasing intensity during this time. The pathogenetic relations between the mass-development of the beetles and the development of the patients' disease could clearly be demonstrated by our diagnostic work in connection with a careful anamnesis. However, the pathological importance of this beetle as an inhalant allergen in house dust has yet to be demonstrated by further investigations.

An attempt to irradicate the beetles by insecticides failed.

If one looks at the electron-microscopical pictures one could assume that the vesicles of the cuticula on the antennae and the legs of the beetles could contribute to the allergenicity of house dust. A positive rub-test especially with the legs of the beetles underline this assumption. It is yet to be established whether also the faeces of the beetles can act as allergens. In earlier investigations on the role of insect dust as inhalative allergens we have reported about this phenomenon. Bellas (1) has summarized the literature on this special topic recently. To our knowledge it is not yet completely known whether there are cross reactivities of allergens from different orders or families of the insects which would mean that a sensitization to one species would automatically include a sensitization to related species. We know about such cross reactivities from our own observations with allergies to grasshoppers, crickets and cockroaches (3,4, 5).

In the course of several months with strict avoidance of exposition and with temporary inhalative bronchiolytic therapy the lung function of the patient could be restored to almost normal values (Resistance AW: 2.1 mbar/1/ sec, vital capacity VK = 6.63 1, forced experitory volume FEV: (1.0) 4.89,

PEF: 15.3).

There were no asthma attacks and the patient was able to perform sports again. Only when he is visiting his parents' house he is experiencing again asthma attacks. Another hint that our diagnosis was correct.

The demonstrated case shows once more the importance of a specific anamnesis for the diagnosis of allergies. In some cases only repeated interrogation of the patient leads to the right solution. The cooperation of the patient and the shrewdness of the investigatior are of great importance.

This case is also another example for the high allergenic potency of dust from insects.

References

1) Bellas, T.E. (1981, 2nd Edt. 1982) Insect as a cause of inhalant allergies. CSIRO Division of Entomology. Commonwealth Scientific and Industrial Research Organization, Canberra Australia.

2) Crowson, R.A. (1981) The biology of coleoptera. Academic Press London.

3) Fuchs, E. (1979) Insects as inhalant allergens. Allergol. Immunopathol. 7, 227-230.

4) Rudolph, R., D. Jung, G. Kunkel, G. Diller, C. Baumgarten (1980) Über einen Fall von allergischem Asthma durch häusliche Exposition gegen Trogoderma angustrum. Allergologie 3, 17-20.

5) Schwartz, H.J. (1981) Inhalant insect allergy. In: Monograph on Insect Allergy Edited by Macy, I., Levine, M.D., Richard, F., Lockey, M.D. Library of Congress Pittsburgh.

6) Weiner, H. (1982) Bestimmungstabellen der Vorratsschädlinge und des Hausungeziefers Mitteleuropas. 4. Aufl., Fischer, Stuttgart.

Author's address

Prof. Dr. med. habil. Erich Fuchs, Stiftung Deutsche Klinik für Diagnostik GmbH, Fachbereich Allergologie, Aukammallee 33, D-6200 Wiesbaden (FR-Germany).

Fig. 1

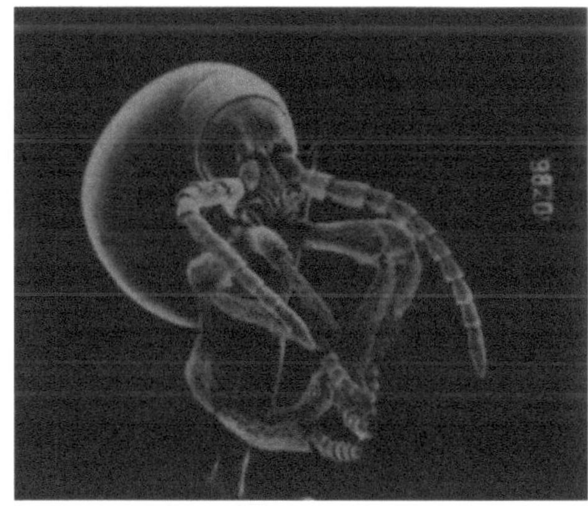

(magnified
24 times)

Fig. 2

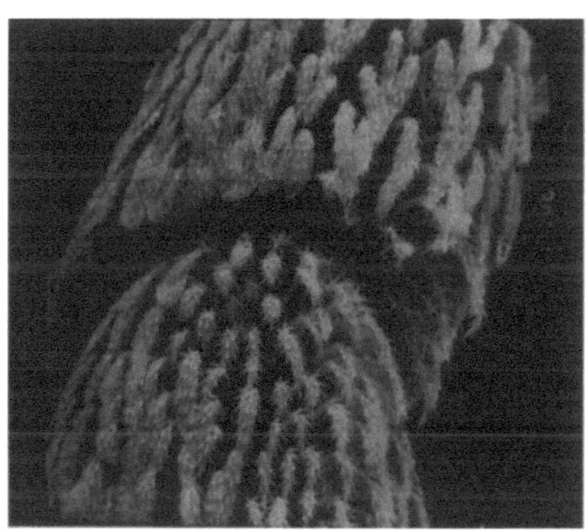

(magnified
580 times)

Fig. 3

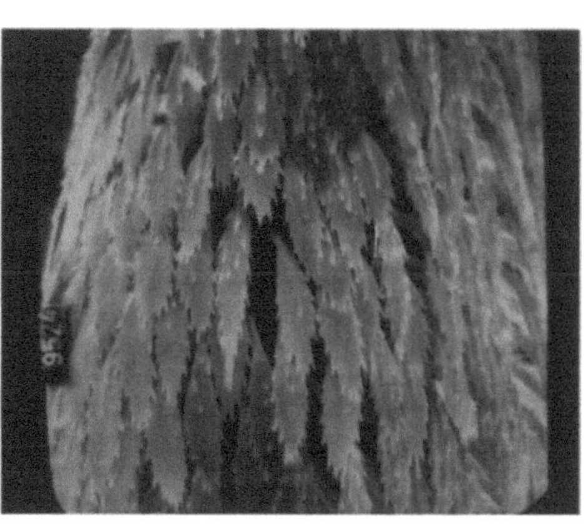

(magnified
670 times)

INVESTIGATIONS OF ALLERGEN-CONTAINING DUST SAMPLES
FROM THE INTERIOR OF THE HOUSE

E. Bischoff and W. Schirmacher
Gesellschaft für hausbiologische Forschung mbH
D-6500 Mainz 1, FRG

1. The problem

Among the contaminants of biological origin of the air in rooms, that
derived from housedust mites is the most important (1-4). It is not the
mites themselves which are present in the inhaled air but their excreta,
which also contain the allergens (5-8). They get into the air as the finest
fraction of the housedust by the activities of people inhabiting the rooms
or by air convection (4-6). There are mite-infested houses almost every-
where.Accordingly, most people in rooms inhale air which contains housedust
mites excreta.

2. Method

The sources of allergen-containing housedust are places, where mite
biotopes are able to develop. This is not only the bed (9).
The dust samples are obtained as usual by vaccum-cleaning. A critical

discussion of this sampling method is given elsewhere (10). The number of live and dead mites in the dust samples is determined using the flotation method (11).

More important is the determination of the excreta content, since these contain the allergens which pass into the air. For this we use the guanine method which was developed by us and is now available under the name ACAREX^R test (9, 12, 13).

The guanine is a product of the metabolic degradation of the mites extracted quantitatively and subsequently detected by formation of an azo dyestuff. A colour scale indicates the gradings 0 "no finding", 1 "slight", 2 "moderate" and 3 "heavy". All guanine levels $\geq$1 % with the colour step 3 indicate that the housedust and thus the air in the room is heavily contaminated with mite excreta.

The allergen contents are determined by RAST inhibition (22).A new evaluation scale MAC (Mite Antigen Content) is used.

3. Biological contamination of air in rooms

3.1 Live and dead mites and mite-excreta in a house

The house near Lake of Constance was inspected in February 1985 and then in June. Six weeks earlier, one of the two mattresses which were heavily infested with mites had been replaced by a new one which had been treated prophylactically with an acaricide (14, 15).

There are live and dead mites on the upholstery. ACAREX^R test shows moderate to heavy infestation. Only dead mites are found on the carpets in the living room, there being no detectable mite excreta (tab.I). This indicates that the source of the contamination of the air in the living room is solely the upholstery. It is merely necessary for this to be treated with an acaricide or removed.

As expected, the old mattress still in the bedroom shows a marked increase in live mites, appropriate for the season, as well as a high concentration of excreta. In contrast, on the new mattress is to be found only a little dust, and this contained only a few live and dead mites.

Table I: Dead and live house dust mites, loading of dust by mite excrements and dust quantity in carpets, upholstery and mattresses found in a house near Lake of Constance (Ba 03.06.85)

Room	object	area m^2	dust g	mites/g dead	live	mites (total) dead	live	ACAREX^R-step
living room	upholstery	25	51,2	145	3	7424	154	2-3
living room	carpet	25	24,5	5	0	123	0	0
living room	carpet	45	54,0	7	0	378	0	0
guest room	mattresses	26	20,5	310	7	6355	144	1-2
guest room	carpet	10	14,0	0	0	0	0	0
bedroom (04.02.)	mattresses (2, old)	18	3,6	248	66	893	238	3
bedroom	mattresses(old)	9	4,1	113	44	463	180	3
bedroom	mattresses(new)	9	0,6	1	2	-	-	-
bedroom	carpet	16	13,3	6	0	80	0	0

3.2 Sources of biological contamination of the air

The excreta determinations by the ACAREX^R test for eight households investigated in 1985 are reported in tab.II for ACAREX^R steps above grading 1.

In every case the beds contain large amounts of excreta. The upholstery in the living rooms is hardly less contaminated. Only in one of the houses (We.) a heavy contamination is found solely in the bedroom. Upholstery in childrens rooms is also a relatively common source of air contamination. On occasion, carpets are also heavily contaminated.

Table II: Biological charge of dust samples by mite excrements in upholstery, mattresses and carpets of 8 houses in South Germany (guanine values) - ACAREX^R-Test steps > 1.

House Date		Ba 6/85	Bö 6/85	Fe 8/84	Ga 7/84	Le 6/85	Pö 6/86	We 7/84	Wo 7/84
Room:	object:								
liv.room	upholstery	2-3	3	2	1-2	3	3	-	1-2
liv.room	carpet	-	2	-	-	3	-	-	-
childr.room/ workroom	upholstery	-	-	2-3	?	-	-	-	3
bedroom	bed (mattress)	3	3	2	2-3	3	3	2-3	2
bedroom	carpet	-	-	-	2	-	-	1-2	-

3.3 Dust contamination and mite population

The examples show that the contamination of dust samples can be determined using the ACAREX^R test. The question is whether there is a relation between dust contamination and mite population.

The excreta content in dust may be regarded as relatively fixed since it is the "result" of the presence of a mite population, beginning with the first colonization, perhaps before years. This content is reduced by losses caused by dust being removed, whether by cleaning or by use.

The population of live mites, however, depends on the current microclimatic conditions which promote or inhibit the development of a habitat suitable for mites. It completely or partially disappears when room-heating begins, i.e. when the air becomes dry. Nevertheless, the dust and the air continues to be contaminated.

Thus, there is not expected to be a direct relation between dust contamination and housedust mites. The "correspondence" - sometimes observed - is, however, no more than the finding that places, where viable populations have been able to develop, the habitat must previously have been suitable as well, as shown by the high excreta content of the dust (9, 16, 17).

Example: In a house in lower Bavaria (tab.III) the dust in some rooms is significantly contaminated although no live mites are to be found. Sometimes, there tends to be a rather loose relation between dust contamination and the number of dead mites.

Table III: Dead and live house dust mites, loading of dust by mite excrements and allergens and dust quantity in textiles in a house in Low Bavaria (Wo 24.07.84 etc.)

Room	object	area m^2	dust g	mites/g dead	live	mites (total) dead	live	ACAREX[R] step	Guanine (%) $(10^{-2}g/$ g dust)	Allergen P 1 $(10^{-9}g/$ g dust)
liv.room	upholstery	19	59	225	0	13 275	0	1-2	-	-
liv.room (04.05.)	upholstery	19	-	210	0	-	-	1	0,15	5 980
liv.room (14.08.)	upholstery	19	36	500	0	18 000	0	0-1	0,07	2 280
liv.room	pillow	-	23	148	0	3 404	0	1	-	-
liv.room (14.08.)	pillow	-	19	313	0	5 813	0	0-1	0,05	1 380
liv.room	carpet	6	44	17	0	748	0	0-1	-	-
kitchen	upholstery	11	60	40	0	2 400	0	0-1	-	-
kitchen (14.08.)	upholstery	11	42	35	0	1 481	0	0	0,05	460
child.r.	upholstery	28	39	103	85	4 017	3 315	3	1,71	125 500
child.r.	carpet	4	21	27	3	567	63	0-1	-	-
bedroom	mattresses	26	58	163	3	9 454	174	2	0,55	16 800
bedroom (04.05.)	mattresses	26	-	35	0	-	-	2-3	0,55	19 800

4. The content of guanine and allergen in dust

For samples from the same house (tab.III) the content of allergen P 1 is determined by RAST inhibition (22), and the guanine content with ACAREX[R] test, as well as quantitatively by direct colorimetry and with enzymatic breakdown of guanine.

It can be seen (fig.1) that the guanine and the allergen contents in the dust samples behave in a parallel fashion. This confirms preliminary results of this kind (18, 19), as well as recent medical findings (20, 21). (In these investigations only the antigen P 1 is determined.)

Whereas allergen content and guanine content are correlated, this is not the case for allergen content and numbers of mites (fig.2), although, on occasion, there is apparently a temporary relation (4, 5, 21).

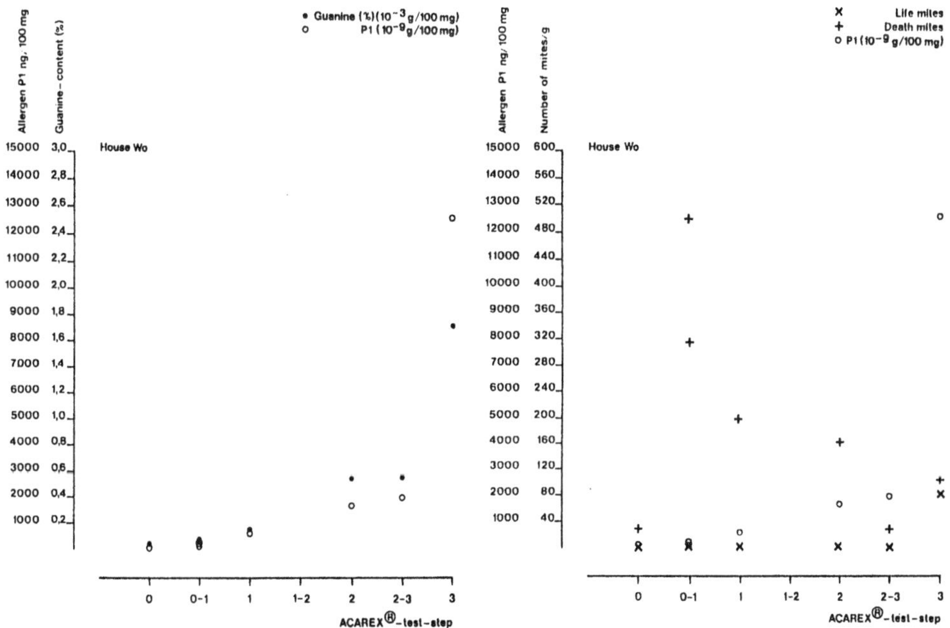

Fig.1: Guanine- and allergen-P1-content in dust-samples.

Fig.2: Allergen-P1-content and number of housedust mites.

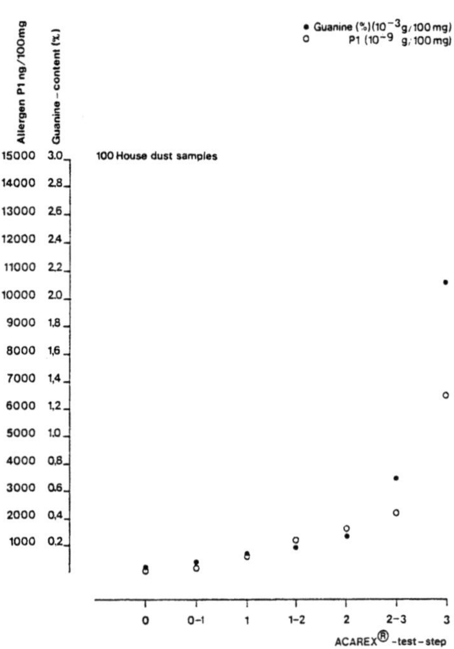

Fig.3: Mean values of guanine- and allergen-P1-content of 100 dust samples from different houses.

On the other hand, a survey of the findings of the guanine and antigen P 1 determination on 100 dust samples (tab.IV) confirms the correlation which was to be seen in fig.1. Fig.3 shows the mean contents of guanine and allergen found for each step in the ACAREX[R] test (9, 22).

In addition tab.IV provides a new classification of the antigen P 1 content determined by RAST inhibition, being based on a suggestion by Platts-Mills et.al.(1985) (21).

This MAC classification (Mite Antigen Content) serves for evaluating the medical significance of the allergen content of dust samples and is - in this manner - comparable with the RAST classes for the evaluation of the relative contents of antibodies (spec.IgE) in a serum. The MAC-scale shows for the same dust samples analogical results to the guanine scale of ACAREX[R] test, although the MAC scale was conceived independently (tab.IV).

Table IV: ACAREX[R]-Test, colour indication, steps and guanine classes
Mite Antigen Content (MAC), classification, RAST-Inhibition, mean values of allergen P 1 and corresp. MAC-classification
100 dust samples from houses in the Netherlands and in South Germany

| | | ACAREX[R]-Test | | | MAC (Mite Antigen Content)* | | | | |
Colour finding	classi-fication	ACAREX[R] step	guanine class theoret. range ($\%$) (resp. 10^{-2}g/g dust)	guanine content mean value found ($\%$) (resp. 10^{-2}g/ g dust)	antigen content classification (ng/g dust)	MAC classi-fication	MAC step	allergen P 1 content (10^{-9} g/ g dust) mean value found RAST-Inhib.	MAC classi-fication mean value found RAST-Inhib.
yellowish	no	0	0,0	0,0 - 0,04	0,04	0 - 500	very low	0	260	very low
		(0-1)	0,03	0,01 - 0,08	0,07	500 - 2 000	low	(0-1)	1 240	low
light red	slight	1	0,06	0,03 - 0,16	0,14	2 000 - 10 000	significant	1	5 490	signif.
		(1-2)	0,12	0,06 - 0,24	0,17	10 000 - 20 000	rath.high	(1-2)	11 390	rath. high
red	mean	2	0,25	0,12 - 0,36	0,35	20 000 - 35 000	high	2	15 550	rath. high
		(2-3)	0,50	0,25 - 0,75	0,68	35 000 - 60 000	very high	(2-3)	21 020	high
intensive red	strong	3	$\geq 1,0$	0,50 - >1,0	2,50	$\geq$ 60 000	extr.high	3	64 660	extr.high

* amplified proposal of Platts-Mills et al. (1985) (21)

5. Summary

Using ACAREX[R] test, it is straightforward to determine the allergen-containing excreta of housedust mites in dust samples and to detect the sources of a contamination.

Investigation of the antigen content by RAST inhibition shows that there is a correlation between guanine and allergen content in a dust sample.

There is no direct relation with the number of live (or dead) housedust mites, nor is this to be expected.

References

1) Berrens (1975) Naturwissenschaften 62, 29-33.
2) Bronswijk,van, Sinha (1971) J. Allergy 47, 31-52.
3) Bessot, Pauli (1986) Bull. Eur. Phys. Resp. 22, 1-8.
4) Tovey,Chapman,Wells,Platts-Mills (1981) Am. Rev. Resp. Dis. 124,630-635.
5) Bischoff (1986) Allergol. Fortbildung Bd. 1, in press.
6) Tovey, Chapman, Platts-Mills (1981) Nature 289, 592-593.
7) Tovey, Chapman, Platts-Mills (1980) Clin. Allergy 10, 466-467.
8) Chapman, Heymann, Sutherland, Platts-Mills (1985) Int. Archs Allergy appl. Immun. 77, 166-168.
9) Bischoff (1986) Swiss Med 8, 39-44.
10) Bischoff, Bronswijk, van (1986) Allergologie 9, 375-378.
11) Bronswijk, van (1973) J. Med. Ent. 10, 63-70.
12) Bischoff,Schirmacher(1984)Allergol.7,446-449,and (1985)Allergol.8,36-38.
13) Bischoff, Schirmacher, Schober (1985) Allergol. 8, 97-99.
14) Bischoff, Fischer, Wetter (1986) Allergol. 9, 171-176.
15) Bischoff, Liebenberg, Wetter (1986) Allergol. 9, 287-292.
16) Bischoff, Krause-Michel, Nolte (1986) Allergol. 9, in press.
17) Mitchell, Wilkins, McCallum Deighton, Platts-Mills (1985) Clin. Allergy 15, 235-240.
18) Bronswijk, Bischoff, Schirmacher, Berrens, Schober (1986) J. Med. Entomol. 23, 217-218.
19) Bronswijk, van, Exp. and appl. Acarology, in press.
20) Pauli, Tenabene, Bessot, Hoyet (1986) Rev. fr. Allergol. 26, in press, and Pauli, Tenabene, Bessot (1986) 3rd Internat. Conf. on Aerobiology, Basel.
21) Platts-Mills, Heymann, Chapman, Smith, Wilkins (1985) Int. Archs Allergy appl. Immun. 77, 163-165.
22) Aalberse, Reerink-Brongers, Vermeulen (1973) Int.Arch.Allergy 45, 46-49.

Dr. E. Bischoff, Gesellschaft für hausbiologische Forschung mbH, Rheinallee 96, D-6500 Mainz 1, FRG.

HOUSE DUST MITE IN DIFFERENT ALTITUDES OF GRISONS

G. Menz[1], E. Petri[1], P. Lind[2], Chr. Virchow[1]
Hochgebirgsklinik Davos-Wolfgang/Switzerland[1] and University of Copenhagen
Protein-Laboratory[2]

KERN (4) and COOKE (3) were the first who connected the pathogenesis of house dust asthma with an immunological process. "House dust contains an allergen" they proclaimed. "Is a mite the producer of the house dust allergen?" asked VOORHORST and coworkers 1964 (10). They found that mites of the genus Dermatophagoides could be recognized to be an important source of house dust allergen. They also noticed that dust samples of houses in Davos (high mountain area in Grisons-CH) contained less allergen than dust from houses in the Netherlands. (8) The optimum environment for Pyroglypidae is a temperature of 20-25° Celsius and a humidity of 70-80%. (7) Using a major allergen purified from Dermatophagoides pteronyssinus TOVEY et al reported an assay suitable for measuring mite derived allergen in house dust. This major allergen was called antigen P1. (9) Using this radioimmunoassay they demonstrated that the allergen accumulates in mite culture as fecal pellets and also becomes airborne in form of fecal pellets. These studies suggested that the relevant form of mite allergens was almost certainly the fecal pellet.

Techniques used in mite allergen analysis are: microscopy, cultivation and immunochemical analysis. Counting mite numbers in dust is tedious and requires some flotation procedure prior to microscopic examination. In order to get a better understanding of the importance of dust mites it was clearly nessessary to have easier and better methods to quantitate mite allergens. (9, 6, 5) TOVEY et al used RIA. The sensitivity for P1 was 5 ng P1/g dust that means less than 1 mite/g dust. Species-specific ELISA for major excrement allergens of Dermatophagoides pteronyssinus (D.pt.), Dermatophagoides farinae (D.f.) and Dermatophagoides microceras was established by P. LIND (5). He used immunabsorbed monospecific rabbit antibodies coupled to horse radish peroxidase. The limit of detection is 13 ng/ml for D.pt. that means less than one mite per gram dust. Acarex®-Test is a test for Guanine which is a product of the mite. (1, 2)

<u>Results:</u> The house dust samples were collected from houses in different altitudes of Grisons (530-2383 meter a.s.L.) and a house of an asthmatic patient in the Saarland (West Germany). Dust was sampled by vacuum cleaner in the last days of June 1986. <u>Normal values and range</u> (tab. I): We consider that 10.000 ng of antigen Dp 42 = P1 equivalent per gram should be regarded as high, more than 2.000 is significant while less than 2.000 is low and less than 400 is very low. (5, 6) (The Enzyme-linked immunosorbent assay was performed in the protein laboratory of the University of Copenhagen.) Guanine-test-graduation (Tab. 4) varies from negative (0), low (1), moderately (2) to high (3). (1, 2) Obviously these figures are only a guide comparable to a pollen count. (6)

Table I

Mite Antigen P_1 (ng/g)

ng/g dust	degree of concentr.
10.000	high
> 2.000	significant
< 2.000	low
< 400	very low

Guanin-Test (Acarex^{o})

% Guanin	Graduation
0	negative (0)
0,06	low (1)
0,25	moderately (2)
1	high (3)

<u>Bedrooms</u> (tab. II): In the dust sample of the bedroom in Neunkirchen/Saar there was less than 2.000 ng/g dust of Dp 42 = P1 (1.740 ng/g dust). That means a low to borderline significant level. This bedroom of an asthmatic patient had been changed earlier in order to reduce and avoid mite allergen. (7) The concentrations of P1 (Dp 42) in samples taken from bedrooms in houses in different altitudes of Prättigau-Valley were 372 ng/g, 318 ng/g and 384 ng/g that was very low! In the patient's room in the Hochgebirgsklinik Davos-Wolfgang there was not found any mite allergen. In a sample of the bedroom of a private flat in Davos we measured 1.440 ng/g which is <u>low</u> but higher we had expected for this altitude. In this room there was a wall to wall carpeting and in addition an oriental woolen carpet. The average temperature in this room was 20-22° C constantly.

<u>Living rooms</u> (tab. III): Significant to high levels of D.pt.'s major allergens in the asthmatic patients' privat flat in Saarland was found (4.320 ng/g dust). An especial high concentration was found in a sample of upholstered furniture (16.800 ng/g). In the different altitudes of Prättigau valley there were very low concentrations between 300 (Klosters) and 372 (Landquart). Major allergen of D.pt. could not be detected in the Hochgebirgsklinik (0 ng/g dust), low concentration in the Davos-private-flat (480 ng/g) and very low in the Flüela-Pass-Hospiz (2.383 m) (192 ng/g).

Discussion: Comparing the results of the different methods we are able to state:

1. The results of the <u>ELISA</u> and of the <u>Guanine-test</u> and if done <u>mite count</u> correspond in a sufficient manner. (There was only one exception. In the dust sample of the private bedroom in Germany Guanine-test was negative while ELISA titer was 2. In this special case the amount of the fine dust sampled for Guanine-test was less than the whole test spoon which was recommended for the test procedure).

2. We found much higher levels of mite antigen in the dust samples in Neunkirchen/Saar than in the samples in Grisons. According to the increasing altitude the number of mites and content of mite antigen decreases. Low but higher than expected was the concentration in a sample of the private bedroom in Davos. These results show that not only the altitude but also microclimatic factors and the biotops (wall to wall carpeting, cerpets, high room-temperature) are of importance; even in the higher altitude.

3. The ELISA method for the 3 mites used in this study is a very sensitive and specific assay.

4. Guanine-test is easily to perform and could be regarded as a good screening method.

Obviously these figures are only a guide. Nonetheless use of the assays will make it easier to estimate the significance of mite sensitivity in asthmatic patients. In addition it will be possible to evaluate <u>both</u>: a) avoidance measures in houses and the effects of acaricidal chemicals in a quantitative manner, (6) especially with the ELISA method for the 3 mite

Content of mite major allergen (D.pt)

Table II — Bedrooms

Location	Altitude	ng/g dust	Titer	Guanine-Test	Mite-Count
Neunkirchen (Saar) BRD	265 m	1.740	2	0	
Landquart (GR)	530 m	372	1	1	1 D.f.
Küblis (GR)	800 m	318	1	1	
Klosters (GR)	1.180 m	384	1	1	
Davos-Wolfgang a) Hochgebirgs-klinik klinik Patient's Room b) Private flat	1.600 m	0 1.440	0 2	0 2	0 17
Flüela-Pass (Hospiz)	2.383 m	192	0	0	

Table III — Living-Rooms

Location	Altitude	ng/g dust	Titer	Guanine-Test	Mite-Count
Neunkirchen (Saar) BRD	265 m	4.320 (16.800 x upholstered furniture)	0 16	2	
Landquart (GR)	530 m	372	1	1	
Küblis (GR)	800 m	360	1	1	
Klosters (GR)	1.180 m	300	0	1	
Davos-Wolfgang a) Hochgebirgs-klinik klinik Patient's Room b) Private Living-Room	1.600 m	0 480	0 1	0 0	0
Flüela-Pass (Hospiz)	2.383 m	192	0	0	

species. It is planned to continue this study by collecting samples in a
different period of the year to detect seasonal changes in the mite concentrations in this area.

Literature:

1. Bischoff, E., Schirrmacher, W.: Farbnachweis für allergenhaltigen
   Hausstaub, 1. Mitteilung, Allergologie 7, 446-449 (1984)

2. Bischoff, E., Schirrmacher, W.: Farbnachweis für allergenhaltigen
   Hausstaub. 2. Mitteilung: Weitere Gesichtspunkte. Allergologie 8,
   36-38 (1985)

3. Cooke, R.A.: Studies in specific hypersensitiveness. IV. New etiologic
   factors in bronchial asthma. J. Immunol. 7, 147 (1922)

4. Kern, A.: Dust sensitization in bronchial asthma. Med. Clin. N. Amer. 5,
   751 (1921)

5. Lind, P.: Enzyme linked Immunosorbent Assay (ELISA) for determination
   of major excrement allergens of the house dust mite species D. pteronyssinus, D. farinae and D. microceras. Allergy (1986) (in press)

6. Platts-Mills T.A.E., Heymann P.W., Chapman, M.D., Smith, T.F., Wilkins,
   S.R.: Mites of the Genus Dermatophagoides in Dust from the Houses of
   Asthmatic and Other Allergic Patients in North America: Development
   of a Radioimmunoassay for Allergen Produced by D. farinae and/or
   D. pteronyssinus. Int. Archs. Allergy appl. Immun: 77, 163-165 (1985)

7. Petri, E., van Bronswijk, J.E.M.H., Rijckaert, G., van de Lustgraf, B.,
   Trendelenburg, F.: Prophylaxe bei Hausstauballergie. Allergologie 5,
   109-119 (1982)

8. Spieksma, F. Th., Zuidema, P., Leupen, M.J.: High Altitude and House-
   Dust Mites. Brit. Med. J. 82-84 (1971)

9. Tovey E.R., Chapmann, M.D., Platts-Mills, T.A.E.: Mite faeces are a
   major source of house dust allergens. Nature 289, 592-593 (1981)

10. Voorhorst, H., Spieksma-Boezeman, M.I.A., Spieksma, F.Th.M.: Is a mite
    (Dermatophagoides sp.) the producer of the house-dust allergen?
    Allergie und Asthma 10, 329-334 (1964)

Author's address:

Dr. G. Menz, Hochgebirgsklinik, CH-7265 Davos-Wolfgang/Switzerland

EXS 51:
Advances in Aerobiology
©1987 Birkhäuser Verlag Basel

GUANINE DOSAGE IN HOUSE DUST SAMPLES
AND QUANTIFICATION OF MITE ALLERGENS

G. Pauli, A. Tenabene, J.C. Bessot, C. Hoyet
Service de Pneumologie, Centre Hospitalier Universitaire,
Strasbourg (F)

Introduction

Mites represent the most important aeroallergen. Since first studies
of VOORHORST et al. (7), the role of pyroglyphid mites was clearly established
in the etiology of respiratory allergic diseases. Identifying house dust mites
through visual count methods is exceedingly cumbersome. Major house dust mite
allergens have been purified and it is now possible to measure these allergens
accurately in house dust or air samples. TOVEY et al. (6) demonstrated that
the antigen P1, the major allergen from dermatophagoïdes pteronyssinus (Dpt)
was mainly identified in mite fecal pellets in which P1 concentration can
reach 0.1 ng / particle. Recently, PLATTS-MILLS et al. (5) have developed a
radioimmunoassay (RIA) for antigen P1 measurement. However visual count methods as well as RIA cannot be easily used on large numbers of samples and
without elaborated technicity.

In 1984, BISCHOFF and SCHIRMACHER (1) reported a colour test allowing
a semi-quantitative dosage of the guanine found in house dust samples. This
technique allows simultaneous extraction and quantification of guanine,
through the formation of an azo dye ; the results can be classified into four
groups depending on increasing amounts of guanine. The test is based on the
fact that guanine is the main component of the nitrogenous waste of cheliciferous arthropods. Whereas in other arthropods i.e. in insects whose waste products can be found in house dust, the nitrogenous waste is uric acid and not
guanine.

The present study was undertaken to assess the correlation between the
mite allergenicity of a house dust extract and its content in guanine. For

this purpose we performed two different studies : - measurement of the mite allergenicity of H.D. extracts variable in their guanine concentration by skin tests, in volunteer individuals sensitive to Dpt, - assessment of different allergenicities of H.D. samples selected on clinical observations.

<u>Material and methods</u>

* <u>Guanine level detection (ACAREX-TEST[R])</u> : H.D. samples were collected by use of a vacuum cleaner. Dust samples were stored in sealed plastic bags until samples were sieved to obtain fine dust. 0.4 ml of sieved dust i.e. about 150 mg are added to 2 ml of the alkaline alcoholic extraction liquid. A test strip is dipped into the supernatant and the intensity of the colour is read on a standardized colour four step scale. Results are expressed in 4 classes : negative (guanine content : < 0.01 %), slightly positive (0.06 %), moderately positive (0.25 %), strongly positive (guanine content : at least 1 %).

* <u>Skin test study</u> :

   - 10 patients were selected with perennial asthma with or without rhinitis. All were skin tested prior to the study and were found to be skin reactive to Dpt extract, with a positive cutaneous test after intradermal injection of 0.04 ml of a 1/100.000 Dpt extract (Institut Pasteur, details of the preparation have been described previously (3)). All were found to have negative skin tests to cat and dog extracts. The 10 patients selected had high Dpt specific IgE serum levels, but had no detectable serum specific IgE for cat and dog allergens.

   - House dust previously selected according to its guanine concentration was extracted in a sodium phosphate saline buffer solution at pH 6.8 and mixed at the same time with a mechanical stirrer for 24 h at 4°C. After centrifugation (for 5 minutes at 1000 g), the different extracts were stocked (1 : 10, W : V) for prick and intradermal tests, in the presence of 50 % glycerol and 5 ‰ phenol. Four crude H.D. samples with increasing amounts of guanine were used for scratch tests.

   After 15 minutes the wheal reactions were outlined with a ball point pen and the markings transfered by means of tape. The wheal figures were reproduced on homogeneous paper, cut out, and weighted. The weight of each paper is proportional to the wheal area and reflects the intensity of the skin test reaction.

* <u>Analysis of house dust samples</u> collected in areas known to be highly bene-
fical to mite allergic patients : symptoms improvement in altitude being
related to mite number decrease, we quantified guanine contents in dust
samples obtained from high altitude homes (CHAMONIX VALLEY 1200-1450 m :
15 houses visited from November to March 1986). Many studies support the
view that hospital beds are free of mites or contain small numbers only.
We measured guanine concentrations in 6 house dust samples collected in our
hospital where matresses are incased in plastic.

- Analysis of house dust samples from homes where Dpt allergic patients
  showed exacerbated symptoms : 7 patients (4 asthma, 3 rhinitis) exhibit
  symptoms only when they move to another home. H.D. samples from the home
  where symptoms occured and H.D. samples from the usual home were investigated for guanine concentration, at the same time of year.

<u>Results</u>

* <u>Skin test studies</u> : with various skin test methods we could confirm a concordance between the mite allergenicity of a given H.D. sample and its
guanine concentration. For 3 guanine free H.D. samples no significant cutaneous reaction was observed among Dpt allergic patients. The slight
wheal that can be observed in certain patients with scratch tests is non
specific as it was also observed with saline control solution (table I).
In contrast increasing guanine amounts in H.D. extracts gave increasing
cutaneous reactions. We evaluated the multiplicative factor between the
wheal area obtained with a slightly positive H.D. extract and a strongly
positive H.D. extract. This multiplicative factor varied from 1.6 to 3.6
in the ten patients tested. In 3 patients the wheal observed after scratch
tests with a strongly positive crude H.D. sample was similar to the wheal
caused by the positive control test performed with a commercial Dpt extract (Dpt 1/1.000 Bencard).

* <u>House dust samples analyses</u> : results are reported in table II and III.
The results of analyses carried out in 21 samples of H.D. reputedly poor
in mites showed complete absence of guanine for 15 samples, and a very
slight content in 5 samples. In 7 samples of dust from homes where allergic patients presented with symptoms, the guanine concentration was higher
than in the samples from homes where the same patients were symptom free
(table II). We performed positive controls by applying the same colour
test to 3 H.D. mite culture and to one sieved H.D. mite culture containing
mostly faeces (table III).

WEIGHT IN mg OF THE WHEAL AREA OBTAINED BY SKIN TESTING WITH H.D. EXTRACTS CONTAINING VARIABLE GUANINE CONCENTRATION

Guanine concentrations \ Patients	SCRATCH TESTS					INTRACUTANEOUS TESTS			PRICK TESTS	
	BO..	OG..	BO..	MA..	DE..	MA..	WI..	SCH..	RE..	HE..
NEGATIVE	15.7	5.1	3.3	0	4.1	0	0	0	0	0
SLIGHTLY (+)	22	6.8	6.2	4.3	6.4	2.3	2.9	3.6	1	1.2
MODERATELY (++)	42.4	7.6	9.1	8.6	8.5	3.8	5.1	4.3	2	1.9
STRONGLY (+++)	47.1	10.9	13.1	15.7	13.6	8	9.7	9.4	3	3.7
MULTIPLICATIVE FACTOR BETWEEN + AND +++	2.1	1.6	2.1	3.6	2.1	3.4	3.3	2.7	3	3.1
POSITIVE CONTROL (Dpt 1/1000 BENCARD)		14.2	11.4	14.6						

Table I

GUANINE CONCENTRATION IN PATIENTS WHO NOTICED WORSENING SYMPTOMS IN PARTICULAR HOMES

	AGE	SYMPTOMS	$\frac{D + d}{2}$ 1/100.000	RAST	HOME 1 (without symptoms)	HOME 2 (with symptoms)
ZAE ♀	24	Rhinitis	14 mm	4	Slightly (+)	Strongly (+)
GIL ♂	35	Asthma	48 mm	4	Moderately (+)	Strongly (+)
HER ♂	22	Asthma	30 mm	N.D.	Negative	Strongly (+)
WAR ♂	20	Rhinitis	17 mm	4	Negative	Strongly (+)
MOU ♂	16	Rhinitis	14 mm	4	Bed (1) Slightly (+)	Bed (2) Moderately (+)
MAN ♂	17	Asthma	18 mm	3	Slightly (+)	Moderately (+)
BOU ♂	20	Asthma	15 mm	4	Slightly (+)	Moderately (+)

Table II

GUANINE CONCENTRATION IN HOUSE DUST SAMPLES			
GUANINE CONCENTRATION	HIGH ALTITUDE (n = 15)	HOSPITAL H.D. SAMPLES (n = 6)	HOUSE DUST MITE CULTURE (n = 3)	SIEVED H.D. MITE CULTURES i.e. FAECES++ (n = 1)
NEGATIVE ($\leq$ 0.01 %)	11	4		
SLIGHTLY POSITIVE (0.06 %)	3	2		
MODERATELY POSITIVE (0.25 %)	1			
STRONGLY POSITIVE (at least 1 %)			3	1

Table III

Discussion

The correlation between the intensity of the colour test and the guanine concentration of a given H.D. sample has been verified by BISCHOFF and SCHIRMACHER (1). However in order to assess the ability of this test to detect H.D. allergenicity and to disclose its full applicability, it is necessary to establish a good agreement between guanine concentration and mite allergenicity. The preliminary results reported, only used H.D. RAST inhibition (2). Skin testing is another reliable procedure for measuring allergen potencies (3), while offering the advantage of being closer to clinical reality. We obtained an excellent concordance between the intensity of H.D. cutaneous reactions and guanine concentrations. All our patients were exclusively sensitized to Dpt allergens and were not sensitized to other H.D. major allergenic components i.e. cat and dog allergens (4). A carefull patient selection is indeed necessary in order to eliminate the interference of other offending allergens present in a crude H.D. extract. Possible interferences with the test must also be considered : excrements of other cheliciferous arthropods such as spiders in our area, containing guanine too, could contaminate H.D. samples ; however interference with the colour test is unlikely because sentization to such arthropods appears to be rare.

The results of our H.D. analyses confirm that the ACAREX TEST gives reliable information about the presence of mite allergens in H.D. In this study, we did not evaluate the mite population through visual counts but we were able to correlate exacerbations of symptoms in mite allergic patients with a higher guanine level in H.D. samples ; we could also confirm previous data which showed that high altitude and hospital H.D. samples do not contain high mite allergen levels.

References

1. Bischoff E., Schirmacher W. (1984) Farbnachweiss für allergenhaltigen Hausstaub. Allergologie, 11, 446-449.

2. Bronswijk J.E.M.H., Bischoff E., Schirmacher W., Berrens L., Schober G. (1986) A rapid house dust allergen test : preliminary results. J. Med. Entomol., 23, 217-218.

3. Le MaO J., Weyer A., Pauli G., Lebel B., David B. (1980) Studies on Dpt allergens measurement of the relative potencies of Dpt purified extracts by in vitro and in vivo methods. J. Allergy Clin. Immunol., 65, 381-388.

4. Pauli G., Bessot J.C., Hirth C., Thierry R. (1979) Dissociation of house dust allergies. J. Allergy Clin. Immunol., 63, 245-252.

5. Platts-Mills T.A.E., Wilkins S.R., Heymann P., Heyden M.L. (1985) Dust mite allergen in the house of asthmatic patients. Annals of Allergy, 55, 310 (abst.).

6. Tovey E.R., Chapman M.D., Platts-Mills T.A.E. (1981) Mite faeces are a major source of house dust allergens. Nature, 289, 592-593.

7. Voorhorst R., SpieksmaF.T.M., Spieksma-Boezeman M.I.A. (1964) Is a mite (Dermatophagoïdes sp) the producer of the house dust allergen ? Allergy and Asthma, 10, 329-334.

Professeur G. Pauli, Service de Pneumologie, Pavillon Laennec, Centre Hospitalier Universitaire, 67091 STRASBOURG Cédex (FRance)

IV Air, our environment

1. Lectures of general importance

EXS 51:
Advances in Aerobiology
©1987 Birkhäuser Verlag Basel

THE QUANTITATIVE SIGNIFICANCE OF ASBESTOS FIBRES IN THE AMBIENT AIR

Richard Doll

ICRF Cancer Epidemiology & Clinical Trials Unit, Oxford, UK

Medical effects of asbestos

The evidence relating to non-occupational exposure to asbestos is too fragmentary to provide any quantitative estimates of risk and we can estimate these only by extrapolation from the observed effects of the larger doses in the asbestos industry. The effects include asbestosis, lung cancer, mesothelioma, and some benign conditions of the pleura that are seldom of lasting importance. It seems likely that asbestos can also cause cancer of the larynx, but the reported excesses of other non-respiratory cancers are probably due to the misdiagnosis of peritoneal mesothelioma and cancer of the lung (7).

Difficulties in assessing quantitative effects

Estimates of the quantitative relationship between the amount of asbestos to which individuals are exposed and the extra risk of developing any specific disease are difficult to make. First, asbestos is not a single chemical but a family of compound chemicals that have crystallised in nature as long thin separable fibres with useful mechanical properties in common. Secondly, the biological effects are due partly to the chemical constitution of the material and partly to the physical configuration of the fibres, both of which vary. Thirdly, the proportions of fibres with specific configurations vary with the way in which the material is handled. Fourthly, asbestos is commonly mixed with other materials which may modify its effect. Fifthly, knowledge of the mechanisms of human carcinogenesis is insufficient to tell precisely what epidemiological observations are most appropriate for measuring the biological response.

Fibre type

Four types of asbestos have been used - the common chrysotile and the three amphiboles (crocidolite, amosite, and anthophyllite). The laboratory evidence relating to the effects of these types is clear, but this has not prevented it from occasional misinterpretation (see 1,7,8,16). All four produce pulmonary fibrosis, lung cancer, and pleural and peritoneal mesothelioma in animal experiments with much the same frequency, when the effects of equal numbers of fibres are compared; none produce cancers of the gastrointestinal tract or other non-respiratory cancers (other than peritoneal mesothelioma).

The human evidence, however, suggests a different conclusion. The most comparable observations, extracted from 22 studies of 30 populations with exposure in many different operations, are summarized in table I. These show that exposure that was solely or principally to chrysotile led to fewer excess lung cancers and fewer mesotheliomas in proportion to the total number of deaths and a smaller proportion of mesotheliomas in comparison with the excess lung cancers than exposure that was principally, or solely to amphibole asbestos.

Differences in the site of origin of the mesotheliomas are even more marked, relatively few originating in the peritoneum when the exposure was due principally to chrysotile. Indeed no peritoneal tumours at all have been validated as occurring from exposure to chrysotile alone (7).

The difference between the human and the animal evidence may be due to the configuration of the curly fibres, which make them more likely to be trapped in the upper air passages than the straight fibres, and the greater rapidity with which chrysotile is cleared from the lungs. This could be less important in rats that live for only two or three years than in humans.

Fibre size
Knowledge of the significance of fibre size depends almost entirely on animal experiments, which have consistently shown that the pathogenicity of asbestos dust correlates better

Table I. Excess lung cancer (a) and mesothelioma (b) deaths by fibre type (30 populations of asbestos workers)

Sex	Type of fibre to which exposed	No. of deaths observed	Per cent of total		Ratio (a) to (b)
			(a)	(b)	
M	Solely or nearly all chrysotile	8319	1.90	0.53	0.28
	Chrysotile & substantial amphibole	4123	10.11	5.02	0.50
	Solely or mainly amphibole	1426	8.72	2.88	0.33
F	Solely or nearly all chrysotile	653	1.58	1.07	0.68
	Chrysotile & substantial amphibole	200	11.90	10.50	0.91
	Solely or mainly amphibole	388	3.40	6.19	1.82

with the number of long thin fibres than with the total mass to which the animals are exposed. The evidence suggests that the hazard is greatest from fibres between 5 and 100 μm in length with diameters less than 1.5 or 2 μm and ratios of length to diameter of more than five to one. Very short fibres, 1-2 μm long, may not be carcinogenic at all (15,16), but there is no evidence of any diminution in carcinogenicity with reduction in diameter down to at least 0.05 μm. These conclusions are imprecise partly because of the difficulties in determining the proportion of fibres of different sizes used in the experiments and partly because there are no sharp boundaries between hazardous and non-hazardous configurations. They are, moreover, mainly based on experiments in which fibres were injected into the pleural or peritoneal spaces or were instilled intratracheally, and such conditions are far removed from the human situation.

Measurement of exposure

The need to convert data collected in the old form of millions of particles per cubic foot (mppcf) to fibres of regulated size per ml creates further difficulties. Fibre counts ranging from 3% to more than 50% of the particle counts have been obtained for different processes and substantial variation has been observed within a single plant. Even with simultaneous counts, the correlations have been weak and the conversion factor has diminished as the particle count has increased. In our study of textile workers in the UK (12), a comparison of the average of the results obtained by the two methods for the same areas in two successive years gave a conversion factor of 34 particles per fibre and of 1 mppcf to 1 regulated fibre per ml.

Relationship between environmental measurements and risk of disease in industry

Asbestosis

Asbestosis presents the special problem of clinical recognition. Defined as "fibrosis of the lungs caused by asbestos dust" (1) it is radiologically indistinguishable from cryptogenic fibrosing alveolitis. Diagnosis is seldom difficult with a clear history and advanced disease, but may be difficult in the early stages as there is no sharp point in the development of signs and symptoms at which it can be said that a change has occurred from healthy to diseased. The precautions that have been taken in some countries have eliminated the gross disease that led to early death in the past, but it remains possible for asbestosis to cause disability or to increase the risk of death from other respiratory or circulatory disease.

Whether there is a threshold dose below which the fibrotic process will not advance to the point of clinical manifestation is still unsettled. If there is, it is below a cumulative dose of 100 f/ml-yrs, at which level the cohort of asbestos textile workers studied by BERRY et al. (3) experienced an incidence of "certified asbestosis" of 0.5% a year. Epidemiologically we cannot be sure that any dose is so small

that the resultant fibrosis could not increase the disability
attributable to some other condition, but it is difficult to
believe that minute doses could produce sufficient disability in
an otherwise healthy person for a diagnosis of clinical disease
to be made. No clinical case of asbestosis has been described in
the general public, nor was one found in over 600 members of the
families of asbestos workers in the USA (2) and I see no reason
to disagree with the ONTARIO ROYAL COMMISSION (14) that "The
lifetime occupational exposure to asbestos at which the fibrotic
process cannot advance to the point of clinical manifestation
.... is in the range of 25 f/cc-yrs and below."

Cancer
        The difficulty in predicting the risk of cancer is in
deciding how to take account of the effect of biological and
environmental differences. In common with most others we have
assumed that the risk of cancer following exposure relative to
that in its absence is independent of age at exposure, linearly
proportional to duration of exposure, constant after exposure has
ceased, and independent of cigarette smoking.
        BROWNE (4) has argued that a threshold exists for lung
cancer but we should not postulate different mechanisms for the
production of bronchial and pleural tumours and the latter can
certainly be produced by very small doses. The argument that
fibrosis is the underlying cause of lung cancer is unimpressive
as there is little if any excess risk of lung cancer associated
with gross fibrosis in other occupational pneumoconioses. The
epidemiological evidence is compatible with a linear dose-
response relationship without a threshold and such a relationship
should be accepted until it is disproved.
        Some dozen sets of data have been used to provide
estimates of the risks that may be expected from different levels
of exposure. These have differed 100 fold and there is no just-
ification for assuming that the best estimate is a weighted mean
of all the results. It is better to examine the separate sets of
data to see which, if any, are adequate for the purpose. Three
stand out as providing the largest amounts of environmental data
combined with large numbers of cancer deaths in groups of men who
have been adequately followed up. One concerns the experience of
chrysotile miners in Quebec (10) and gives results that are very
different from the others. The environmental conditions in mines
are, however, very different from those observed elsewhere in the
asbestos industry. It is better, therefore, to use only the two
sets of data obtained for chrysotile textile factories in S.
Carolina (9) and Rochdale UK (11). Both provided estimates in
terms of particles per unit volume of air, but the instruments
used to count the particles were different and gave different
conversion factors, the most representative being respectively 6
and 1 for converting mppcf to f/ml. The results give best
estimates of an increase in the Standardized Mortality Ratio of
approximately 1.25% and 0.54% for each year's exposure to 1
regulated fibre per ml. I doubt, therefore, whether we can
estimate the risk of lung cancer from exposure to chrysotile more
precisely than by using the rounded off intermediate value of an

increase of 1% per regulated f/ml-yr.

No estimate of the risk of mesothelioma can be made in the same way, as the numbers of cases in the two cohorts are too few. We can, however, derive a model for the relationship between the incidence of mesothelioma and the age at starting and stopping exposure from the much larger studies that have not provided measurements of dose (12) and use the two cohorts with measurements of dose to estimate the constants in the model. This leads to the equation

$$I(t) = 0.62 \times 10^{-10} L[(t-t_1)^4 - (t-t_2)^4]$$

where $I(t)$ is the incidence at age t years due to exposure to L f/ml, from age t_1 to age t_2.

These two equations enable us to calculate the numbers of asbestos induced deaths from lung cancer and mesothelioma . that would be expected to occur from a given exposure which are shown in table II. From this it appears that the risk of death from cancer, attributable to exposure to 1 f/ml at work for 20 years, starting exposure at 20 years of age, is about 2%.

Asbestos in the general environment

If it is assumed that there is no threshold for the induction of cancer and that the risk is linearly proportional to dose, these results can be used to estimate the risk from exposure to asbestos in houses and in the general environment. Pollution measurements made in terms of the mass of asbestos per m^3 air are useless, as they include thick non-respirable and minute non-carcinogenic fibres, and counts of fibres of regulated size made by optical microscopy are no better, as they do not distinguish asbestos fibres from others of the same configuration. This is unimportant in the asbestos industry, where the majority of fibres are asbestotic, but is crucial in the general environment where the vast majority may consist of innocuous materials. At the very low levels that are encountered in ordinary life only

Table II. Risk of death from lung cancer or mesothelioma before 80 years of age from exposure to chrysotile asbestos in asbestos textile factories at level of 1 f/ml (7)

Age at first exposure (yrs)	Cause of death	Duration of exposure (yrs)		
		5	20	35
20	lung cancer	3.7	14.8	25.2
	mesothelioma	2.1	4.9	5.6
30	lung cancer	3.7	14.5	24.0
	mesothelioma	0.9	2.0	2.2
40	lung cancer	3.7	13.0	17.0
	mesothelioma	0.3	0.6	0.7

Table III. Asbestos pollution in buildings containing asbestos

Transmission electron microscope counts: regulated fibres per ml	No. of buildings	
	Canada (14)	England (6)
*0.0	14	11
less than 0.0009	–	4
0.001 to 0.003	3	1

*below limit of quantification

counts of fibres that have been identified as asbestos by transmission electron microscopy (TEM) are of any value. These then have to be converted back to the numbers of "regulated fibres" that would be counted by the optical microscope in industry. Counts in two asbestos textile factories suggest that for this purpose the TEM counts should be halved (5,12).

Two sets of observations in houses have been reported (6,14). These are summarized in table III. The 19 Canadian buildings all contained asbestos insulation while the British buildings included 5 with asbestos insulation, 7 with asbestos sprayed on ceilings, and 4 with other uses. The pooled material for all 16 British sites gave a fibre count of 0.0002 f/ml. It seems, therefore, that people within asbestos containing buildings are seldom likely to be exposed to concentrations equivalent to more than 0.0005 regulated f/ml as observed by optical microscopes in the asbestos textile industry and that by extrapolation the corresponding lifetime risk of death for a man with average smoking habits from exposure for 40 hours a week for 20 years would be 1 per 100,000.

Conclusion

In this review I have outlined the scientific problems that have to be faced when trying to make a quantitative assessment of the risks to health from exposure to small amounts of asbestos in industry and in buildings. The estimates could certainly be improved by further epidemiological and laboratory research and by more attention to the measurement of environmental pollution. They should not, however, be more than an order of magnitude out and without some quantitative estimate it is easy to spend Society's limited resources in unrewarding ways and even to create new hazards that are greater than the old.

References

1. Advisory Committee on Asbestos. (1979) Asbestos: final report of the Advisory Committee, vol 1 (HMSO, London).
2. Anderson, H.A., Lilis, R., Daum, S.M., Selikoff, I.J. (1979) Asbestosis among household contacts of asbestos factory workers. Ann. N.Y. Acad. Sci. 330, 387-399.

References cont'd

3. Berry, G., Gilson, J.C., Holmes, S., Lewinsohn, H.C., Roach, S.A. (1979) Asbestosis: a study of dose-response relationships in an asbestos textile factory. Brit. J. Industr. Med. **36**, 98-112.
4. Browne, K. (1986) Is asbestos or asbestosis the cause of the increased risk of lung cancer in asbestos workers? Brit. J. Industr. Med. **43**, 150-157.
5. Dement, J.M., Harris, R.C. (1979) Estimates of pulmonary and gastrointestinal deposition for occupational fiber exposures. DHEW (NIOSH) Publication No. 79-135 (US Dept. of Health, Education and Welfare, National Institute for Occupational Safety and Health, Cincinnati, Ohio).
6. Department of the Environment. (1985) Asbestos materials in buildings (HMSO, London).
7. Doll, R., Peto, J. (1985) Effects on health of exposure to asbestos. A report to the Health and Safety Commission. (HMSO, London).
8. International Agency for Research on Cancer. (1977) IARC monographs on the evaluation of carcinogenic risk of chemicals to man. Asbestos, vol 14, (IARC, Lyon).
9. McDonald, A.D., Fry, J.S., Woolley, A.J., McDonald, J.C. (1983) Dust exposure and mortality in an American chrysotile textile plant. Brit. J. Industr. Med. **40**, 361-367.
10. McDonald, J.C., Liddell, F.D.K., Gibbs, G.W., Eyssen, G.E., McDonald, A.D. (1980) Dust exposure and mortality in chrysotile mining, 1910-75. Brit. J. Industr. Med. **37**, 11-24.
11. Peto, J., Doll, R., Hermon, C., Glayton, R., Goffe, T., Binns, W. (1985) Relationship of mortality to measures of environmental asbestos pollution in an asbestos textile factory. Ann. Occup. Hyg. **29**, 305-335.
12. Peto, J., Seidman, H., Selikoff, I.J. (1982) Mesothelioma mortality in asbestos workers: implications for models of carcinogenesis and risk assessment. Brit. J. Cancer **45**, 124-135.
13. Rood, A.P., Streeter, R.R. (1984) Size distributions of occupational airbourne asbestos textile fibres as determined by transmission electron microscopy. Ann. Occup. Hyg. **28**, 333-339.
14. Royal Commission on Matters of Health and Safety Arising from the Use of Asbestos in Ontario. (1984) Report of the Royal Commission on matters of health and safety arising from the use of asbestos in Ontario, (Ontario Ministry of the Attorney General, Toronto).
15. Wagner, J.C. (1986) Personal communication.
16. Walton, W.H. (1982) The nature, hazards and assessment of occupational exposure to airborne asbestos dust: a review. Ann. Occup. Hyg. **25**, 117-247.

Sir Richard Doll, Imperial Cancer Research Fund Cancer Epidemiology and Clinical Trials Unit, Gibson Laboratories, Radcliffe Infirmary, Oxford OX2 6HE, UK.

THE INDOOR ASBESTOS PROBLEM: FACTS AND QUESTIONS

M. Guillemin, G. Litzistorf, P. Madelaine, F. Iselin and Ph. Buffat

Institute of Occupational Medicine and Industrial Hygiene, University of Lausanne, Switzerland
Swiss Federal Institute of Technology, Lausanne, Switzerland

1. Introduction

The spray application of asbestos fiber insulation was introduced in 1932 in Great Britain and within a few years was adopted in other countries. In Switzerland, this process has been used from 1936 to approximately 1975 for more than 4000 buildings. This friable material may deteriorate with time or be damaged during maintenance operations, transformations or by the users themselves.

In 1971, in the Yale University Art and Architecture building (USA), the ceiling deterioration became a subject of concern (SAWYER 1977). The first measurements carried out in this case revealed that the students' exposure to asbestos, which is recognized as a carcinogen, could be significant. Since that time public concern has continued to increase rapidly, which prompted regulatory agencies to begin evaluation of the problem and determine appropriate corrective actions.

To properly assess the exposure to asbestos fibers and therefore the related health risk, one must be able to continuously measure, by personal sampling, the airborne fiber concentration with respect to the species, the lengths and the diameters of the asbestos fibers. Such an assessment is impossible, due to many technical limitations. Actually, the best approach is to measure (as well as identify and size) the fibers by the transmission electron microscopy (TEM) method. But this method is extremely expensive and time consuming. Therefore other approaches and acceptable compromises have to be found. The purpose of this presentation is to critically comment on these approaches in the light of our own research.

2. Exposure assessment

To better define the dimensions of this public health problem, a valid and reliable risk assessment system is needed. In the USA, the Environmental Protection Agency (EPA) proposed, in order to avoid the costs and the difficulties related to the air measurements, an Asbestos Exposure Algorithm for non occupational settings (DHEW 1980). This algorithm utilizes eight factors to predict the exposure to asbestos within buildings. This simple and attractive method has been extensively used to determine priorities relative to corrective actions. However this approach has been shown to be unreliable (FINDLEY 1983) and not correlated to the real exposure of building occupants (CONSTANT 1983). Moreover the relative weights given to each factor seem inappropriate. For instance the asbestos content of the material is a multiplicative factor and is in fact not very important compared to friability or "releasibility" (CONSTANT 1983). Our results confirm this lack of correlation between the EPA index and the airborne fiber concentration, including the fact that the eight factors used in the EPA method are probably not the most relevant.

It became obvious that the only way to make a reliable health risk estimate was to measure the real exposure to asbestos fibres. In the early seventies, the phase contrast microscopy (PCM) method was used to collect data on indoor pollution but it was realized within a few years that this method was inappropriate, mainly due to the fact that it cannot differentiate between asbestos and non asbestos fibers. Other methods utilizing the electron microscope have been used, but currently no uniform (standardized) method has been agreed to at the international level. This makes the comparison of the published results almost impossible.

The Federal Republic of Germany introduced a standardized method (Verein Deutscher Ingenieure 1985) based on scanning electron microscopy, which is considered in this country as an acceptable compromise based on cost and performance considerations. This method has a detection limit of 0.2 µm, which is the same as the PCM method but has the advantage of allowing the identification of the fibers by X ray fluorescence. It must be stressed that this identification may be erroneous in some situations. We observed with the TEM method that in some buildings up to 70% of long (> 5 µm) and thin (< 0.2 µm in diameter) chrysotile fibers will not be detected by the SEM method. By using such a method, it is essential to be aware of its limitations (BURDETT 1986, ALTREE-WILLIAMS 1985, SPURNY 1984, LOHRER 1983).

The TEM method we have developed (STEEN 1983) is similar to the one proposed as an international standard (CHATFIELD 1983). It has been used in different buildings, table I summarizes the results obtained for long asbestos fibers.

Table I. Indoor asbestos fiber concentrations measured by transmission electron microscopy

No	Type of building	Asbestos type	F/L ≥ 5 µm	F/L ≥ 2.5 µm	Potential sources
1	Research lab.	chrysotile	n.d.	1.95	unknown
2	Class-room	chrys.-amos.	8.58	17.17	unknown
3	Offices	chrysotile	1.67	3.35	ACFM
4	School hall	chrysotile	7.93	23.80	ACFM
5.a	Factory offices	amosite	1.41*	2.66*	ACFM
5.b	Fact. computer-r.	amosite	4.85*	7.05*	ACFM
6.a	School kitchen	chrys.-crocid.	0.12*	0.83*	ACFM
6.b	School hall	chrysotile	n.d.*	0.19*	ACFM
7.a	School theater	chrys.-crocid.	2.27*	4.38*	ACFM
7.b	School hall	chrysotile	0.19*	0.56*	ACFM
8	Shopping center	chrysotile	0.12*	0.36*	ACFM
9	Research lab.	chrysotile	2.49	5.97	unknown

n.d. : not detected
ACFM: asbestos containing friable material (insulation)
*   : counting carried out at 3'700 x ; all the others at 30'000 x

224

In table I it is interesting to observe that buildings with unknown potential source of asbestos (chosen as control buildings) may have significant concentrations. The contribution of the outdoor ambient levels have been considered but do not influence to a large extent the indoor levels.

Our results appear somewhat higher than other published data (BURDETT 1986, DUPRE 1984). Reasons for this may be the result of the ashing and ultrasonic treatment or of the different sampling strategies used. Other results published in the literature cannot be used for comparison since most of the TEM results are expressed in mass per volume (ng/m^3) and not in fibers/m^3.

In parallel to this TEM reference method, we have used the instrumentation shown in figure 1.

INSTRUMENTATION

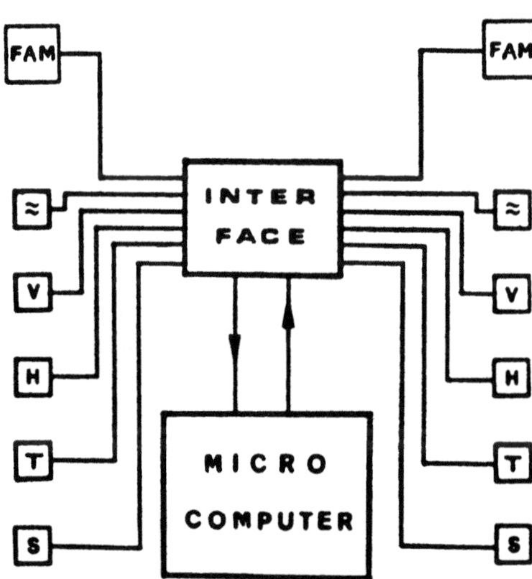

Figure 1 : Monitoring system for five parameters related to the indoor environment under study

FAM : Fibrous Aerosol Monitor (GCA Inc., USA)
≈   : vibration   V : air velocity   H : relative humidity
T   : temperature   S : sampling pump

This monitoring system allows the simultaneous measurement of the airborne fiber levels, the floor vibration, the air velocity and the climatic factor (T,H). A sampling pump may be activated at given times.

The use of such a system in different indoor environments allowed us to detect the relevant periods of use of the room under study during which the exposure assessment should be done (LITZISTORF 1985).

It appeared that the floor vibration is a good index of room occupancy and could be used to drive the sampling pump.

3. Results interpretation

What do the results of table I mean ? With the actual body of knowledge available it is almost impossible to answer this question. Many attempts have been made to assess the risk of such low exposures. Although most conclude that the estimated risk is very low and quite acceptable, one must admit that the determination of a reliable method of risk assessment is prevented by numerous problems (i.e. analytical methods, physiology and toxicology questions). Even the epidemiological studies done to extrapolate the risk of occupational exposures to low non-occupational exposures are unreliable, in many respects. One example of a problem area is the size and species distribution of the fibers in different environments. It has been shown that these distributions differ greatly from one place to another. Therefore it is impossible to predict for a given environment which is the proportion of the fibers longer than 5 μm detectable by the light microscopy method (PCM) used to assess the occupational exposure in the above mentioned epidemiological studies. Moreover it has been shown in animals that fibers shorter than 5 μm but longer than 2.5 μm may also be toxic, and that mineral fibers other than asbestos but with the critical sizes may also be carcinogenic and should be considered in any exposure assessment (SPURNY 1984, LE BOUFFANT 1985).

MCDONALD (1985) proposed a new approach to assess the risk of non occupational exposure to asbestos by looking at actual deaths due to mesothelioma in these situations. This would imply the further investigation of the indoor environment.

Aknowledgement

The Swiss National Research Fund supported this research.

4. References

Altree-Williams, S. (1985) Asbestos and other fibre levels in buildings. Ann. occup. Hyg. 29 (3), 357-363

Burdett, G.J. (1986) Airborne asbestos concentrations in buildings. Ann. occup. Hyg. 30 (2), 185-199

Chatfield, E.J. (1983) Short and thin mineral fibres. Identification, exposure and health effects. Proc. Symp. Nat. Board Occup. Saf. Health Res. Dept, Solna, Sweden

Constant P.C. Jr. (1983) Airborne asbestos levels in schools. EPA 560 (003), 186 p.

DHEW (1980) Asbestos detection and control, local educational agencies, asbestos detection and state plan. Fed. Reg. 61950-61954

Dupré J.S. (1984) Report of the Royal commission on matters of health and safety arising from the use of asbestos in Ontario (Queen's Printer, Toronto, Canada)

Findley M.E. (1983) An assessment of the Environmental Protection Agency's asbestos hazard evaluation algorithm. Am. J. public Health 73 (10), 1179-1181

Le Bouffant L. (1985) Pouvoir carcinogène des fibres de chrysotile de longueur <5 μm. Cah. Notes Doc. 118, 83-89

Litzistorf G. (1985) Influence of human activity on the airborne fiber level in paraoccupational environments. J. Air Pollut. Control Assoc. 35 (8), 836-837

Lohrer W. (1983) Asbestbelastete Innenräume - Analyse und Bewertung des Gefahrenpotentials. Staub Reinhalt. Luft 43 (11), 434-438

McDonald J.C. (1985) Health implications of environmental exposure to asbestos. Environ. Health Perspect. 62 (Oct.), 319-328

Sawyer R.N. (1977) Asbestos exposure in a Yale building. Environ. Res. 13, 146-169

Spurny K. (1984) Die Konzentration von faserigen Stäuben in der atmosphärischen Umwelt. Staub Reinhalt. Luft 44 (10), 456-458 (1984)

Steen D. (1983) Determination of asbestos fibres in air - Transmission electron microscopy as a reference method. Atmos. Environ. 17 (11), 2285-2297

Verein Deutscher Ingenieure (1985) VDI-Richtlinie 3492: Messen anorganischer,

faserformiger Partikeln in der Aussenluft: rasterelektronenmikroskopisches Verfahren. (VDI-Verlag, Düsseldorf, B.R.D.)

EFFECTS OF SHORT-TERM EXPOSURE TO AMBIENT NITROGEN DIOXIDE CONCENTRATIONS
ON HUMAN BRONCHIAL REACTIVITY AND LUNG FUNCTION.

G. Bylin[1,2], T. Lindvall[2], T. Rehn[2] and B. Sundin[1]

[1]Division of Allergology, Department of Medicine, Huddinge University
 Hospital, Sweden
[2]The National Institute of Environmental Medicine, Stockholm, Sweden

Nitrogen dioxide (NO_2) in high concentrations is a well-documented airway
irritant. The effects of short-term exposure to ambient concentrations
(0 to about 1000 $\mu g/m^3$) of NO_2 are, however, not so well known. Small, but
statistically not significant, increases in specific airway resistance mean
values are reported (OREHEK 1976, HAZUCHA 1981) both for asthmatic (+8-10%)
and non-asthmatic (+5%) subjects after short-term exposure to 200 $\mu g/m^3$ NO_2.
Bronchial reactivity was increased in asthmatic subjects after exposure to
200 $\mu g/m^3$ NO_2 in one study (OREHEK 1976), but in a similar study (HAZUCHA
1981) no significant effects were seen either in asthmatic or non-asthmatic
subjects.

The acute effects of short-term NO_2-exposure at ambient concentrations seem
to be small, at least in healthy persons, but they are nevertheless important
to detect. Small changes after short-term exposure may proceed to greater
effects after months´ or years´ of continous exposure. In real life, effects
of other pollutants, temperature and humidity may be imposed on the NO_2
effects. A laboratory finding of moderate or even small changes after short-
term NO_2-exposure may therefore be of considerable clinical importance.

Bronchial obstruction is an effect observed after exposure to high concentrations of NO_2, and narrowing of airway calibre is an effect also expected at exposure to NO_2 in ambient concentrations. Measurement of airway resistance by body plethysmography is a method often preferred to detect airway obstruction. The method is sensitive for small changes in the calibre of at least central airways and the performance of the measurement does not affect the state of the airways (as does methods based on forced expiratory manoevres). A drawback with the method is however that airway resistance varies with lung volume. This may give difficulties in interpreting small changes in airway resistance as exposure to irritant gases also can affect lung volume. To detect even rapidly occurring changes in airway resistance we have modified a whole body plethysmograph, so it can both serve as a dynamic exposure chamber and as an instrument for measurement of airway resistance and lung volume during on-going exposure. (Fig. 1)

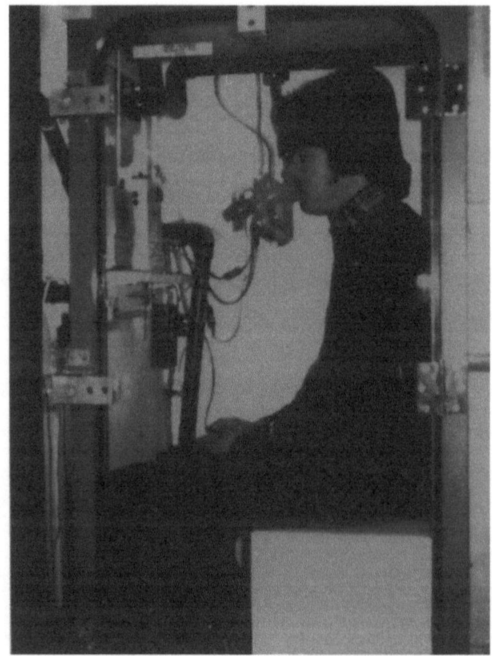

Fig. 1. The subject is sitting in a constant-volume whole-body plethysmograph, which is modified to a dynamic exposure chamber by installation of inlet and outlet systems for the testgas. This technique permits repeated measurements of airway resistance during exposure.

Increased respiratory rate has been observed in mice and guinea-pigs after long-term exposure to NO_2 at 1500 $\mu g/m^3$ and higher concentrations (FREEMAN ET AL 1966, MURPHY ET AL 1964). In humans NO_2-induced changes in breathing frequency have not been reported. However, alterations in respiratory rate and tidal volume might reflect defence mechanisms against NO_2-induced injuries, and the variables are therefore important to measure during NO_2-exposure.

In a study comprising both asthmatic and non-asthmatic subjects exposed to 0, 230, 460 and 910 $\mu g/m^3$ NO_2 during 20 min (BYLIN ET AL 1985, a) the subjects experienced no discomfort (except odor) during or after the exposure. However, statistically significant changes were seen in airway resistance with an increase (+11%) after exposure to 460 $\mu g/m^3$ NO_2 and a <u>decrease</u> (-9%) after 910 $\mu g/m^3$ NO_2 in the non-asthmatic group. In the asthmatic group the trend in airway resistance was the same but not statistically significant. These data give some support to the hypothesis that there is a non-monotonous dose-effect relationship for NO_2 induced airway resistance. The trend is increased specific airway resistance (SR_{aw}) at low NO_2 concentrations, decreased SR_{aw} at moderately high concentrations and at still higher NO_2 concentrations (above 3000-4700 $\mu g/m^3$) SR_{aw} is again increased.

In the asthmatic group the lung volume was significantly decreased (-0.45 1) after exposure to 910 $\mu g/m^3$ NO_2. As a change of 1 1 in lung volume in itself may give about 10% change in SR_{aw} (BYLIN ET AL 1985, b) the SR_{aw}-data are to be interpreted with some caution. By constructing an airway conductance - lung volume curve for each subject it is possible to compare preexposure and exposure airway resistance values referring to the same lung volume and thus correct for the airway resistance dependence of lung volume (BYLIN ET AL 1986).

The changes in SR_{aw} at 460 $\mu g/m^3$ NO_2 developed between 10 and 20 min exposure and were normalized 10 min after exposure. At exposure to 910 $\mu g/m^3$ NO_2 the changes also emerged between 10 and 20 min exposure but persisted even 10 min after exposure. These findings show that NO_2-induced effects on SR_{aw} at ambient concentrations may vary during the exposure-time and be of quite short duration.

The bronchial reactivity of the asthmatic subjects increased significantly by 20 min exposure to 910 $\mu g/m^3$ NO_2. For 4 of the 5 asthmatic subjects showing increased bronchial responsiveness the changes were insignificant from a clinical point of view but 1 subject showed a pronounced increase in reactivity which must be judged as clinically important.

The respiratory rate tended to decrease with increasing NO_2-concentrations in both asthmatic and non-asthmatic subjects, but the differences were not statistically significant.

Short-term NO_2-exposure in concentrations even below 1000 $\mu g/m^3$ seems to have effects on human lung function and bronchial reactivity. The changes found in the laboratory are small but may nevertheless, especially for asthmatic persons, be of clinical importance in real life.

References

Bylin, G., Lindvall, T., Rehn, T. and Sundin, B. (1985, a) Effects of short-term exposure to ambient nitrogen dioxide concentrations on human bronchial reactivity and lung function. Europ. J. Resp. Disease 66, 205-217

Bylin, G., Hedenstierna, G., Rehn, T. and Sundin, B. (1985, b) A study comparing the use of specific airway conductance (SG_{aw}) and airway conductance - lung volume curves (G_{aw}/TGV) in determination of changes in airway conductance. Report no 6, The National Institute of Environmental Medicine (Stockholm, Sweden).

Freeman, G., Furiosi, N.J., Haydn, G.B. (1966) Effects of continuous exposure to 0.8 ppm NO_2 on respiration of rats. Arch. Environ. Health 13, 454-456

Hazucha, M.J., Ginsberg, J.F., Mc Donnell, W.F., Haak, E.D. Jr, Pimmel, R.C., Salaam, S.A., House, D.E., Bromberg, P.A. (1983) Assessment of 0.1 ppm nitrogen dioxide effects on the airways of normal and asthmatic subjects. J Appl. Physiol. 54 (3), 730-739

Murphy, S.D., Ulrich, C.E., Frankowitz, S.H., Zintaras. C. (1964) Altered function in animals inhaling low concentrations of ozone and nitrogen dioxide. Am. Ind. Hyg. Assoc. J. 25, 246-253

Orehek, J., Massari, J.P., Cayrard, P., Grimaud, C., Charpin, D. (1976) Effects of short-term, low-level, nitrogen dioxide exposure on bronchial sensitivity of asthmatic patients. J. Clin. Invest 57, 301-307

MUTAGENIC AND CARCINOGENIC EFFECTS OF AIRBORNE PARTICULATE
MATTER FROM POLLUTED AREAS ON HUMAN AND RODENT TISSUE CULTURES

N.H.Seemayer, W.Hadnagy and R.Tomingas

Medizinisches Institut für Umwelthygiene, Düsseldorf, F.R.G.

Introduction

Air pollutants are generated by various combustion processes
from power plants, factories, households and from automobile
traffic by exhaust.
Among these numerous air pollutants, airborne particulate matter
or city smog represents a special hazard to human health. In
airborne particulate matter more than 500 chemical substances
have been detected, among them known carcinogens and mutagens,
particularly polycyclic aromatic hydrocarbons, acridines, traces
of heavy metals etc.. Airborne particulates with a diameter
smaller than $5\,\mu m$ represent respirable dust which can reach
broncho-alveolar space in human lung affecting cell functions.
For assessment of noxious effects cell and tissue cultures
offer special advantages. As compared to the in vivo animal
model, cell and tissue cultures grant shorter testing periods
and exhibit a better approach to study mutagenic and carcinogenic effects of environmental noxae on the cellular and molecular level (SEEMAYER et al., 1984).

Material and Methods
Collection and extraction of airborne particulate matter
City smog was collected in the city of Duisburg during the
winter period of 1979/80 using a "Draeger Box Micron Filter
MB 1700" - apparatus for sampling. The dust sample was extracted by cyclohexane and the global extract (GEX) was quantita-

tively transferred to dimethyl sulfoxide (DMSO) for tissue
culture experiments. Used dosage was calculated according to
its benzo(a)pyrene - content and is expressed as benzo(a)-
pyrene - equivalent. A benzo(a)pyrene-equivalent of 0.01 ug/ml
is related to airborne particulate matter from 4 cbm of air.

<u>Detection of "Sister Chromatid Exchanges"</u>

We employed human lymphocyte cultures prepared as whole blood
cultures stimulated by phytohemagglutinine and the Chinese
hamster cell line V 79. Cells were exposed to global extract in
presence of 5-bromodeoxyuridine. After defined incubation
periods we used standard cytogenetic techniques.

<u>HGPRT - Point Mutation Assay</u>

V 79 cells were incubated with various concentrations of global
extract for 24 hours. After 8 days expression period cells were
cultivated in presence of 8-azaguanine. 11 days later the
number of resistant colonies was estimated.

<u>Cell Transformation Assay</u>

Logarithmically growing cultures of Syrian hamster kidney cells
were exposed to various concentrations of global extract for
18 hours. Thereafter cell cultures were infected with Simian
virus (SV-) 40. After 4 weeks of cultivation number of transformed colonies was estimated macroscopically and verified
microscopically as described earlier (SEEMAYER, 1978; SEEMAYER
et al., 1980).

<u>R e s u l t s</u>

Global extract of city smog induced a dose-dependent increase of
"sister chromatid exchanges" in cultures of human lymphocytes.
A significantly increased rate of "sister chromatid exchanges"
was already observed at a amount of extracted particulates from
1 cbm of air.
The dose-effect curve showed a steep increase at low concentrations and went more flat a higher concentrations, indicating a
saturation type.
A similar dose-dependent induction of "sister chromatid exchanges was obtained with exposed Chinese hamster cell line V 79.

To discover "point mutations" induced by global extract of
city smog we determined the 8-azaguanine resistant mutants in
exposed cell cultures of V 79 cells. Mutants were produced with
concentrations of the global extract of 0.01 and 0.02 /ug/ml of
benzo(a)pyrene-equivalents corresponding to particulates contained in sampling air volumes of 4 and 8 cbm, respectively.
In reconstruction experiments, by cocultivation of wild type
V 79 cells and azaguanine resistant mutants in presence of
8-azaguanine we confirmed the phenotypic stability of mutants
induced by global extract. "Enhancement" of malignant cell
transformation was shown by a dose-dependent increase of transformation frequency of Syrian hamster kidney cell cultures pre-
treated with global extract and consecutively infected with
Simian virus 40. Already small amounts of global extract,
equivalent to particulates from 0.25 - 0.5 cbm of air volumes
led to significant "enhancement" of cell transformation. We
could demonstrate even a 600% increase of transformation frequency when exposing cell cultures to a benzo(a)pyrene-equi-
valent of 0.01 /ug/ml which is related to particulates from
4 cbm of air.

Summary and Conclusion
Global extract of airborne particulate matter from an industrialized area contains chemical substances which exert a dose-
dependent mutagenic and carcinogenic activity on mammalian
cells in vitro. Mutagenic activity was demonstrated by induction of "sister chromatid exchanges" in human lymphocyte
cultures and in Chinese hamster lung cell line V 79. Furthermore, "point mutations" at the HGPRT-locus were induced in
V 79 cells. Carcinogenicity was shown by "enhancement" of malignant cell transformation of Syrian hamster kidney cultures exposed to global extract and thereafter infected with Simian
virus 40.
When considering a human respiratory volume of 12 - 14 cbm of
air per day we have to emphasize that amounts of global extract

equivalent to particulates from O.5 - 1 cbm clearly showed
mutagenic and carcinogenic activity. This relationship is
visualized on Fig. 1.

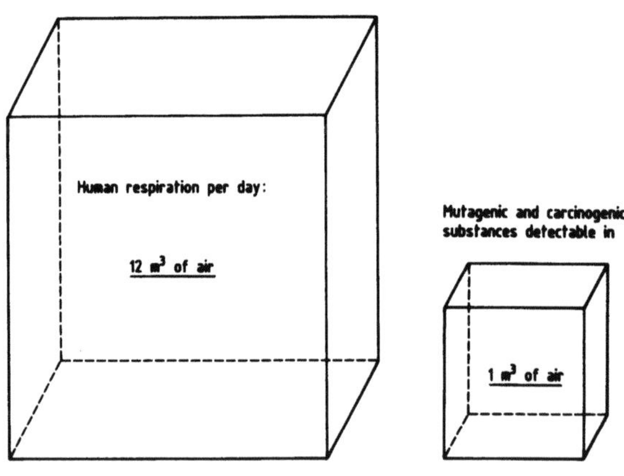

Seemayer, N.H. (1978) Zelltransformation von Hamsternieren-
zellen in vitro unter dem Einfluß von atmosphärischen Feinstaubextrakten und dem Papovavirus SV40. Staub-Reinhalt. Luft
38, 254-258

Seemayer, N.H., N.Manojlovic u. R.Tomingas (1980) Die Wirkung
von kanzerogenen Stoffen, insbesondere von atmosphärischen
Feinstaubextrakten im Zelltransformationstest. In: Polycycli-
sche aromatische Kohlenwasserstoffe, VDI-Verlag Düsseldorf
VDI-Berichte No. 358, 285-291

Seemayer, N.H., N.Manojlovic, C.-C.Schürer and R.Tomingas
(1984) Cell cultures as a tool for detection of cytotoxic,
mutagenic and carcinogenic activity of airborne particulate
matter. J.Aerosol Sci. 15, 426-430

Prof. Norbert H. Seemayer, Medizinisches Institut für Umwelthygiene an der Universität, Auf'm Hennekamp 50, D-4000 Düsseldorf, F.R.G.

EXS 51:
Advances in Aerobiology
©1987 Birkhäuser Verlag Basel

MODULATION OF PULMONARY DEFENSE MECHANISMS BY ACUTE EXPOSURES TO NITROGEN DIOXIDE

George J. Jakab

Aerobiology Laboratory, Department of Environmental Health Sciences, The Johns Hopkins School of Hygiene and Public Health, Baltimore, Maryland, USA.

Introduction

Atmospheric pollutants such as nitrogen dioxide, ozone and sulfur dioxide have been shown to impair pulmonary antibacterial defenses resulting in increased susceptibility to respiratory infections (Ehrlich 1980). Herein, nitrogen dioxide (NO_2) was used to determine its effect on the individual components of the co-ordinated biocidal mechanisms of the lung to elucidate the mechanism by which NO_2 suppresses pulmonary antibacterial defenses.

Bronchopulmonary defense mechanisms against bacterial infections depend primarily on the integrated activity of the phagocytic and immune systems (Green et al 1977). In the normal lung, the alveolar macrophage serves as the surveillance phagocyte (Goldstein et al 1974). This resident phagocytic system can be augmented by the intraalveolar influx of polymorphonuclear leukocytes (PMNs) to provide the lungs with additional defense capabilities (Pierce et al 1977; Rehm et al 1979 & 1980). Specific immune mechanisms augment the biocidal defenses of the lung by enhancing phagocytic activity (Jakab 1976). It is possible to study the individual components of the co-ordinated biocidal mechanisms of the lungs by using different challenge organisms to probe specific defense parameters. Herein, Staphylococcus aureus was used as a probe to study the functional integrity of the alveolar macrophage phagocytic system and the gram-negative bacterium Proteus mirabilis to examine the dual phagocytic system of the lungs (Rehm et al 1979). Finally, Pasteurella pneumotropica, a gram-negative rodent respiratory bacterium (Brennan et al 1969) was used to examine the effect of NO_2 on pulmonary antibacterial defenses against an organism endogenous to the host. The results show that the individual components of the defense mechanisms of the lungs are adversely affected by different concentrations of NO_2.

Materials and Methods

Animals: White female Swiss mice weighing 20-23 gm were used. Serologic tests demonstrated serum antibodies against P. pneumotropica but not against S. aureus or P. mirabilis.

Bacterial Challenge and Bactericidal Assay: S. aureus (FDA strain 209 P, phage type 42D), a laboratory strain of P. mirabilis and P. pneumotropica (isolated from the nasopharynx of a mouse) were prepared for aerosol inhalation challenge by previously described methods (Jakab and Green 1972). Animals were challenged by aerosol inhalation with each of the bacteria for 30 min during which time 1 to 5×10^5 of each of the bacteria were deposited in the lungs.

Pulmonary bactericidal activity was assessed by methods previously described (Ruppert et al 1976). Briefly, animals were sacrificed by luxation of the neck immediately and at 4 hrs after cessation of bacterial challenge. The lungs were aseptically removed, homogenized and appropriately diluted for quantitation by standard bacteriologic pour plate methods. Intrapulmonary bacterial killing in each animal was calculated as the percentage of the initially deposited bacteria killed at 4 hours following bacterial challenge.

Characterization of the Pulmonary Cell Populations: Total and differential cell counts of free pulmonary cells were quantitated 4 hrs after inhalation challenge with bacteria (Warr and Jakab 1983). The animals were killed, bled by cardiac puncture and the lungs surgically removed in toto and lavaged 3 times with 1.5 ml of lavage fluid. Total and differential cell counts were performed by standard methods.

Nitrogen Dioxide Exposure: NO_2 exposure was accomplished in stainless steel horizontal flow exposure chambers (Hemenway and MacAskill 1982) utilizing methods previously described (Hemenway and Jakab 1986). Two exposure chambers were used in these studies. In one the mice were exposed to various levels of NO_2 whereas the other served as a control chamber to expose animals to an identical flow of HEPA filtered room air without the NO_2. During the exposure periods the animals were housed in individual stainless steel wire cages.

Results

The data for the experiments in which mice were challenged with either S. aureus, P. mirabilis or P. pneumotropica and then exposed for 4 hrs. to various concentrations of NO_2 are presented in Figures 1.

Pulmonary bactericidal activity against S. aureus decreased progressively with exposure to increasing concentrations of NO_2. These differences were significant for comparison of control and mice treated with 4.0 ppm of NO_2 or greater. When P. mirabilis was used as the challenge organism exposure to NO_2 concentrations of 5.0 ppm had no effect on the intrapulmonary killing of this organism. In contrast, 10.0 ppm of NO_2 significantly enhanced the bactericidal activity of NO_2 exposed mice and the trend toward this enhancement was still evident at 15.0 ppm. Increasing the NO_2 exposure concentrations to 17.5 ppm resulted in equivalent rates of intrapulmonary bacterial killing. Finally, at exposure concentrations of 20.0 ppm a significant suppression of bactericidal activity of the lungs against P. mirabilis was observed. This impairment continued in a dose-dependent manner at 25.0 ppm; at 30.0 ppm mice began to die if exposed to NO_2 after P. mirabilis challenge.

When P. pneumotropica was used as the challenge organism yet another pattern of intrapulmonary bactericidal dysfunction was observed

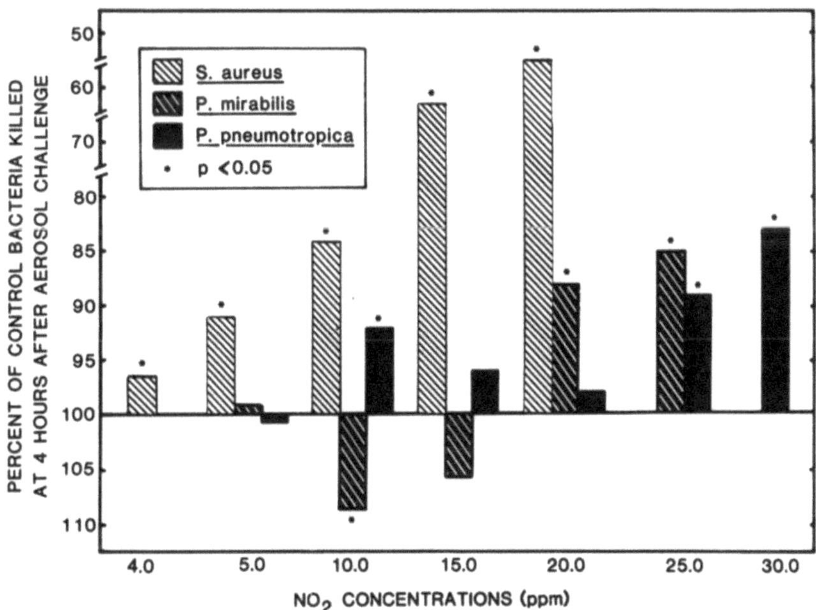

Figure 1. Intrapulmonary killing of bacteria in mice exposed to increasing concentration of NO_2. The data for the animals exposed to NO_2 for 4 hours is normalized to the values obtained formt he animals that were allowed to breath ambient air (value = 100%). Each value represents the mean of 15 to 25 individual determination.

upon exposure to increasing concentrations of NO_2. NO_2 levels of 5.0 ppm had no effect on bactericidal activity whereas exposure to 10.0 ppm of NO_2 caused a significant suppression. Thereafter, increasing the exposure levels of NO_2 to 15.0 20.0, 25.0 and 30.0 ppm induced a significant bactericidal dysfunction only at the highest 2 concentrations.

Inhalation challenge with _S. aureus_ did not result in any appreciable change in the phagocytic cell population recovered from the lungs (Figure 2). In contrast, bacterial challenge with _P. mirabilis_ and _P. pneumotropica_ significantly increased the number of lavageable pulmonary cells from control values of $7.0\pm0.3\times10^5$ to $32.7\pm1.2\times10^5$ and $43.6\pm9.2\times10^5$

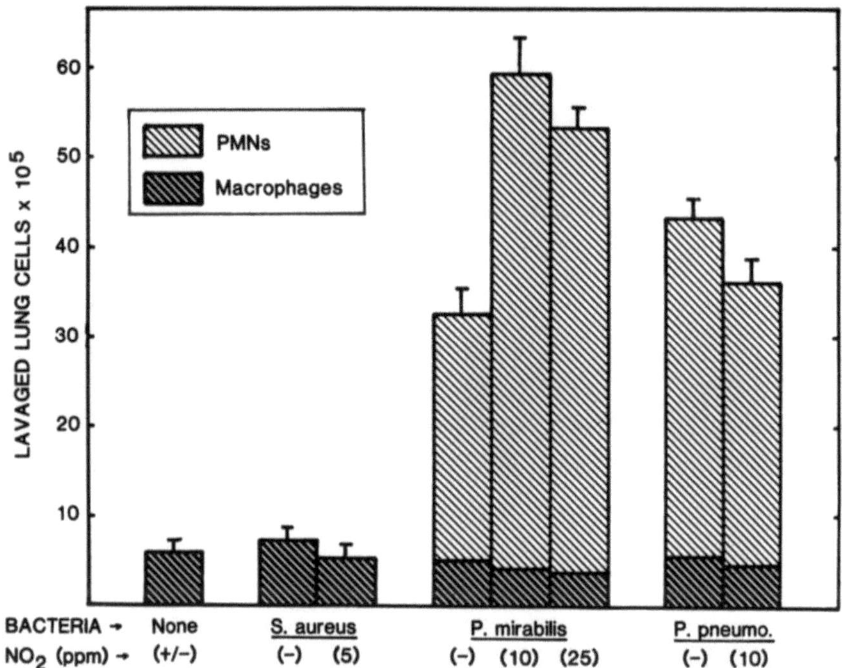

Figure 2. Lavaged phagocytic cell population from the lungs of animals 4 hours after bacterial challenge with or without subsequent exposure to NO_2. Each value represent the mean of 6 to 12 individual determinations.

respectively with the dramatic increase in the cell population consisting primarily of PMNs.

NO_2 exposure of mice challenged with **P. mirabilis** significantfly increased the number of cells retrieved. For example, Figure 2 shows that alveolar macrophages increased from $2.7\pm0.4\times10^5$ to $4.7\pm0.9\times10^5$ and PMNs from $29.7\pm1.4\times10^5$ to $53.6\pm4.2\times10^5$ upon exposure to 10.0 ppm of No_2. The 10.0 ppm concentration of NO_2 however appeared to have the opposite effect when the challenge organism was **P. pneumotropica**. Fewer cells were recovered from the lungs of **P. pneumotropica** challenged and NO_2 exposed mice than those only challenged with the bacterium.

Discussion

The studies performed herein clearly demonstrate that a 4 hr. acute exposure to NO_2 adversely affects the bactericidal activity of the murine lung. However, the concentration of NO_2 at which this dysfunction occurs is dependent on the challenge organism. Herein, _S. aureus_ and _P. mirabilis_ were used respectively as indicator organisms to quantitatively probe the effect of NO_2 exposure on the functional integrity of the alveolar macrophage and auxiliary (PMN) phagocytic systems of the lungs. In addition, bacterial challenge with _P. pneumotropica_ was also used to probe the effect of NO_2 exposure on the phagocytic defense mechanisms of the lungs against a pathogen endogenous to the respiratory tract of the experimental host (Brennan et al 1960; Jakab and Dick 1973).

Pulmonary antibacterial defenses against _S. aureus_ is primarily dependent on the alveolar macrophages (Goldstein et al 2974) whereas lung defenses against gram-negative bacteria are dependent on a dual phagocytic system consisting of both alveolar macrophages and PMNs (Pierce et al 1977; Rehm et al 1979, 1980). The peripheral phagocyte is not a constituent of normal murine lung cell populations but can immigrate rapidly to the lung in response to gram-negative bacteria and thus provide additional phagocytic defense capabilities to the lung (Pierce et al 1977). Failure of PMNs to immigrate to the lungs is associated with a decrease of pulmonary defenses against gram-negative bacteria but not against _S. aureus_ (Rehm et al 1979; Astry et al 1983).

NO_2 exposure after staphylococcal challenge caused a progressive dose-dependent suppression of bactericidal activity with increasing concentrations of the gas. This defect was present in animals exposed to levels of NO_2 of 4.0 ppm or greater.

The initial enhancement followed by the suppression of pulmonary bactericidal activity against _P. mirabilis_ by exposure to increasing concentrations of NO_2 was unexpected. To gain insights into the possible mechanisms of this modulating effect the lungs of mice were lavaged to quantitate the phagocytic cell populations. Exposure of _P. mirabilis_ challenged mice to 10.0 ppm and 20.0 ppm of NO_2 increased the auxiliary (PMN) phagocytic cell population of the lungs nearly two-fold over that observed with the bacterial challenge alone (Fig. 2). The results herein demonstrate that NO_2 acts as a concommitant inflammatory agent in _P. mirabilis_ challenged lungs. Therefore, a possible explanation for the enhanced bactericidal activity against _P. mirabilis_ at an NO_2 exposure concentration of 10.0 ppm may be that the augmented recruitment of PMNs into the lungs increased the intrapulmonary killing capabilities. This contention is supported by the observation that recruitment of PMNs into the lungs by inhalation of endotoxin (Hudson et al 1977) and gram-negative bacteria (Beseler et al 1983; Rehm et al 1980). Prior induction of intra-alveolar PMNs enhances the intrapulmonary killing of organisms (Rylander et al 1975). Following this line of reasoning, the enhancing effect in bactericidal activity brought about by augmented recruitment of phagocytic cells to the lungs at exposure to 10.0 ppm of NO_2 was finally overcome at 20.0 ppm most likely through impairments in the intrinsic phagocytic process itself (Acton et al 1972; Amoruso et al 1981; Vassalo et al 1973).

The modulating effect of exposure to increasing concentrations of NO_2 was again different when the challenge organism was _P. pneumotropica._

This organism is endogenous to the respiratory tract flora of many rodent colonies (Brennan et al 1969) including the mice used in this study. Exposure to 10.0 ppm of NO_2 significantly suppressed the intrapulmonary killing of this organism. This defect was not noted at 15.0 ppm and 20.0 ppm yet appeared again at exposure concentrations of 25.0 ppm and greater. Pulmonary lavages showed that NO_2 exposure concentrations of 10.0 ppm impaired the recruitment of the auxiliary PMN phagocytic defenses to the lung upon challenge with this bacterium. Following the line of reasoning established previously with **P. mirabilis**, fewer intra-alveolar phagocytic cells may offer an explanation for the bactericidal defect against **P. pneumotropica** at NO_2 exposure levels of 10.0 ppm. In addition, compared to **S. aureus** and **P. mirabilis** another host defense parameter undoubtedly plays a role in pulmonary resistance against **P. pneumotropica**; the role of antibody. Active or passive systemic immunization is known to enhance the intrapulmonary killing of gram-negative bacteria in normal animals (Jakab 1976) and mitigate the bactericidal defects when pulmonary antibacterial defenses are suppressed (Jakab and Green 1973). Serum antibody transudated into the lungs during the inflammatory process may account for the observation that the final bactericidal defect against **P. pneumotropica** was at exposure levels of 25.0 ppm whereas the defect against **P. mirabilis** (in these mice that possessed no demonstrable antibody against this organism) occured at 20.0 ppm.

The studies detailed above clearly demonstrate that an acute exposure to NO_2 for 4 hrs following bacterial challenges causes dysfunctions in pulmonary antibacterial defenses but that the level at which this defect occurs depends on the organism used to probe a specific defense parameter. When **S. aureus** was used to probe the functional activity of the alveolar macrophage phagocytic system the bactericidal defect was initially observed at exposure concentrations of 4.0 ppm and a clear dose-response relationship was observed with increasing concentrations. When gram-negative bacteria were used to probe the functional activity of both the resident and auxiliary phagocytic defenses of the lung the results depended whether the challenge organism was exogenous or endogenous to the respiratory tract of the host.

References

Acton, J.D., and Myrvick, Q.N. (1972). Nitrogen dioxide effects on alveolar macrophages. Arch. Env. Hlth. **24**, 48-52.

Amoruso, M.A., Witz, G., and Goldstein, B.D. (1981). Decreased superoxide anion radical production by rat alveolar macrophages following inhalation of ozone or nitrogen dioxide. Life Sciences **28**, 2215-2221.

Astry, C.L., Warr, G.A., and Jakab, G.J. (1983). Impairment of polymorphonuclear leukocyte immigration as a mechanism of alcohol-induced suppression of pulmonary antibacterial defenses. Amer. Rev. Respir. Dis **128**, 113-117.

Baseler, M.W., Fogelman, B., Burrell, R. (1983). Differential toxicity of inhaled gram-negative bacteria. Infect. Immun. **40**, 133-138.

Brennan, P.C., Fritz, T.E., and Flynn, R.J. (1969). Murine pneumonia. A review of the etiologic agents. Lab. Anim. Care 19, 360-371.

Ehrlich, R. (1980). Interaction between environmental pollutants and respiratory infections. Env. Hlth. Perspect. 35, 89-100.

Goldstein, E., Lippert, W., and Warshauer, D. (1974). Pulmonary alveolar macrophage: Defender against bacterial infection of the lung. J. Clin. Invest. 54, 519-528.

Green, G.M., Jakab, G.J., Low, R.B., and Davis, G.S. (1977). Defense mechanisms of the respiratory membrane. Amer. Rev. Respir. Dis. 115, 479-514.

Henemway, D.R. and Jakab, G.J. (1986). Nitrogen dioxide inhalation studies: problems and solutions. Appld. Ind. Hyg. In Press.

Hemenway, D.R. and MacAskill, S.M. (1982). Design, development and test results of a horizontal flow inhalation toxicology facility. Amer. Ind. Hyg. Assc. J. 43, 874-879.

Hudson, A.R., Kilburne, K.H., Halprin, G.M., and McKenzie, W.N. (1977). Granulocyte recruitment to airways exposed to endotoxin aerosols. Amer. Rev. Respir. Dis. 115, 89-95.

Jakab, G.J. and Green, G.M. (1972). The effect Sendai virus infection on bactericidal and transport mechanisms of the murine lung. J. Clin. Invest. 51, 1989-1998.

Jakab, G.J., and Dick, E.C. (1973). Synergistic effect in viral-bacterial infection: Combined infection of the murine respiratory tract with Sendai virus and Pasteurella pneumotropica. Infect. Immun. 8, 762-768.

Jakab, G.J., and Green, G.M. (1973). Immune enhancement of pulmonary bactericidal activity in murine virus pneumonia. J. Clin. Invest. 52, 2878-2884.

Jakab, G.J. (1976). Factors influencing the immune enhancement of intrapulmonary bactericidal mechanisms. Infect. Immun 14, 389-398.

Pierce, A.K., Reynolds, R.C., and Harris, G.D. (1977). Leukocytic response to inhaled bacteria. Amer. Rev. Respir. Dis. 116, 679-684.

Rehm, S.R., Gross, G.N., Hart, D.A., and Pierce, A.K. (1979). Animal model of neutropenia suitable for the study of dual phagocyte systems. Infect. Immun. 25, 199-303.

Rehm, S.R., Gross, G.N., Pierce, A.K. (1980). Early bacterial clearance from murine lungs: Species-dependent phagocytic response. J. Clin. Invest. 66, 194-199.

Ruppert, D., Jakab, G.J., Sylvester, L.L., and Green, G.M. (1976). Sources of variance in the measurement of intrapulmonary killing of bacteria. J. Lab. Clin. Invest. <u>87</u>, 544-558.

Rylander, R., Snella, M.C., and Garcia, I. (1975). Pulmonary cell response patterns after exposure to airborne bacteria. Scan. J. Respir. Dis. <u>59</u>, 195-200.

Vassalo, D.L., Domm, B.M., Poe, R.H., Duncombe, M.L. and Gee, B.L. (1973). NO_2 gas and NO_2 - Effect on alveolar macriphage phagocytosis and metabolism. Arch. Env. Hlth. <u>26</u>, 270-274.

Warr, G.A., Jakab, G.J. (1983). Pulmonary inflammatory responses during viral pneumonia and secondary bacterial infection. Inflammation <u>7</u>, 93-104.

Acknowledgement

Research described in this article is conducted under contract to the Health Effects Institute (HEI), an organization that supports the conduct of independent research and is jointly funded by the United States Environmental Protection Agency (EPA) and automotive manufacturers. Publications here implies nothing about the view of the contents by HEI or its research sponsors. HEI's Health Review Committee may comment at any time and will evaluate the final report of the project. Additionally, although the work described in this document has been funded in part by the U.S. Environmental Protection Agency under assistance agreement X808859 with HEI, the contents do not necessarily reflect the views and policies of the Agency; nor does mention of trade names or commercial products constitute endorsement or recommendation for use.

Prof. George J. Jakab, Department of Environmental Health Sciences, The Johns Hopkins School of Hygiene and Public Health, 615 North Wolfe St., Baltimore Maryland, 21205, USA.

LYSOSOME RESPONSE AND CYTOSKELETON ALTERATION IN CELL CULTURES EXPOSED TO AIRBORNE LEAD

E.V. Orsi, M. Bavlsik, R. Viera, M. Petersheim[+] and O.F. Baturay[#]
Seton Hall University, Departments of Biology and Chemistry[+], South
Orange, N.J. and Technion Inc.[#], Belleville, N.J., U.S.A.

We present here the delayed leakage of lysosome acridine orange
in cell cultures exposed to lead acetate. Delayed leakage was accompanied by loss of actin. These responses were not elicited by
sodium acetate but could be duplicated by extracts of air sample
filters from a high air pollution area but not a cleaner air site.

<u>Materials and Methods</u>:Air samples were collected at the same high
particle count (New Jersey-urban) and low particle count(northern
New York-Adirondack mountains)sites. Particle counts and metal
assay by Zeeman Effect atomic absorption methods were performed
as previously reported[3] with the following major modifications.
I.Air samples:a)A more remote Adirondack site was added utilizing
a solar powered vacuum pump system for the filter manifold.b)All
filters were 25 mm diameter. c) Only outside air was sampled. The
larger filters provided multiple samples which were minced,ground
with sterile saline,sonicated and centrifuged to provide debris
free supernatants. Replicate samples of each were used for metal
assay and inoculated in the cell culture medium.
II.Cell cultures: Cell lines of mouse(3T3) or monkey(VERO)origin
were the main target cells. Others were L-929(mouse),Wi-38(human)
and B95-8(marmoset)lines. All cells were grown in MEM with 10%
fetal bovine serum and conventional antibiotics.
III.Acridine orange fluorescence:Dye accumulation and retention
in cell lysosomes was monitored and evaluated by two methods.
a)Fluorescence Densitometry. A photomicrograph was taken,followed
by a second of exactly the same field after a measured time of ir-
radiation by UV light of about 400 nm and with the same exposure
time as the first. The red intensity of the same regions in the

two slides were measured by projecting a 7x enlargement and track-
ing across with a light sensitive point probe of a silicon diode
photometer provided with analog readout Fig.1(Oriel Corp.,Stamford
Conn.). With this method decreased image red fluorescence was correlated with well defined and reproducible decreased amplitudes on
the chart recorder as seen in Fig. 2.
b)Continuous cytophotometry of dye fluorescence. Half-life fluorescence calculations for intralysosome acridine orange retention
were measured with the same microscope and filter combination used
for the fluorescence densitometry photomicrographs. The cells were
irradiated continuously without changing the field and the emission measured at 30 seconds intervals with a photomultiplier(AMINCO
Silver Spring,MD). The amount of dye diffusing out of the lysosomes
depends on the duration of exposure to UV. In Fig. 3 the red intensity is plotted against duration of UV exposure. Since the rate of
intensity loss is roughly exponential the rate of color shift can
be expressed in terms of half life. Using data typified in Fig.3,a
linear regression was carried out and half life values obtained.
IV.Cytoskeleton distribution: a)Cytophotometric optical density procedures were done after extraction of cytoplasmic proteins using
tungsten light and the same microscope-photometer unit as above.
b)Fluorescent-antibody methods utilizing first antibodies against
actin,myosin,prekeratin and tubulin (ICN,Lisle,Illinois) were done
with the same exciter and barrier filters used for acridine orange
fluorescence but with a dark field condenser.
V.Cell exposure to lead: a) Lead acetate and sodium acetate were added to about one million cells in 3 ml growth medium following 24
hours of incubation. A final concentration of 0.30 mg/ml was usually employed. Following appropriate incubation periods,cells were
used for the various procedures. In some cases lead assays of the
media components,cells and overlying growth media were carried out.
b)Filter extract concentrations were adjusted so that the above lead
concentrations would not be exceeded by lead in the urban filters.

Results:

I.Air quality: The typical high particle(condensation nuclei)count
associated with high vehicular traffic (table I)was again found[2]

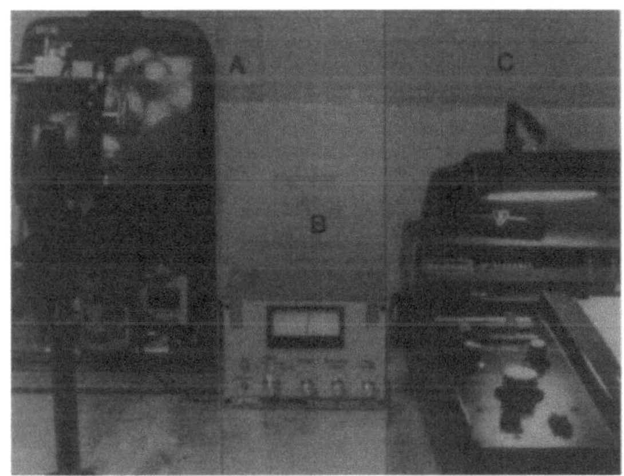

Fig. 1. A,B,C indicate enlarged image light probe control and chart recorder.

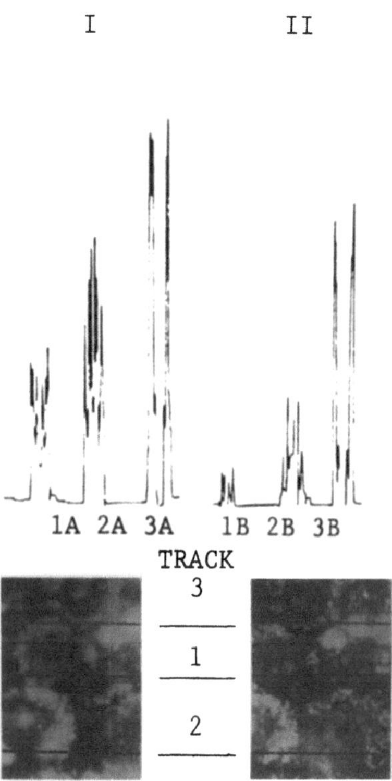

Fig.2. Red color intensity of the same tracks(A and B). I and II are first and second photomicrographs of a control 3T3 cell culture.

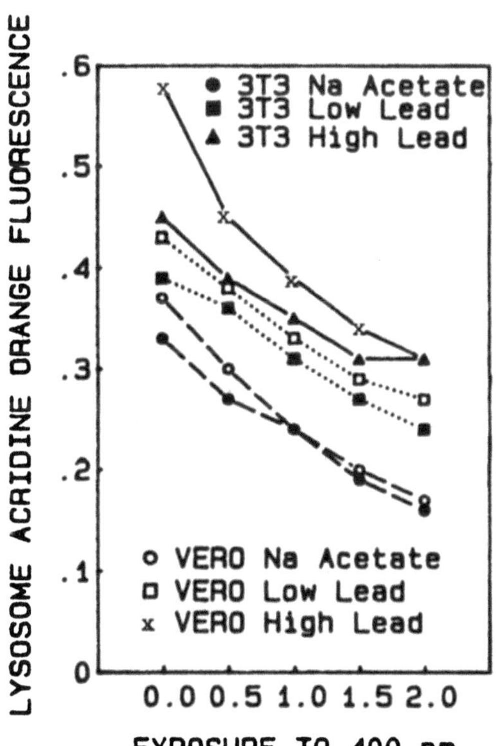

Fig.3. Loss of dye fluorescence at half minute intervals.

Table I. Vehicle traffic and regional particle (CN) counts

Site	Vehicles/hr.		CN/cc air
N.J. (urban)	lowest	120	14,000
	highest	624	54,000
N.Y. (mts)	usual	2	740
	unusual	18	1,850

throughout all the sampling periods.

II. Airborne metal burden:The consistently[3] higher lead levels typical of the urban site are shown after a 30 day collection period
in table II.

Table II. Metal from filters at low and high air pollution sites

Location	Lead/ppm	Copper/ppm	Other	CN/cc of air
Control*	1.5	1.2	#	----------
B-W Laboratory**	1.8	2.2	#	1200--1500
SHU Laboratory***	20.5	30.3	#	23000-84000

*Unexposed filter
**Adirondack Mts.(Wilmington,NY)
***New Jersey,urban area(So.Orange,NJ)
#Levels of cadmium,zinc,nickel and titanium did not exceed 1.0 ppm

III.Acridine orange retention:Fluorescence densitometry scans seen
in Fig.4 was greater with both species when exposed to lead acetate. The mouse cells usually showed a greater response to lead confirmed by comparing the ratio of the amplitudes before and after
UV exposure. There was no difference between 3T3 and VERO sodium
acetate controls but 3T3 showed less decrease (P=0.05) with lead.
Enhanced acridine orange retention by lead particularly with cells
of the 3T3 line was also seen with half life measurements. Again
the mouse cells showed greater retention (P=0.05) than the monkey
line. Filter extracts from the SHU site increased the half life of
3T3 cells in contrast to control or the low particle count site.
The results are summarized in table III.
IV.Cytoskeleton response: Regardless of the methods employed exposure to lead acetate or SHU filter extracts disclosed loss of the
actin components of the cytoskeleton. These results to be presented elsewhere also showed a consistent relation between rodent
cell enhanced dye retention and greater loss of actin at least in
the cell lines tested. These included 3T3 and L-929 (both mouse)
origin and the VERO and Wi-38 lines both(primate origin).

3T3 CELL CULTURES   VERO CELL CULTURES

sodium acetate lead acetate   sodium acetate lead acetate

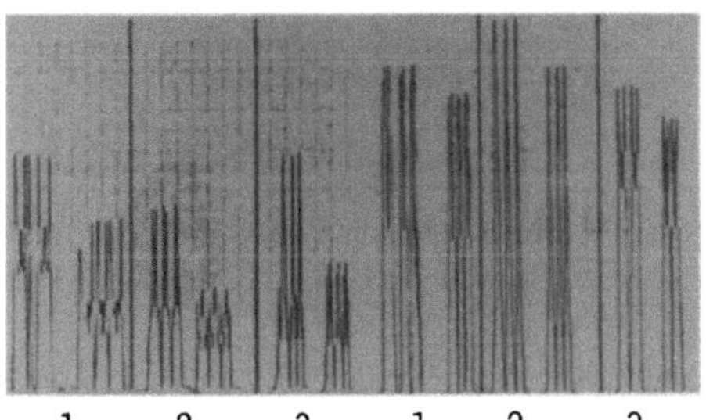
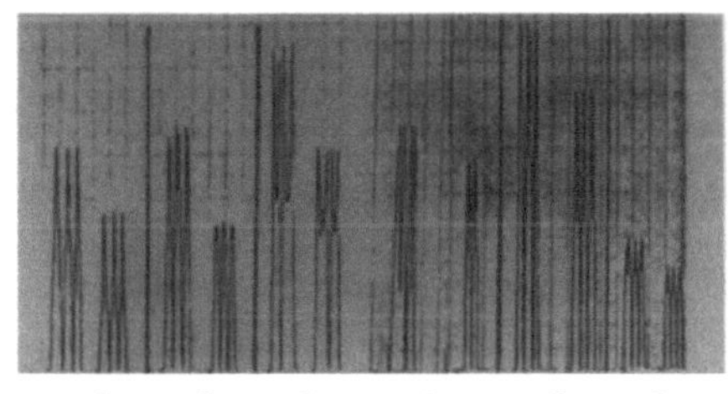

$A^1B \quad A^2B \quad A^3B \quad A^1B \quad A^2B \quad A^3B \qquad A^1B \quad A^2B \quad A^3B \quad A^1B \quad A^2B \quad A^3B$

Fig.4. Fluorescence intensity scans along identical tracks of replicate photomicrographs. A is the first photomicrograph and B is the second. The number denotes the scanning track

Table III. Acridine orange half life of 3T3 and VERO lysosomes

| | 3T3 cultures | | VERO cultures | |
Source of lead	Half life#	P	Half life#	P
sodium acetate(exp.a)	1.11	-----	1.14	-----
lead acetate	2.11	0.05*	1.65	0.05*
saline	4.50	-----	inc	
control filter(exp.b)	4.97	NS**	inc	
B-W filter	4.86	NS***	inc	
SHU filter	8.38	0.16***	inc	

*compared to sodium acetate control   inc denotes in progress

**compared to saline control   NS= P greater than 0.20

***compared to control filter extract

Summary and Conclusions:

The results indicate that the delay in acridine orange loss by lysosomes exposed to UV can serve as a sensitive probe to the lead content of the cell milieu and the induced responses. This lysosome stabilization is in agreement with the reported decrea-

sed release of lysosome enzymes by lead in rat cerebral tissue[1] .
The lead induced lysosome stabilization reported here is undoubtedly pathologic in view of the cytoskeleton alteration that usual-
lly accompanied the introduction of lead in the cell media either
from laboratory sources or natural conditions.

<u>References</u>

1. Nakagawa, K., Asami, M. and Kuriyama. K.(1983) Some factors
affecting enzyme release from cerebral lysosomes: inhibitory effects of lead. Japan J. Pharmacol. <u>33</u>,9-15.

2. Orsi, E.V., Levins,P.,Ames, G., Buonincontri, L., Holyk, N.and
Balciunas, R. (1977) Human cell cultures as sentinal systems for
monitoring air quality. pp. 637-650. in "Cell Culture and Its Applications", R.T. Acton and J.D. Lynn (eds.) Academic Press, New
York .

3. Orsi, E.V., Zois, W.J., and Baturay, O.F. (1983) Human cell
cultures response to simulated and natural levels of air pollution. pp. 85-98. in "In Vitro Toxicity Testing of Environmental
Agents, Part B", A. Kolber, T.K. Wong, L.D. Grant, R.S. DeWoskin
and T.J. Hughes (eds.) Plenum Pub. Co., New York and London.

Dr. Ernest V. Orsi, Department of Biology, Seton Hall University
So. Orange, N.J. 07079, U.S.A.

2. Outdoor air

HEALTH RISKS FROM RESPIRABLE DUSTS PRODUCED DURING THE OPERATION OF BIG HARVESTERS

K. Stalder and M. Kundel
Department of Occupational and Social Medicine, University
of Göttingen, D-3400 Göttingen, FRG

In modern agriculture the use of big machines may be the cause
of considerable amounts of dust. We have investigated in particular the health risks from dusts generated during the operation of combine harvesters. These dusts are highly complex
mixtures of inorganic and organic materials depending on the
type of soil and the kind of crop (7). Four special situations
could be included in our study: There were fields of loamy
soil and fields of sandy soil. On both types of soil wheat was
grown; furthermore there were fields of sandy soil with winter
barley and with spring barley. Dust was collected by a sampler
with a preseparation device (2) so that the respirable dust
fraction was obtained separately. 15 mg aliquots of this fraction in 0,3 ml saline were administered to female rats (Wistar
strain, weight: 18o-2oo g) by intratracheal injection. After
three months the lungs of the animals were examined histolo-
gically and the degree of fibrosis was quantified by determination of the hydroxyproline content after hydrolysis in 6N
hydrochloric acid at $105^{\circ}C$ for 22 hrs (5). As hydroxyproline
is a biochemical marker of collagen the amount of this sclero-
protein could be calculated from the amount of the amino acid.
Dust samples collected during the harvest of wheat or spring
barley in sandy soil elicited an almost threefold increase of
collagen in the lungs of the eperimental animals (Table I). It
has to be noticed that these dust samples also have the highest
percentages of SiO_2 (determined by a colorimetric method (4)

Table I: Lung collagen of rats 3 months after intratracheal application of dust samples in percent of untreated controls

| Source of dust | | SiO_2-content | Lung Collagen |
Soil	Crop	of dust (%)	(% of control)
Loamy	Wheat	9	154
Sandy	"	12	29o
"	Winter barley	7	1o1
"	Spring "	25	274
Reference dusts	Quartz (DQ 12)	98 (86% Quartz)	6o4
	Anatase		139

after solubilization in molten Na_2CO_3 an K_2CO_3). This method measures not only quartz which may be the cause of fibrosis. It gives information about the total SiO_2-compounds including silicates. However, the presence of quartz was confirmed in several specimens by X-ray diagrams. As fibrogenic reference dust Dörentrup Quartz No.12 (DQ 12) was included; this is used as standard in silicosis research. As inert reference dust Anatase was chosen because it is known to be practically neither cytotoxic nor fibrogenic. Particle diameter of all dusts was about 3 /um.

The fibrogenicity of the collected dusts indicates a certain risk that exposed persons may develop silicosis. This risk may be limited by the relatively short total exposure time compared with the situation e.g. in mining or tunneling. Dusts generated in agriculture may contain allergens and these may even add to the risk of lung fibrosis as a possible result of allergic alveolitis. Evidence for an immune response after the administration of some of the dust samples was obtained in experiments with guinea pigs and rats. In these experiments 1o mg samples of dust were injected intraperitoneally in complete Freund's adjuvant. 45 days later the animals were

challenged by intratracheal injection of 3 mg of the same
dust in o,3 ml saline. After 6 hrs. the animals developed
dyspnoa. Free Alveolar cells were increased in the broncho-
alveolar lavage. Histological examination revealed a predominant polymorphonuclear infiltration of the interstitium
of the lung.

A special problem is the influence of the mineral components
of the dust samples on the immune response. Several mineral
particles, in the first place quartz, have an immune adjuvant effect (1, 6). In the present context, model experiments
were performed using the standard quartz DQ 12 and isolated
allergens produced by the actinomycete Micropolyspora faeni,
known to cause the occupational disease "farmer's lung" (3).

Soluble allergen was prepared from cultures of M. faeni in
Synthetic Broth AOAC (Difco Laboratories Detroit). Supernatants of the cultures were purified by ultrafiltration,
the substances of the broth removed by dialysis, and the
retentate freeze dried. Rats (as above) were sensitized
intraperitoneally either by 4 mg allergen in 5 ml saline or
by 4 mg allergen and 1o mg DQ 12 in 5 ml saline. After 28
days the animals were bled and 2 days later boostered intraperitoneally with 2 mg allergen alone or with 1o mg DQ 12 in
5 ml saline. On day 58 the animals were boostered again with
2 mg allergen. They were bled a second time on day 68.
Precipitating antibodies were determined in dilutions of the
serum by the Ouchterlony technique. As shown in Table II at
the end of the immunisation schedule there were only low
antibodies titers when the allergen was administered alone
whereas much higher titers could be observed in the sera of
animals that received quartz particles in addition to the
allergen. These results indicate that there is a strong
influence of the mineral fraction, especially of quartz
particles, on sensitizing processes by allergens that may be

present in respirable dusts caused by modern agricultural procedures.

Table II: Precipitating antibodies in the sera of rats after immunization by M. faeni antigens with and without Quartz dust

Immunization	Titers	
	after 28 days	68 days
Antigen + Quartz	0/0/0/0/1/0/0/	16/32/32/32/32/32/16
Antigen	0/0/0/2/1/0/0	0/1/2/0/4/1

<u>References</u>

(1) Antweiler, H. (1961) Über die Adjuvanswirkung von Quarz und anderen Stäuben beim Kaninchen und bei der Ratte. Naunyn-Schmiedeberg's Arch. exp. Path. Pharmak. <u>241</u>, 215 - 216

(2) Batel, W. (1975) Messung zur Staub-, Lärm- und Geruchsbelastung an Arbeitsplätzen in der landwirtschaftlichen Produktion und Wege zur Entlastung. Grundl. Landtechn. <u>25</u>, 135 - 157

(3) Pepys, J., P.A. Jenkins, G.N. Festenstein, P.H. Gregory, M.E. Lacey, F.A. Skinner (1963) Farmer's lung: thermophilic actinomycetes as a source of farmer's lung hay antigen. Lancet II, 6o7 - 711

(4) Stegemann, H., J. Fitzek (1954) Die mikroanalytische Bestimmung von Silizium in quarz- und silikathaltigen Staubproben. Beitr. Silikoseforsch. <u>31</u>, 29 - 4o

(5) Stegemann, H., K. Stalder (1967) Determination of hydroxyproline. Clin. Chim. Acta <u>18</u>, 267 - 273

(6) Stalder, K. (1985) Das Immunsystem als Angriffspunkt gewerblicher Noxen. Verh. Dt. Ges. Arbeitsmed. 23. Jahrestagung, Gentner Verlag, Stuttgart, 31 - 37

(7) Stalder, K., W. Batel (1985) Staubexposition der Beschäftigten in der landwirtschaftlichen Produktion und Daten zu damit verbundenen Gesundheitsrisiken. Verh. Dt. Ges. Arbeitsmed. 25. Jahrestagung, Gentner Verlag, Stuttgart, 595 - 599

Prof. Dr. med. Karlheinz Stalder, Abteilung Arbeits- und Sozialmedizin, Universität Göttingen, Windausweg 2, D - 34oo Göttingen

EXS 51:
Advances in Aerobiology
©1987 Birkhäuser Verlag Basel

THE INFLUENCE OF METEOROLOGICAL AND AIR POLLUTION FACTORS ON
ACUTE DISEASES OF THE AIRWAYS IN CHILDREN AS ILLUSTRATED BY
THE BIEL REGION (SWITZERLAND)

H. Marty, Department of Medicine, Obersimmental District
Hospital, CH-3770 Zweisimmen, Switzerland

1. Introduction

In Biel, an interdisciplinary working group on meteorology and
air pollution has existed since 1979. This small town at the
southern foothills of the Jura mountains is especially suitable
for investigations of air pollution because of inversion
situations in winter. The objective of this study was to
investigate the influence of meteorological and individual air
pollution factors on acute airway diseases in children.

2. Method

The climatological measurements comprised the following
parameters: Temperature, humidity, cloud cover, precipitation,
barometric pressure, wind speed, wind direction and weather
type as well as the occurrence of fog and mist. The measurements
of pollution were restricted to sulphur dioxide and dust. With
ten measurement points, deposition values could be registered.
In the months October, November and December of 1981, additional
measurements of SO_2 concentration were carried out. The
registration of medical data was carried out at the
Wildermeth Children's Hospital and in two pediatrician's offices
over 17 consecutive months from November 1st 1980. The following
diseases were registered: Croup, epiglottitis, pneumonia, acute
bronchitis, extrinsic (allergic) asthma and intrinsic
(infectious) asthma.

3. Results

3.1 Meteorological and climatological measurements

They show that in every weather type, there exist less or more
local circulations caused by the topography. In winter, there
are often pronounced inversion situations, which may persist
for several days.

3.2 Measurements of pollution

Fig. 1 shows the 24-hour mean values of SO_2 concentration
measured in early winter 1981. The highest value measured was
180 $\mu g/m^3$, and the mean value of all SO_2 concentration
measurements was 73 $\mu g/m^3$.

The SO_2 deposition values are presented in Fig. 2.
The highest values were registered during the autumn and winter
months. Dust measurements revealed a slight to moderate dust
exposure which greatly depended on the weather situation. On
certain days episodes of air pollution were observed which are
typical for Biel and are due to the weather-dependence of the
pollutant level (strong influence of local wind systems).

3.3 Cases of illness

During the 17 months study, 563 cases of acute airway conditions
were registered in children. Table I shows the distribution
of the cases and of the individual illnesses.

3.4 Seasonal distribution of illnesses

Fig. 3 shows the seasonal distribution of the cases of disease.
An increased occurrence of illness in the autumn and winter
months is apparent. The same annual curve applies to croup and
acute bronchitis.

3.5 Correlations between meteorological/air pollution data and of illness

The climatic data are correlated singly with the air pollution
values and the various diseases. By means of simple statistical
tests, it was checked whether the individual parameters showed

significant differences on days with different numbers of cases. The following correlation was shown: Airway diseases were significantly more frequent at lower temperatures and at a high relative and absolute humidity. The correlation between humidity and acute bronchitis was especially pronounced. In the analysis of the weather types, the evaluation of the overall patients investigated showed an unequivocally raised occurrence of acute airway conditions in northerly wind situations. This correlation is especially pronounced in croup.

On the air pollution side, the values for SO_2 deposition show a significant correlation with the occurrence of the cases of illness. This may possibly involve a parallel, but independent annual curve with maxima for both events in the winter months. In the measurements of SO_2 concentration, there were no significant correlations, but only a certain trend could be detected, as the number of cases increased with increased SO_2 exposure and lowered air temperature. There was no significant correlation between dust deposition and concentration values with the number of cases of illness.

4. Discussion

We can thus summarize the results of our investigations as follows: We found an unequivocal influence of certain meteorological parameters on acute airway conditions in children. Low temperature, high humidity and northerly weather types proved to be unfavorable. This applied above all to croup and acute bronchitis. On the other hand, we could not detect any statistically significant correlations with SO_2 and dust exposure, which is only slight to moderate for Biel. However, continuous measurements of these pollutants are largely lacking and other very much more important noxae such as nitrogen oxides and ozone were not measured. However, our study shows that in future investigations the meteorological factors must be given the necessary attention besides the parameters of air pollution.

<u>Fig. 1</u>: 24-hour mean values of SO_2 concentrations in µg/m^3

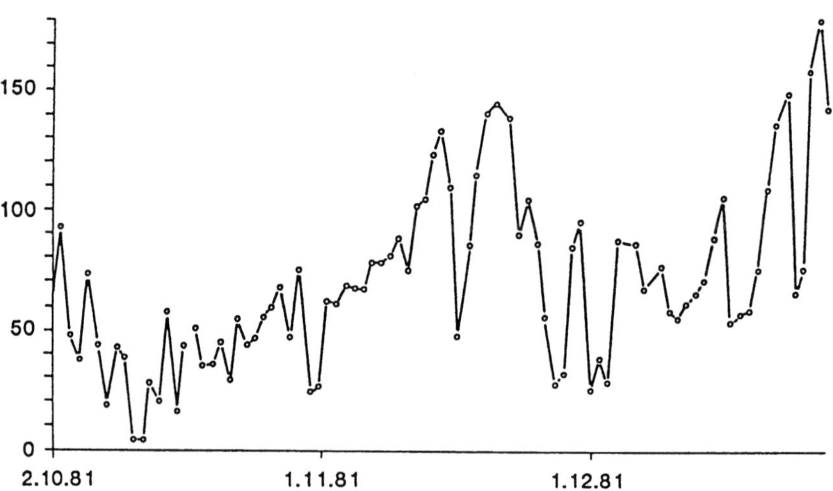

<u>Fig. 2</u>: SO_2 deposition in mg/m^2/100 hours

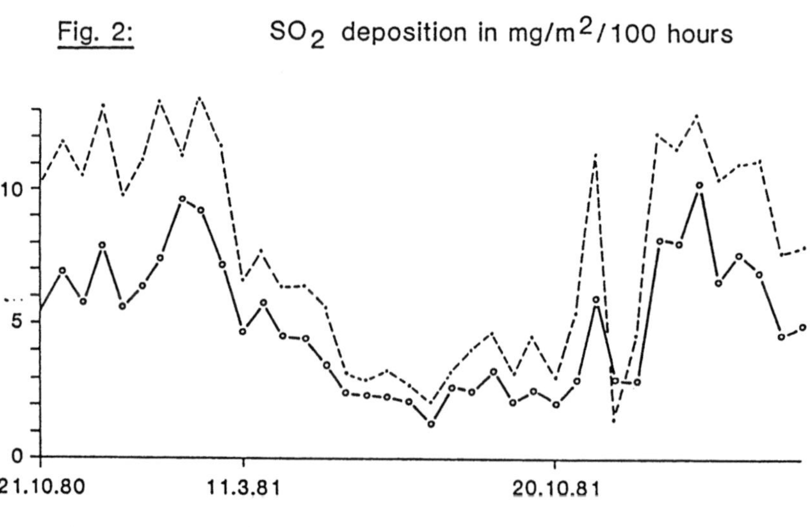

——— Average values of all 10 stations
----- values at central square

Tab. I

	City of Biel	Environs	Total
Croup	141	89	230
Epiglottitis	6	0	6
Acute bronchitis	109	31	140
Extrinsic asthma	19	10	29
Intrinsic asthma	53	20	73
Pneumonia	44	35	79

Fig. 3: Seasonal distribution of all acute airway conditions from November 1st 1980 to March 31st 1982

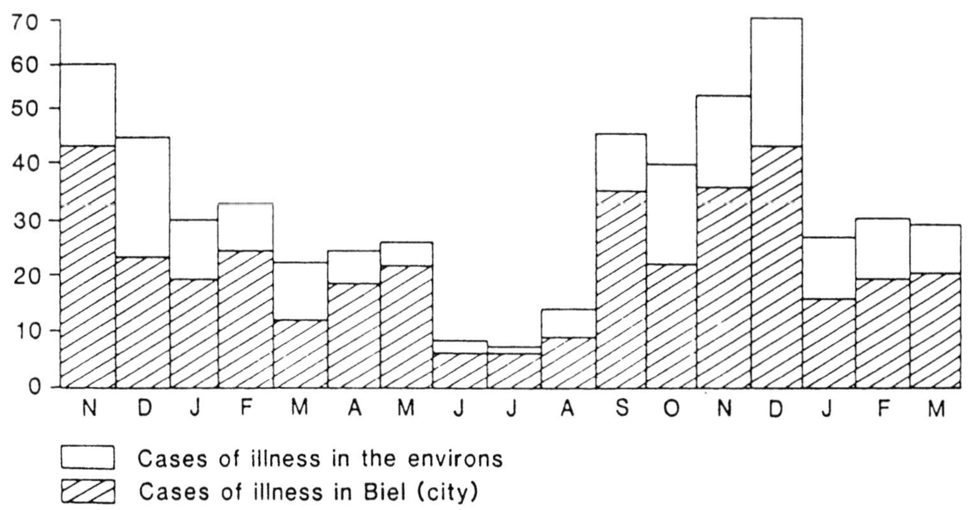

EXS 51:
Advances in Aerobiology
©1987 Birkhäuser Verlag Basel

AIRBORNE CRYSTALS OF ANHYDROUS CALCIUM SULPHATE - A NEW AIR POLLUTANT

H. Morrow Brown, F.A. Jackson
Midlands Asthma and Allergy Research Association, Derby, England.

In July 1983 an outbreak of acute asthma during a thunderstorm in Birmingham was reported.[10] The authors suggested that sudden weather changes were the cause of the epidemic, but when the storm reached Derby, 60 Km. away, we observed a very high spore count which correlated closely with the remarkable numbers of asthma admissions the previous evening, as shown in Fig.1.

In Derby patients with allergic symptoms closely linked to weather changes were first noted in 1960. We have sampled air spora since 1965, and shown [3,4] that dramatic fluctuations in airborne fungal spores frequently account for allergic symptom changes, and that Derby spore and pollen counts are broadly applicable to Birmingham.[2] We suggested [5] that a sudden rise in wet weather spores was the most likely of this asthma epidemic.

Fig 1 - Total daily spore counts from Derby (solid line) superimposed on daily asthma admissions in Birmingham (histogram) June 27 to July 10 1983. Dashed line represents Derby spore counts displaced backwards by one day.

The Morrow Brown spore trap as previously described [1] is modified to
improve trapping efficiency for small spores by having a slit which is
0.5 mm. wide instead of the usual 2.0 mm., and a distance of only 0.5 mm.
from the slit to the Vaseline surface of the slide, which is made
accurately by the use of a special jig. Unpublished studies sampling
simultaneously with spore traps using 2 mm. and 0.5 mm. slits showed no
difference as regards grass pollen or cladosporium, but an obvious increase
in the catch for smaller particles as shown (Fig.2)

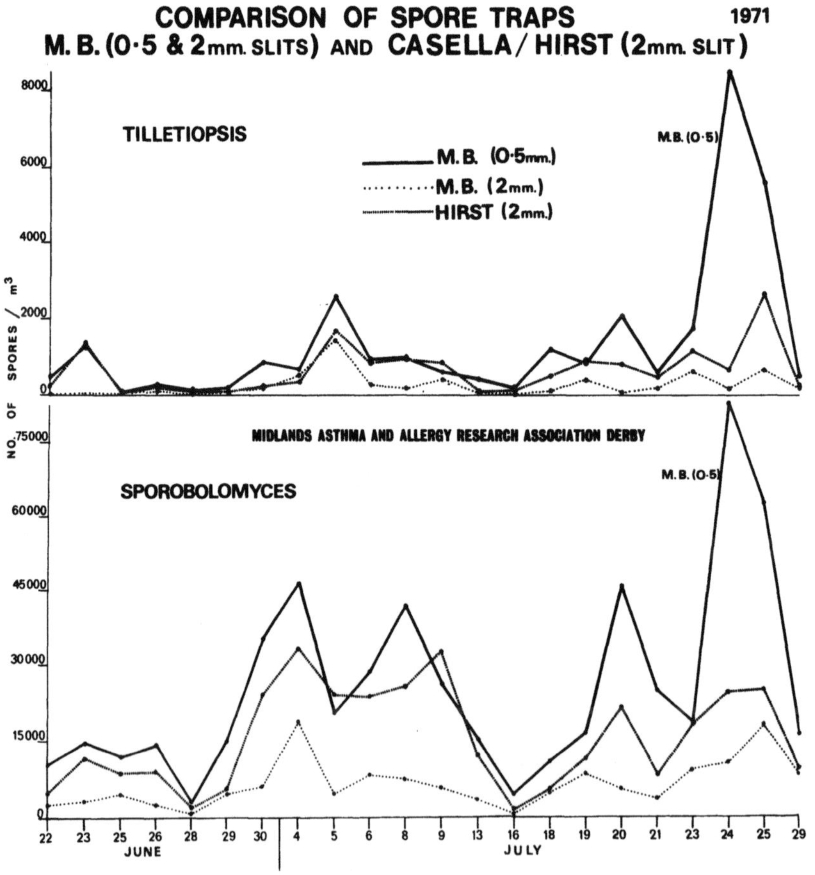

Fig.2. Results of simultaneous air sampling at 10 1/mm.
showing how small spores are more efficiently trapped
by using a narrow slit.

In 1982, while experimenting with alternative adhesives for spore trap
slides, it was found that this modification had increased the impact
velocity to the extent that many spores would stick to a bare slide
without any adhesive. In the summer of 1983 we repeated this experiment
during the period of the epidemic of asthma in Birmingham.
Examination of simultaneously exposed spore trap slides <u>without</u> adhesive
revealed not only large numbers of spores, but many crystalline structures
of various shapes and sizes. (Figs. 3 and 4) Further plain slide exposures
have confirmed that crystals appear frequently during wet or humid
conditions and also at a very rural site, but the identify of these
crystals was quite unknown. They had not been observed previously because
they dissolve in the poly-vinyl lactophenol mountant usually used for spore
trap slides, but they can be seen on Vaseline-coated slides before mounting
by using a long working distance objective lens. (Fig.5)

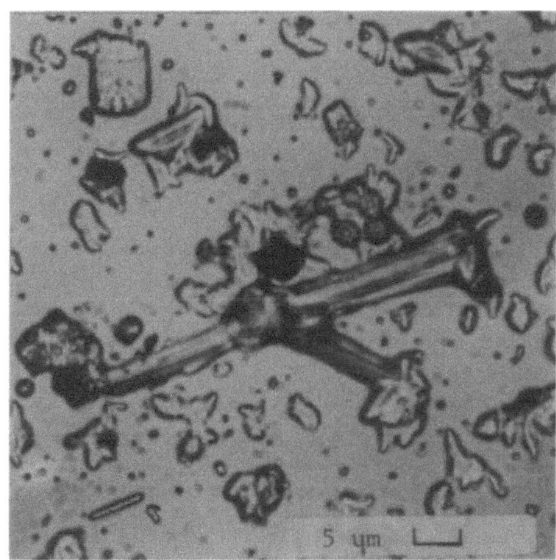

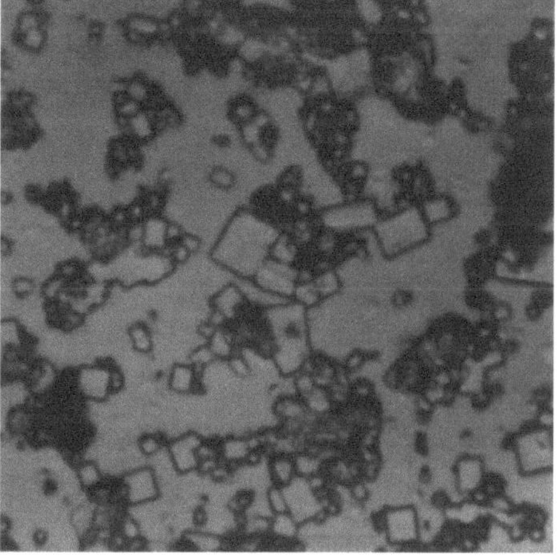

Fig.3. Light Microscope X600.
A feathery crystalline
structure superimposed on a
profuse deposit of wet
weather spores on a bare slide.

Fig.4. Light Microscope X600.
Crystal impacted on a bare
slide during wet weather.

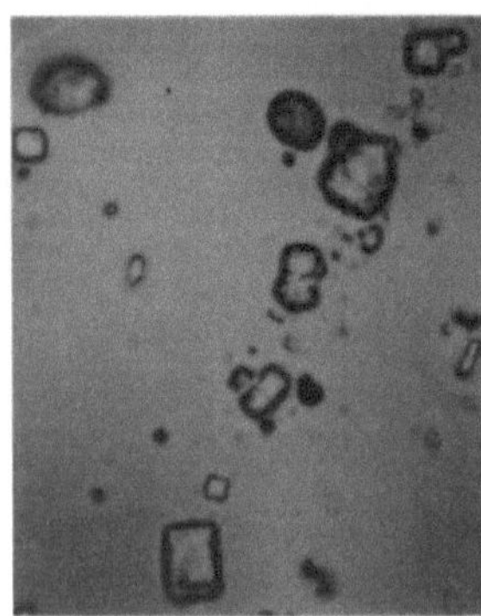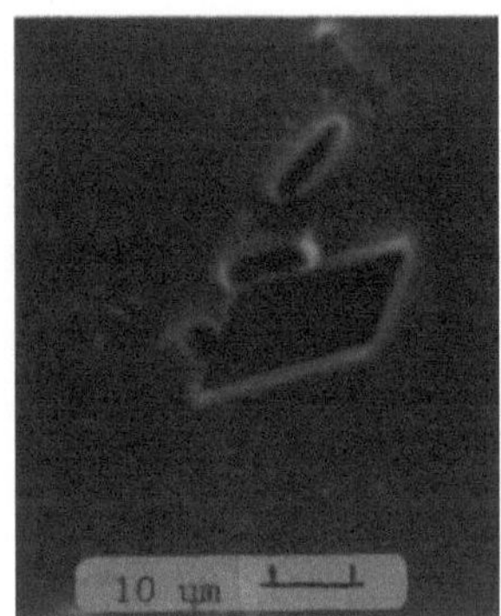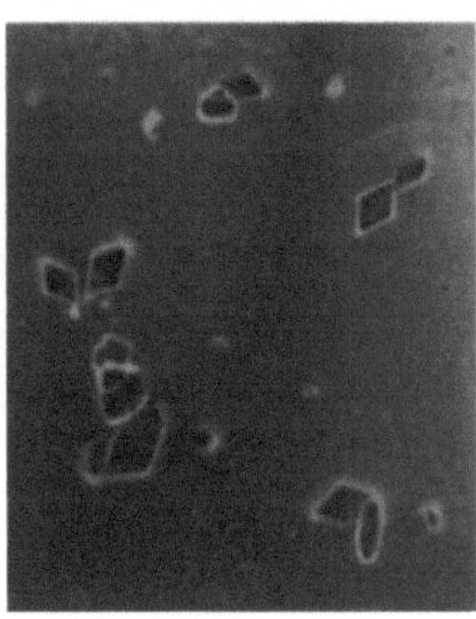

Fig.5. Light Microscope X600. Crystals observed on
Vaseline surface before mounting.

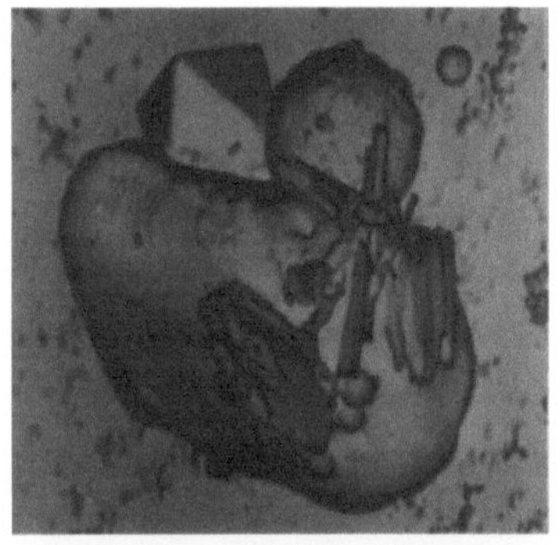

X 5500 1 µm   X 2800 1 µm

X 4700 1 µm   X 11000 1 µm

Fig.6. S.E.M. studies of crystals of calcium sulphate
illustrating pleomorphism. Some are combined with
spores and fly-ash.

In 1984 scanning electron microscopy and anlysis of the crystals by energy
dispensive x-ray microanalysis and x-ray diffraction, was carried out at
Loughborough University. To our surprise all crystals examined, whatever
their shape, were identified as a pleomorphic form of anhydrous calcium
sulphate. The most likely source seems to be the many coal fired electric
power stations within 25 Km. of Derby, which lies in the largest area of
sulphur emissions in England, by combination of Ca and SO_2 during
combustion.[8,9] We now suggested[6] that these crystals might trigger spasm
in hyper-reactive bronchi, thus contributing to asthma epidemics, in agreement with further studies of the outbreak of asthma in 1983.[11]

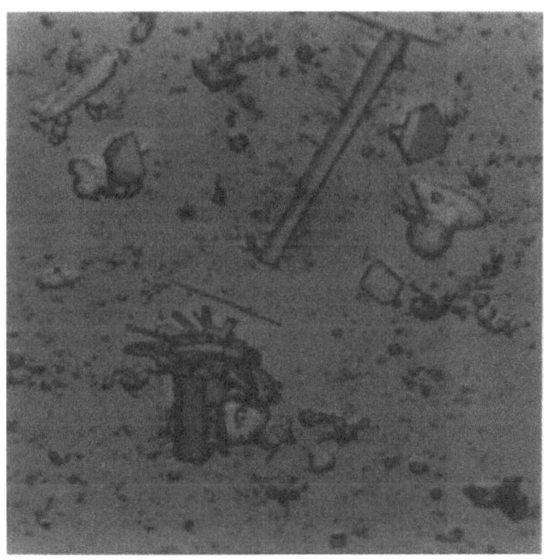

Fig.7. S.E.M. X2200 showing
extreme pleomorphism and
fly-ash.

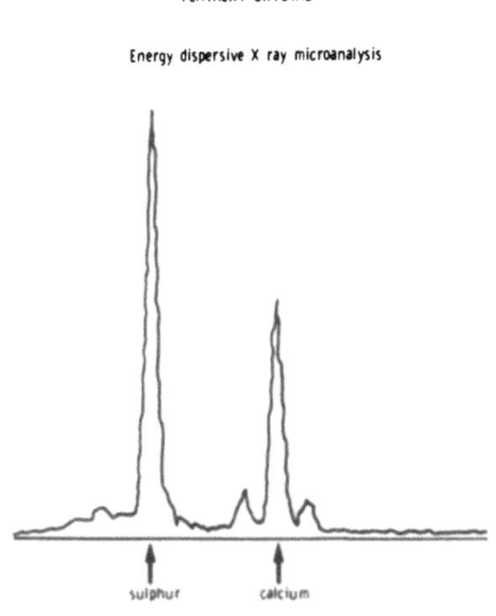

Fig.8. All crystals examined
produced the same result, and
were identified as Calcium
Sulphate by x-ray diffraction
analysis.

The significance of our findings remains unknown, as we have been unable
to produce calcium sulphate crystal aerosols with which to challenge
asthmatic patients. Studies have been made on particulate sulphates in
rural air,[8] and exposure of human volunteers to sulphate aerosols[7] has
been found to have no effect, but there is no data on calcium sulphate.

Our observations have not been reported by others so far, and further
investigation of the origin and possible importance of this new form of

pollution is clearly necessary. This summer we plan to obtain airborne samples to demonstrate clearly if these calcium sulphate crystals are really derived from the coal fired power stations with which Derby is surrounded.

Dr. H. Morrow Brown, Midlands Asthma and Allergy Research Association, 12 Vernon Street, Derby DE1 1FT, England.

1.   Brown, H.M., Jackson, F.A. (1978) Aerobiological studies based in Derby I: a simplified automatic volumetric spore trap. Clin. Allergy 8; 589-97.

2.   Brown, H.M., Jackson, F.A., (1978) Aerobiological studies based in Derby II: Simultaneous pollen and spore sampling at eight sites within a 60 Km. radius. Clin. Allergy 8; 599-609.

3.   Brown, H.M., Jackson, F.A. (1981) Allergy to seasonal mould spores. Allergol. Immunopathol; suppl IX: 137-42.

4.   Brown, H.M., Jackson, F.A. The value of aerobiology to an allergy treatment centre. In: Federal Environment Agency, ed. Proceedings of First International Conference on Aerobiology (Munich Aug 13 - 15 1978). Berlin: Erich Schmidt Veriag, 1980: 278-86.

5.   Brown, H.M., Jackson, F.A. (1983) Asthma and the weather. Lancet ii 630.

6.   Brown, H.M., Jackson, F.A. (1985) Asthma outbreak during a thunderstorm. Lancet ii:562.

7.   Kulle, T.J. et al. (1984) Sulfur dioxide and ammonium sulfate effects on pulmonary function and bronchial reactivity in human subjects. Am.Ind.Hyg.Assoc.J. 45: 156-161.

8.   Martin, A., Barber, F.R. (1978) Some observations of acidity and sulphur in rainwater from rural sites in central England and Wales. Atmosph Envir 12: 1481-87.

9.   Martin, A., Barber, F.R. (1985) Particulate sulphate and ozone in rural air: Preliminary results from three sites in central England. Atmosph Envir 19: 1091-1102.

10. Packe, G.E., Archer, P.St.J., Ayres, J.G. (1983) Asthma and the weather. Lancet ii: 271.

11. Packe, G.E., Ayres, J.G. (1985) Asthma outbreak during a thunderstorm. Lancet ii: 199-203.

EXS 51:
Advances in Aerobiology
©1987 Birkhäuser Verlag Basel

CONIOMETRIC DUST MEASUREMENTS DURING A SEA VOYAGE FROM THE NORTHERN TO THE SOUTHERN HEMISPHERE, ALONG THE ANTARCTIC COAST, AND AT A COAST STATION IN ANTARCTICA

Schrader, G.

Institut für Allgemeine und Kommunale Hygiene (Direktor: OMR Prof. Dr. sc. med. H. Horn) der Medizinischen Akademie Erfurt, Erfurt, GDR, 5082

Antarctic expeditions pursue the aim to win a better knowledge of the natural conditions of the distant ice-covered continent. Of course geological, glaciological and meteorological research tasks are predominant for economic and surviving interests. Medical researches on this continent aim at finding out how extremely different day and night influences, social isolation and inclement natural conditions may affect human health (Lugg,1986).

Researches concerning the percentage of dust particles in the air were no primary task of our expedition groups. They were initiated by one of the scientists when he saw possibilities for it. This may explain the simple method of coniometry by means of a "Conimeter 10" from VEB Carl Zeiss Jena and the unsystematic localization of the measuring points. Coniometrical measurements were taken daily during a voyage which led very far away from the usual searoutes from the Northern to the Southern Hemisphere and during a voyage along the Antarctic coast as well as during a nine months' winter stay in the Soviet Antarctic station Molodezhnaya indoor and outdoor.
Fig. No. 1 shows measuring points respectively the regions where probes were taken.
Fig. No. 2 shows, how a probe by means of a "Conimeter 10" is taken in front the Antarctic coast. The measuring results should be examined under the following aspects:
1. Where is the original source of the particles of the open sea?
2. Are there remarkable differences between the numbers of the

Fig. 1: Atlantic Ocean and the
positions of meas-
urement

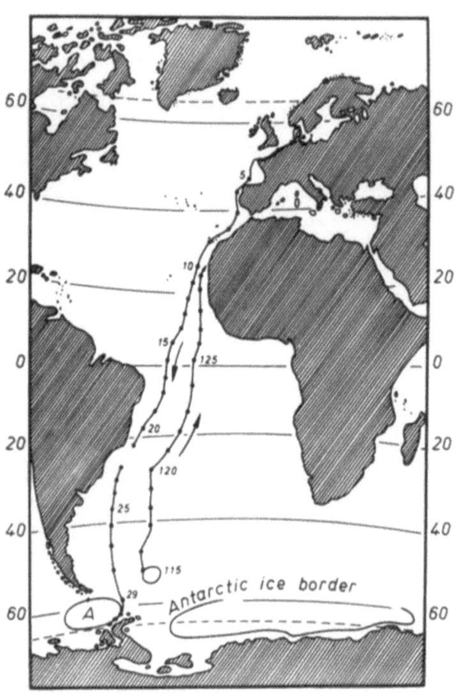

particles at the Antarctic
shelf ice coast compared with
other sea regions?
3. Which relations can be
found between the number of
air particles outdoor and indoor the houses of the expedition members.

In order to examine the influence of the wind velocity to
the concentrations of particles all measuring data were
associated to 4 categories of
wind velocity with the exception of the values from the
west coast of Europe, the original regions of the Leste and
Harmattan wind systems as well as with
the exception of raining days or days
with snow or fog (Table 1).
Together with an increase of wind velocity we could state an increase of
the average number of particles from
about 15.000 to nearly 40.000 per litre
of air. This increase could be statistically secured at a probability of an
error rate of 5 per cent.

Fig. 2: Taking coniometrical data of
particles in front of the
Antarctic coast

Table 1: Medium number of particles in proportion to the wind
velocities in the Atlantic Ocean

	wind velocities (m · s^{-1})			
	1-4	5-8	9-12	>12
number of measuring values (n)	18	36	25	18
medium number of particles ($\bar{x}$)	14.437	20.789	39.376	35.933
dispersion (s)	19.150	18.918	25.989	28.965

When we passed through sea regions lying in influence spheres
of constant continental wind systems, an increase of the number
of the air particles was stated (Fig. No. 3).

Fig. 3: Atlantic Ocean and its
      wind regions
      L = Leste
      H = Harmattan

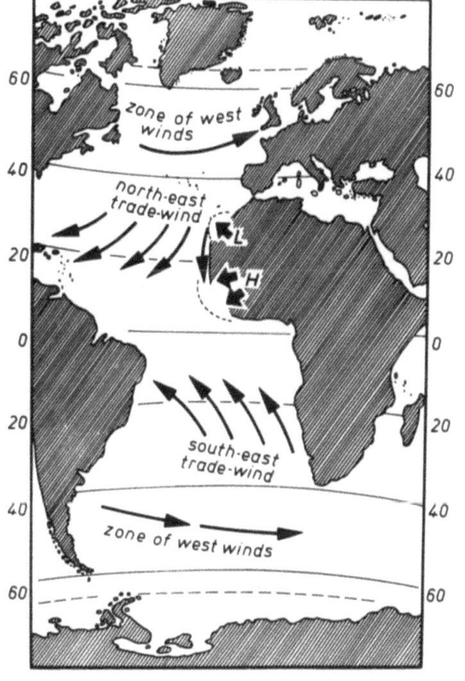

Perhaps this can be explained
by a certain influence of the
above mentioned wind systems.
The average concentrations of
air particles increased in the
neighbourhoud of the Leste
which is a dry eastern wind
loaden with dust particles from
deserts and in the region of
the Harmattan, a dry hot wind
blowing from the north-east
and carrying sand in the beginning and dust particles
later. The particles found
within the reach of the wind
systems of Leste and Harmattan
certainly have their origin on the continental surface.
Particles found above the other sea regions may be all particles of sea salt. The salt percentage of the Atlantic Ocean is

in an average 36 ‰ . A rough calculation of the salt percentage of a (free-falling) little water drop gives a result of
about 0,0018 g of salt. It is easily imaginable that at a high
wind velocity little water drops will stay in the air until at
last all their water has evaporated. Then the remaining salt
particles begin to float in the air. When clashing with other
similar particles even much smaller particles will be produced.
That most particles above the open sea are salt particles is
also shown by the fact that the clothes of seamen are often
covered with fine white salt dust, even when they had no
direct contact with water.

Between the two continents of America and Antarctica, in the
Drake Passage (v. Fig. 1, sea region A) measurements of particles were taken at 70 different points. In the mean 37.998
$P \cdot 1^{-1}$ at a wind velocity between 2 and 17 $m \cdot s^{-1}$ could be
stated. The 43 probes along the shelf ice of "Eastern Antarctica" gave an average value of 5.476 $P \cdot 1^{-1}$. The spreading was
nearly as high as the mean value. This considerable spreading
is apparently a result of the very different orographic conditions. Where there was a very intact cover of snow and ice,
the numbers of particles decreased very much, on the open sea,
however, and especially at a high wind velocity a distinct increase could be observed.

Evaluating the data after such points of view, we get a mean
number of particles of the air of 709 $P \cdot 1^{-1}$ for regions with
an uninterrupted cover of snow and on ice-bergs and a mean-
value of 10.100 $P \cdot 1^{-1}$ for regions on the open sea near the
shelf ice.

In order to find out what is characteristic for the numbers of
particles in an Antarctic coast-station where there lived nearly one hundred people, the measuring data of the air outdoor
the houses were compared with those indoor (Table 2). One can
sea that the increase of the number of dust particles indoor
the houses is very different compared with the outdoor number.
In October the indoor value was only double the number of the
outdoor value, whereas in May the indoor value was 28 times

Table 2: Medium dust content in $P \cdot l^{-1}$ for the indoor and outdoor air at the
station Molodezhnaya in 1975 6.00 GMT
(altogether 296 coniometric measurements)

Characteristic	April	May	June	July	August	September	October	total time
outdoor dust								
$\bar{x}$	1575	3435	3768	1804	6336	5232	3000	3710
s	1177	7294	3659	3599	5961	5034	2000	4958
n	20	20	25	28	25	25	5	148
indoor dust								
$\bar{x}$	9205	98065	16968	5875	29880	22036	5700	27413
s	22050	65933	13091	9977	24562	21576	4807	41201
n	19	20	25	29	25	25	5	148
$q = \dfrac{\text{indoor dust}}{\text{outdoor dust}}$	5,55	28,5	4,5	3,4	4,7	4,2	1,9	7,38

higher than that outdoor. These strong differences are not surprising considering the fact that the number of particles indoor generally differs in a high degree and depends above all from the sources of dust in such rooms, the number and the activities of the persons living and working there. In the mean in the Antarctic houses the concentration of particles were five times higher than outdoor. The indoor values could be compared with those of the outdoor air of Leipzig in the thirties.

Conclusions

The values which are represented in this study should be taken as a contribution to the quantification of natural conditions with the help of the indicator of the particle concentration of air. Especially the values found in the Antarctic station are very low compared with other measurements of numbers of particles by the same methods. The concentrations of particle numbers between outdoor and indoor air with a factor 5 in disfavour of the outdoor air can be used for airtechnical installations which are to be employed in Antarctica. The general trend of an increase of particle concentration in the atmosphere is also observed at the investigation on the Atlantic. In the fifties Effenberger (1958) stated essentially lower values for sea air (200 $P \cdot 1^{-1}$) than in the snow regions of Antarctic shelf ice (700 $P \cdot 1^{-1}$).

Literature:

Effenberger, E.: Z. Aerosolforsch. u. Therapie, 7, 1958, 121-45
Lugg, D.J.: A Review of the SCAR Working Group of Human Biology
            and Medicine, San Diego (California), 16.-27. June
            1986

Oberarzt Dr. med. Georg Schrader, Institut für Allgemeine und
Kommunale Hygiene der Medizinischen Akademie Erfurt,
Gustav-Freytag-Straße 1, Erfurt, 5082

3. Indoor air

EXS 51:
Advances in Aerobiology
©1987 Birkhäuser Verlag Basel

INQUIRIES RECEIVED BY THE CALIFORNIA INDOOR AIR QUALITY PROGRAM ON BIOLOGICAL CONTAMINANTS IN BUILDINGS

J. M. Macher, MPH, ScD
Indoor Air Quality Program, Air & Industrial Hygiene Laboratory
California Department of Health Services, Berkeley, CA, USA

Nonindustrial indoor environments, such as homes, offices, schools, and other commercial and public buildings are now recognized as important environments for exposure to air pollutants. Many urban residents spend 80-90% of their time indoors, especially the elderly, the infirm, and infants, who are indoors almost all the time. It has been found that the concentrations of many pollutants are higher indoors than outdoors. In addition, exposures to most human pathogenic microorganisms occur predominantly indoors, and the concentrations of many aeroallergens are higher inside buildings than outdoors. In order to develop a coordinated state effort to determine the extent and nature of the indoor air quality (IAQ) problem in California, the State Legislature passed a bill in 1982 to form an IAQ program within the Department of Health Services.

To keep track of the large number of telephone inquiries and complaints received by the IAQ program, a telephone log system was instituted. A uniform logging form was developed, on which the respondant recorded the details of phone calls, including the type of complaint, the type of building affected, the health symptoms perceived by the occupants, and the suspected pollutant(s). During the year in which records were kept, 15-20 calls were received each month. This number is a fraction of the total number of calls received, because many could not be responded to by an IAQ staff person. The numbers of complaints about the air quality in office buildings or homes were approximately equal, and represented the major indoor environments in which people experienced difficulties. Other, less frequent calls involved schools, hospitals, and stores. In 70% of the cases, the cause for concern affected five or fewer people, for example, the members of a household or small office. Tens to hundreds of people were or potentially were exposed to indoor air pollution in 24% of the buildings (usually larger office buildings) about which complaints were reported to the IAQ program.

The types of symptoms reported were headache (reported in 19% of all calls in which one or more symptoms were mentioned), eye irritation (17%), nose and throat irritation (14%), nausea (12%), allergy symptoms (watering eyes and running nose) (10%), cough (8%), congestion (7%), dizziness (5%), fatigue (5%), and other respiratory problems (3%). In many instances, the caller had no knowledge of what caused the physical symptoms. Often he or she described an unusual odor or complained that the indoor air was uncomfortable in some way.

Suspected contaminants or potential problems about which calls were received included formaldehyde (16% of all contacts in which a specific topic was identified), biological contamination (16%), ventilation systems (15%), odors (13%), carpeting (11%), asbestos (8%), tobacco smoke (8%), insulation (6%), and radon (3%). Knowing which issues are of concern to the residents of the state of California aids the IAQ program in focusing on areas for research (Sexton & Wesolowski, 1985). To date indoor exposures to formaldehyde and asbestos have received the most attention.

The above statistics indicate that biological contamination, i.e., viruses, bacteria, fungi, algae, and dusts of plant or animal origin represented a significant proportion of the problems that came to the attention of the IAQ program. Although outbreaks of infectious diseases, e.g., Legionellosis, tuberculosis, and measles, are reported to and investigated by the Infectious Disease Section within the Department of Health Services, before 1982 no centralized agency in California had the expertise to respond to complaints about other types of biological contamination. Therefore the IAQ program fills the need for an agency that can respond to complaints of predominantly irritation or allergic reactions to biologically contaminated indoor environments.

According to the IAQ telephone log, the source of airborne allergens in office buildings, schools, and stores was often

the ventilation system, including humidifiers and air con-
ditioners, and contaminated building materials. Additional
sources of contamination, such as pets, house plants, hobby and
craft supplies, and damp basements, bathrooms, and kitchens were
described by the occupants of troubled houses and apartments.

From these data, it is clear that indoor microbiological
contamination in California is often related to problems with the
control of moisture. Some homes in the wetter coastal areas have
seasonal outbreaks of mildew odors and growth of molds, e.g.,
Penicillium, Alternaria, Cladosporium, and Fusarium during the
rainy season, from November to March. The growth of molds on interior windows, walls, ceilings, floors, and furniture can be
simply an annoyance to the housekeeper, or present a severe
problem for an allergic person. Outdoor climatic factors such as
maximum and minimum air temperature and humidity, and amount of
rainfall strongly influence the presence of house dust mites
throughout the state. Dermatophagoides pteronyssinus was the most
abundant species in California in one study (Furumizo, 1975). It
was found in 68% of all samples of house dust that contained
mites. Three other species also occur in the state, D. farinae
(21%), D. microceras (3%), and D. evansi (1%). The remaining 7%
of the samples contained a mixture of the first three species.

The IAQ program does not investigate individual problems
in homes or office buildings. However, information and advice are
provided over the phone, and literature from other agencies and
organizations is often mailed to callers. When appropriate, the
caller is directed to a group that can provide specific assistance. In some cases, the caller is instructed to seek the attention of a physician or allergist. A list of private environmental
consultants and contract laboratories is also made available.

The research efforts of the IAQ program in the area of
indoor biological contamination to date have focused on methods to
collect and identify airborne microorganisms. This work includes
the evaluation of air samplers (Macher & Hansson, 1986), and the
modification of sampling equipment for use in homes and offices.
A project was recently begun to study, at regular time intervals,
the types and numbers of bacteria and fungi in the air, and the

bacteria, fungi, and mites in the house dust of a new apartment building. The overall direction of this research is to become better informed about the biological ecology of buildings, the levels of contamination or infestation that commonly occur and those that represent health hazards, and the best measures for controlling the types of problems seen inside buildings.

The advice given to callers to help them reduce biological contamination can be categorized as source, dilution, or administrative control, using the terms of the National Institute for Occupational Safety and Health. Source control includes such precautions as removing stagnant water, preventing leaks and floods, eliminating sprayer-type humidifiers, and maintaining the indoor relative humidity below 70%. To control pollutants by diluting their concentrations, the amount of outdoor air brought into a building can be increased, and high efficiency filters can be added to a ventilation system to clean recirculated air. Finally, if the other types of control are not practical or do not alleviate the problem, it may be possible to relocate sensitive people to areas that do not adversely affect them.

Indoor air of poor quality can be an inconvenience or a serious health hazard, and office workers and residents are looking to professionals for information on how to improve their environments. Aerobiologists will be called upon more and more often in the future to contribute their specialized knowledge and skills to the understanding and control of biological indoor air pollution.

References:

Furumizo, R. T. 1975. Geographic distribution of house-dust mites (_Acarina_:_Pyroglyphidae_) in California. California Vector Views, Vol. 22, No. 11.

Macher, J.M. and H-C Hansson. 1986. Personal size-separating impactor for sampling microbiological aerosols. Proceedings, 79th Annual Meeting of the Air Pollution Control Association, Minneapolis, MN.

Sexton, K. and J. J. Wesolowski. 1985. Safeguarding indoor air quality. Environ. Sci. Technol., 19(4):305-309.

Dr. Janet M. Macher, Indoor Air Quality Program, AIHL
2151 Berkeley Way, Berkeley, CA 94704, U.S.A., (415)540-2137

MICROBIAL AND DUST POLLUTION IN NON-INDUSTRIAL WORK PLACES.

Gravesen, S., ALK Research Group, Environmental Dept., Copenhagen, Denmark.

During the last decade an increasing number of people with a non-industrial occupation complain of the following symptoms: sensation of dry mucous membranes and skin, erythema, mental fatigue, headache, high frequency of airway infections and cough hoarseness, wheezing, itching and unspecific hypersensitivity, nausea and dizzines (8).
The increase in number of complaints synchronizes with the energy-effecient procedures in the form of tightening of buildings and reduction of ventilation as well as of cleaning procedures. Add to this the increasing industrialization of the building trade including the use of new building materials (6).
Symptoms of the above-mentioned type have often been ascribed to physico/chemical factors such as high temperature, relative humidity, static electricity, vapours and gases. Also psycho/social conditions have been studied in relation to the indoor climate (1).
Very little attention has, however, been drawn to the biological parameters. A few epidemiologic studies report that the prevalence of the "sick" building syndrome is high in institutions for the children (5) and in offices (7).
From field study observations we have noticed that this syndrome is often present in schools and offices which are heavily peopled, use neddle-felt carpets, and are so contructed as to cause dust accumulation and microbial pollution, and furthermore have a reduced cleaning budget. Based on these observations we propose the following hypothesis: Contamination by dust and

microorganisms may be an important factor in the indoor climate
problems since most of the complaints are ascribed to carpeted
rooms. The biological factors should be considered together with
the physico/chemical factors earlier demonstrated.
To elucidate this hypothesis an investigation of the following
parameters has been carried our in a number of schools and offices: accumulated dust on carpets/floors, organic macromolecular
components of dust, dust-bound moulds and airborne moulds and
bacteria (2).

Case Study No. 1: From a school recieving many complaints concerning the indoor environment, 60 (88%) school teachers re-
turned a questionnaire concerning health and subjective feelings
of the indoor climate. The investigation showed that 25-50% of
the teachers suffered from eye-, nose- and throat problems as
well as from headaches and extreme fatigue.These symptoms were
significantly related to work. Physically the school was characterized in that it had thermic problems and water damage due to
bad construction of the roof (flat roof) with wetting of carpets
and subsequent microbial growth as a consequence. A biological
elucidation programme was performed as described in the introduction. Airborne recordings showed: 4000 bacteria colonies/m^3
air in classroom 1 (+carpet) and 6000 bacteria colonies/m^3 air
in classroom 2 (+carpet). Equally high concentrations of bacteria were obtained from cultivation of the carpet dust in both
rooms. In classroom 1, Penicillium sp. and Trichothecium roseum
(a mould infrequently obtainable in the environment) were the
dominating organisms in the carpet dust. The same two species
were recorded in the air of the classroom thus indicating that
the microbial particles from the carpet dust were liberated to
the air. Attemps have now been undertaken to solve both the dust
and microbial problems as well as the thermic conditions: the
roof has been reconstructed, carpets progressively changed to
linoleum and the ventilation system adjusted (3).

Case study No. 2: In 1981 two patients presented themselves at
the Clinic for Occupational Diseases, Copenhagen because of in-
fluenza-like symptoms in the form of extreme fatigue, an occasi-

onal feeling of increased body temperature in the afternoon and general malaise. The patients connected their symptoms to the indoor climate at their working place. Both perform laboratory and clerical work as well as teaching in a teaching- and laboratory unit for engineers.

Examination of building: The building was taken into use in 1964. In 1973 wall-to-wall carpets were installed as part of a cleaning contract. Cleaning has since been limited to sporadic vacuum cleaning 2-3 times a week. During certain periods every year the traffic from students in the building was heavy resulting in accumulation in carpets of dirt and dust brought in by footwaer. The carpets were removed during the summer of 1981 with complete remission of symptoms following cleaning of the working place (4).

In our consequtive study (2) we found higher microbial levels as well as higher levels of dust and macromolecular organic components (proteins) in carpet dust versus floor dust. The higher mean levels of airborne microorganisms from carpeted rooms could not be statistically proven, presumably due to many confounding factors connected to this type of air sampling. From a microbiological point of view, it is logical that carpets are more contaminated than floors. In this investigation we noticed that most of the carpets were more than 5 years old. Carpets stay moist longer after wetting than bare floors. Furthermore, since dust particles are not removed as easily from carpets as from floors, there will be a higher accumulation of especially fine dust particles. Washing floors is a much more efficient way of disinfecting than vacuum-cleaning or shampooing carpets. A bare floor is exposed to the main natural disinfectant, visible light, whereas a carpet is impenetrable.

We are dealing with at least three different modes of exposure: The mechanical influence, with the possibility of irritation of skin and mucosa, the antigen exposure, with possible subsequent immunological reactions, and the possible toxic reactions on metabolic volatiles from the growing microorganisms.

However, as all parameters measured show a higher level in car-

peted room i.e. rooms with most complaints, our findings support
the hypothesis that dust and microorganisms may be an important
factor in indoor climate problems.
Based on this we suggest that many vague complaints concerning
indoor climate may in fact be due to reactions caused by prolonged exposure to organic dust. We further predict that the "dust
disease" will be common in the years to come. This is because
present buildings style involves the use of many factors that
promote microbiological growth and thereby accumulation of organic dust. Examples are use of efficient insolating without
sufficient ventilation and the use of air-conditioning divices
not regularly cleaned. Further use of inadequately cleaned wall-
to-wall carpets in rooms used by large numbers of people coming
in from outside are bad. instead:leave the floors bare, thus preventing needle-felt carpets acting as dust reservoirs or cultivation media, with subsequent release of particles and volatiles.

References:

1. Colligan, M.J. (1981) The psychological effects of indoor air
   pollution. Bull. N.Y. Acad. Med., 57, 1014-1025.
2. Gravesen. S, Larsen, L. Gyntelberg, F. & Skov. P. (1986) Demonstration of microorganisms and dust in schools and offices
   Allergy (in press) Munksgård.
3. Gravesen, S., Larsen, L. & Skov, P (1985) Aerobiology of
   schools and public institutions- part of a study. Ecology of
   Disease, 2, 241-243.
4. Nexø, E., Skov, P. & Gravesen, S. (1985) Extreme fatigue and
   malaise-a syndrome caused by malcleaned wall-to-wall carpets?
   Ecology of Disease, 2, 415-418.
5. Rindel, A.,Bach,E.,Breum,N.O.,Hugod,C.Nilsen,A. & Schneider,T
   (1985) Mineraluulds-lofter i børnehaver. Den sundhedsmæssige
   betydning af at anvende mineraluuldslofteri institutionsbygge-
   ri. Arbejdsmiljøfondets forskningsrapporter.
6. Skov,P.,Valbjørn,O. & The Danish Indoor Climate Study Group.
   The "sick" building syndrome in the office environment. Environmental International(in press).
7. Valbjørn, O. & Kousgård,N. (1986) Headache and mucous membrane irritation at home and at work. SBI-rapport 175, Danish
   Building Research Institute, Hørsholm.
8. World Healthe Organization (1983) Indoor air pollutants: exposure and health effects. Report on a WHO.meeting, Euro Reports
   and Studies No. 78, WHO Regional Office for Europe,Copenhagen

Cand.scient. Suzanne Gravesen, ALK Diagnostic Lab., Environmental Dept., Ved Amagerbanen 23, DK-2300 Copenhagen S, Denmark.

EXS 51:
Advances in Aerobiology

FILTERS OF AN AIRCONDITIONING INSTALLATION AS DISSEMINATORS
OF FUNGAL SPORES

J.H. Elixmann*, W. Jorde**, and H.F. Linskens*

* Dept. of Botany, Catholic University, Nijmegen, The Netherlands
**Asthmakrankenhaus der Kamillianer, Mönchengladbach, F.R.G.

Air-conditioning systems have four functions: They must determine
temperature, humidity, and have to refresh and to clean the
indoor air (STEIMLE, 1969). In order to perform the last
function air-conditioning systems are equipped with filters
which hold back airborne particles. A year ago we began a
study of the microbial community of an air-conditioning system
installed in a new addition to the hospital in Mönchengladbach
in an attempt to examine the effect of the system in dissemina-
ting fungal spores (ELIXMANN et al., 1986). In January 1986,
after 12 months of operation, the filters of this airconditioning system were examined for their contamination with fungi.

The system under investigation consists of two parts. Part I
is a real air-conditioning system combined with an humidifier.
Before leaving the air stream is filtered by coarse and fine
filters. In part II fresh air is only filtered into a coarse
filter (glasswool, Camfil, filterclass EU3, holding back
80% of the airborne particles greater then 0.5 µm Ø) (ANONYMUS),
warmed up and subsequently distributed to four units. Here the
air is filtered again into a fine filter (synthetic material
from Camfil, filterclass EU7, which holds back 95% airborne
particles greater than 0.5 µm) (ANONYMUS) and then brought
to the conditioned rooms. The exhaust air is filtered into
an exhaust filter (coarse EU3).
The coarse filters were vacuumed with a vacuum cleaner and
examined for the dust. The filter material was divided into
an inner and outer layer. These layers were washed in 200 ml
of a 50% saccharose solution; the eluent was diluted serially
and spread on media selective for hydrophilic, mesophilic

and xerophilic fungi. The fine filters were not vacuumed as
the material was too soft for this treatment, but also divided
into two layers.
We also examined the membranes which protect the filter material
from damage from the air flow.

Table 1 shows fungal spore counts in the coarse filter of
part I of the system. It shows that most of the fungi found
in the coarse filter are house dust moulds as *Aspergillus,
Penicillium* and *Wallemia,* and extramural species as *Cladosporium*
and *Aureobasidium.* The high counts of *Cladosporium* in the wintertime are notable. A similar result was found in the other
installation (part II).
In contrast to the high counts of mould spores found in the
coarse filters we found from both layers of the fine filters
only very low counts of fungal spores. Microscopic investigations, however, revealed that there are many spores in each
layer of the filters, but those spores are retained by the
strong absorbing power of the filters, so that they could
not washed off the filters by the technique used.

Table 2 represents the spore counts of the exhaust filters.
It is notable that *Cladosporium* is no longer the dominating
mould in that microbial community. House dust fungi such as
Aspergillus, Penicillium and *Wallemia* are now more present.
Moulds can be found in all exhaust filter layers; notable
are the high counts of *P. citrinum.* We believe that the occurence
of *P. citrinum* is due to the celebration of Christmas. This
mould occurs on orange peels, and as this fruits is often
consumed in wintertime, some of them may be infected by this
mould. The spores get into the air stream where they are
filtered out in the exhaust filters.

The findings of this investigation raise more questions than
providing answers. There is evidence of a mould allergy of
sensitive patients in air-conditioned rooms (SCHATA, 1981).

The filters, especially the fine filters should hold back
not only dust particles but also fungal spores. But we believe
that these spores hold back in the filter system germinate
and produce mycelia which grow into the filter material,
producing spores again. Those spores are distributed into
the air stream and are potential allergen carriers (KREMPL-
LAMPRECHT, 1985). This hypothesis is supported by our high
spore counts found on the inner side of the exhaust air
filters.

Another important question is: Is there a transport of allergenes parallel with the transport of allergen-carriers in
this system? We can put forward the hypothesis that allergens
are extracted from the spores hold back in the filters. Due
to the high humidity in the filters an aerosol is formed
with water, which contains the allergens, and is blown into
the conditioned rooms. If this hypothesis is correct it has
great indications for the whole filter industry.

References

Anonymus: Luchtfiltertechniek. Information leaflet of the
CAMFIL BV, POBox 341, Ede, The Netherlands, 24 pp

Elixmann, J.H., W. Jorde, H.F. Linskens (1986) Apparition
de moisissures dans les appareils d'air conditionné avant
et après la mise en action. Colloque INSERM (ed. C. Molina)
135, 77-85

Krempl-Lamprecht, L. (1985) Bedeutung saisonal auftretender
Schimmelpilze als Allergene. Allergologie 8, 1. Sonderheft,
26-30

Schata, M., W. Jorde (1986) "Air-conditioning Disease"
caused by an allergy to fungal spores? Contribution given
on the 3rd. Int. Conf. on Aerobiology, Basel.

Steimle, F. (1969) Klimakursus, Verlag C.F. Müller, Karlsruhe.

J.H. Elixmann, Department of Botany, Catholic University
of Nijmegen, Toernooiveld, 6525 ED Nijmegen, The Netherlands.

Species	Dust			Inner layer			Outer layer		
	H	M	X	H	M	X	H	M	X
Asp.amstellodami	0.36	0.131	2.15	0	0	0	0	0	0
Asp.flavipes	0	0	0.51	0	0	0	0	0	0
Asp.ruber	0.734	0.192	1.73	0	0	0	0	0	0
Asp.versicolor	0	0.006	0	0	0	0	0	0	0
Aur.pullulans	0	0.021	0	0	0.272	0	0	0	0
Botrytis cinerea	0	0.014	0	0	0	0	0	0	0
Cla.herbarum	10.95	15.4	4.4	0	2.01	0	3.124	0	0
Pen.brevicompactum	0.198	0.822	2.53	0	0	0	0	0	0
Pen.chrysogenum	0.45	2.03	0	0	0	0	0	0	0
Pen.citrinum	0.506	0	0	0	0	0	0	0	0
Pen.frequentans	0.0188	0	0	0	0	0	0	0	0
Wallemia sebi	0.0	0	0.66	0	0	0	0	0	0
Unidentified	2.218	0	2.1	0	0	0	1.03	0	0

Table 1

Counts of fungal spores found in the   filter, part 1 of the
installation, in million spores per filter bag.

H = Hydrophilic   M = Mesophilic   X = Xerophilic

Species	Dust			Inner layer			Outer layer		
	H	M	X	H	M	X	H	M	X
Asp.amstellodami	0	0.002	0.204	0	0	0	0	0	0
Asp.fumigatus	0	0.018	0	0.22	0	0	0.0022	0	0
Asp.glaucus ser.	0	0.104	0.353	0	0	0	0	0	0
Asp.niger	0	0.002	0	0	0.002	0	0	0	0
Asp.ruber	0	0.155	0.207	0	0	0	0	0	0
Asp.versicolor	0	0.003	0.272	0	0	0	0	0	0
Cla.herbarum	1.086	0.457	0.247	1.107	0.704	0	0.884	0.72	0
Pen.brecicompactum	16.28	6.15	2.03	0.3248	0.14	0.78	0	0.67	0.014
Pen.chrysogenum	0	17.5 ·	4.33	0	8.5	0.013	0	0.78	0.002
Pen.citrinum	2.788	0.003	0	0.6836	0	0	0.02	0	0
Wallemia sebi	0	0.326	0.21	0	0	0	0	0	0
Unidentified	0.1088	0	2.1	0.2723	0.009	0.001	0.14	0.967	0

Table 2

Counts of fungal spores found in the exhaust air filter of
Part 1 of the installation, in million spores per filter bag.

H = Hydrophilic   M = Mesophilic   X = Xerophilic

EXS 51:
Advances in Aerobiology
©1987 Birkhäuser Verlag Basel

THE BUILDING ILLNESS SYNDROME: COMPARATIVE STUDIES IN AIR CONDITIONED AND CONVENTIONAL HEATED BUILDINGS

P. Kroeling

1.INTRODUCTION

Complaints about disorders of health and comfort in air-conditioned (AC) buildings are wellknown, but still the knowledge about their quantitative structure -compared to "normal" conditions in conventional buildings- is poor. For some of them there exist no satisfying physical explanations, so mostly psychological factors have been made responsible. By help of a governmental supported research program (1), two different types of studies (A and B) have been performed to get some informations about following questions:

- How many persons in the FRG are working in AC buildings, and are the complaints relevant from an epidemiological point of view?

- Do there exist typical patterns of complaints in AC buildings, compared to those in conventional ones?

- Do there exist suspicious physical, technical or other differences between both types of buildings, which may lead to reasonable explanations for the complaints?

2.MATERIAL AND METHODS

Study A

A representative inquiry, performed by INFRATEST has been included within a monthly performed 'multiple theme' investigation. The inquiry based on 8000 representative persons older than 14 years in the FRG and Berlin. 11 items of a questionnaire comprehended typical health complaints at work, another 12 items dealt with comfort- and other environmental problems. Filterquestions allowed to compare the answers from persons working in AC- and conventional locations (this inqiry A took place at home of each participant, not at the working place itself).

The representative study enables to get some data about the general wellbeing and comfort situation in AC and conventional locations, but gives only little information about the possible causes of the different complaints. Therefore it has been necessary to compare subjective inqiry results with objective measurement data. (study B).

Study B

Some comparative inquiries have been performed in selected, matched buildings with and without AC, by use of an extended questionnaire (comprehending the questions of the representative inquiry, study A). The investigations took place in 6 schools with about 1800 pupils (15 years old) and teachers, and in 5 administration buildings with about 1200 employees (Muenchen, Berlin, Heidelberg). Physical and other boundary conditions simultaneously have been measured, such as:
-Air velocity and turbulance (Institut f. Bauphysik, FhG Holzkirchen)
-Room temperature and humidity (Institut f. Bioklimatologie, Muenchen)
-Low frequency noise (10-100 Hz)
-Carbondioxide
-Weather conditions
-Specifications of AC systems and buildings

3. RESULTS

Study A:

420 persons out of 8000 were working in AC locations; this means, that about 2,5 million (5%) of the population in germany is air conditioned, or about every seventh working place.

In part 1 of the questionnaire the persons evaluated the statement: "During my work I am suffering under following disorders of wellbeing:..."

Fig 1 demonstrates the differences between the AC group and a control group. AC-persons complained nearly double as much about:
"tendency to colds", "dry mucous membranes", "rheumatism", and "lack of energy". Significant were also the items: "headache", "rapid fatigue", and "numbness".

In part 2 of the questionnaire the persons evaluated the statement: "At my working place I feel affected by following conditions:..."
Fig.2 shows, that the AC persons were mostly affected by:
"missing window ventilation", "draught", "dry air", "changing temperatures", "low temperatures "and "stale air".

Study B:
It is not possible in this contribution to give detailed information about the results of each matched investigation. To enable an overlook, however, a summary of some data gained in the administrating buildings and in schools will be given.
Fig.3 bases on about 700 persons in AC buildings and about 500 in conventional ones. Nearly twice as many persons in AC buildings suffered under "rapid fatigue" and "dry mucous membranes." Highly significant different were the items: "tendency to cold", "headache", "loss of energy" and "numbness".
As shown in fig.4 the persons mostly complained about: "missing window ventilation", "dry air", "stale air", "changing temperatures" and "draught".

Similar, if not even worse disorders were found in AC schools, compared to conventional ones, as shown in fig. 5.
Highly significant increased were complaints over: "dry mucous membranes", "lack of concentration", "rapid fatigue" and "tendency to colds", furthermore "headache", "lack of energy" and "numbness".
Fig.6 demonstrates the answers concerning the environmental comfort. Nearly 90% of the children missed the window ventilation. Other highly significant complaints concerned illumination problems, "low temperatures" "changing temperatures" and "dry air".

The physical measurements revealed impressing differences between AC and conventional buildings, especially for the parameters: air velocity and turbulence, air temperature, humidity and low frequency noise. No differences were found for CO_2 concentrations, so that the complaints cannot be related to insufficient air exchange rates. We regret not to be able to go into further details in this contribution.

<u>4. DISCUSSION</u>

Non of the 23 items in the first and second part of the questionnaire in study A had favourable results for persons in AC locations. Considering that these results are gained from persons out of more than 400 different AC buildings, it can be concluded, that the complaints are epidemiological relevant and not limited to some single buildings, for example with malfunctioning technical systems. It can be roughly estimated, that in the FRG about 500.000 out of 2,5 million persons suffer more or less from disorders they would not have, if they worked in conventional heated buildings.

The results of study B varied between the different matched investigations in selected buildings; generally the findings in AC buildings were worse than in conventional ones. At the same time a number of physical conditions were found to be different. This makes it rather improbable, that mainly psychological are responsible for the complaints (such as prejudice).

The described investigation program has to be understood as a first step towards a better knowledge about the so called "building illness" syndrome ("sick building" syndrome). It has not been designed to find definitive answers concerning the causes of the complaints. However, in combination with a systematic literature analysis the results allow to draw some connections between the found effects and possible reasons.

Fig. 7. gives an impression hereof; some of the complaints (left side) may belong together and were put into groups. The arrows indicate, in which way certain connections may occur and overlap. The list is incomplete, and some of the named "possible reasons" (right side) are still rather hypothetical; they may serve as a basis for specific further investigations.

MISSING WINDOW VENTILATION   38,2 / 13,8   δ%=24,4
DRAUGHT   33,9 / 21,1   12,8
DRY AIR   37,8 / 25,7   12,1
CHANGING TEMPERATURES   21,1 / 9,8   11,3
LOW TEMPERATURES   19,0 / 10,6   8,4
BAD ILLUMINATION   25,2 / 17,7   7,5
HIGH TEMPERATURES   24,7 / 18,7   6,0
STALE AIR   31,0 / 26,2   4,8
ROOM PROBLEMS   19,1 / 15,1   4,8
DISSATISFACTION WITH WORK   12,1 / 8,4   3,7
NOISE   30,8 / 27,8   3,0
PROBLEMS WITH KOLLEAGUES   10,6 / 7,8   2,8
0   50   100
AIR-CONDITION (N=420)
NO AIR-CONDITION (N=699)

Fig. 2

TENDENCY TO COLD   32,7 / 18,1   δ%=14,6
DRY MUCOUS MEMBRANES   28,5 / 15,5   13,0
RHEUMATISM   15,4 / 7,9   7,5
LACK OF ENERGY   12,4 / 5,5   6,9
HEADACHE   20,5 / 15,5   5,0
RAPID FATIGUE   20,8 / 15,0   5,8
IRRITABILITY   22,1 / 18,8   4,1
LACK OF CONCENTRATION   14,1 / 18,1   4,8
NUMBNESS   7,8 / 3,7   3,3
TIRED LEGS   25,2 / 22,6   2,6
CIRCULATION PROBLEMS   15,1 / 13,8   1,3
0   50   100
AIR-CONDITION (N=420)
NO AIR-CONDITION (N=699)

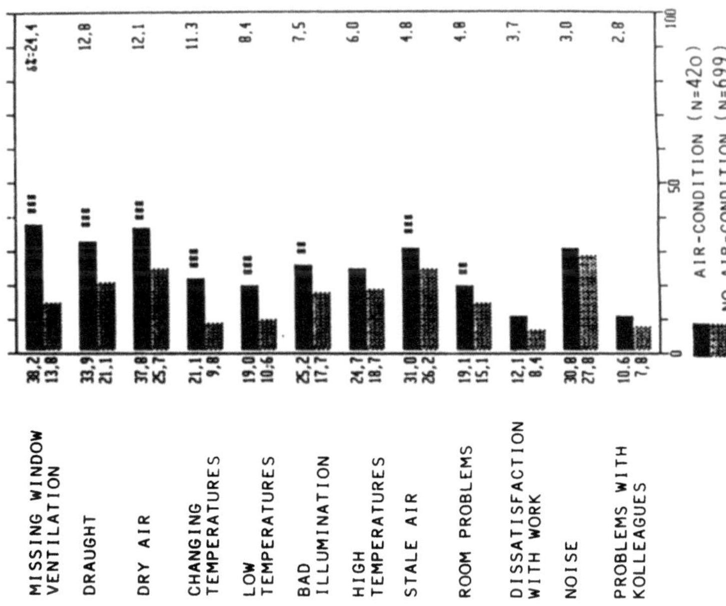

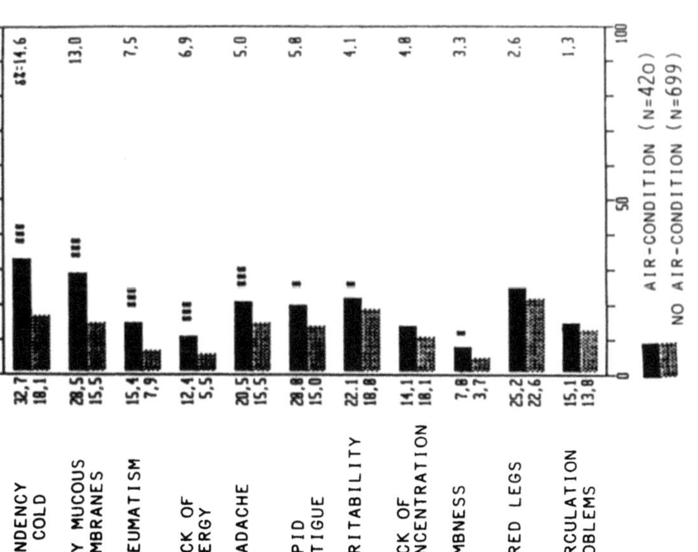

Fig. 1

292

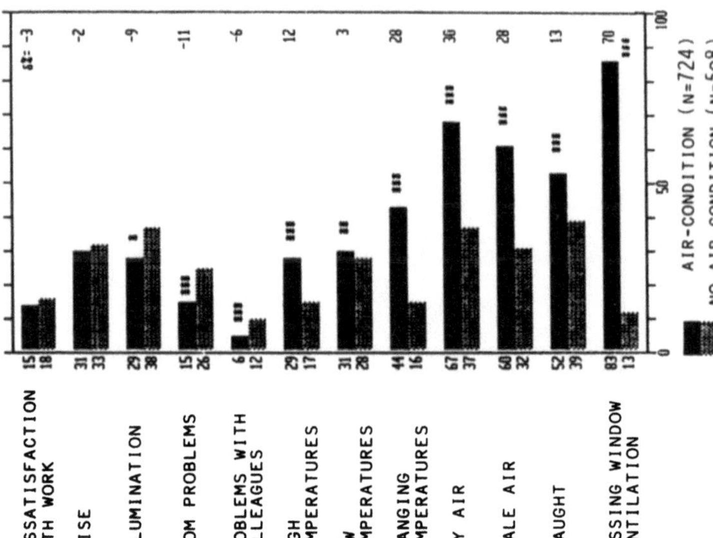

Fig. 4

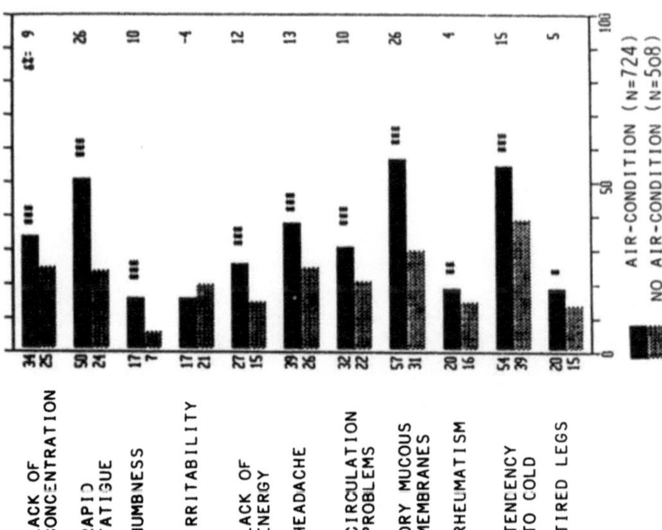

Fig. 3

LACK OF CONCENTRATION 60 42
RAPID FATIGUE 71 54
NUMBNESS 20 11
IRRITABILITY 42 37
LACK OF ENERGY 48 35
HEADACHE 39 23
CIRCULATION PROBLEMS 21 14
DRY MUCOUS MEMBRANES 34 14
RHEUMATISM 8 6
TENDENCY TO COLD 44 28
TIRED LEGS 26 18
42: 18
17
9
5
13
16
7
20
2
16
8
0 50 100
AIR-CONDITION (N=1129)
NO AIR-CONDITION (N= 733)

Fig. 5

ROOM CHANGING TOO OFTEN 39 33
NOISE 50 35
BAD ILLUMINATION 68 38
ROOM DESIGN 56 41
PROBLEMS WITH KOLLEAGUES 21 16
HIGH TEMPERATURES 45 32
LOW TEMPERATURES 51 22
CHANGING TEMPERATURES 49 24
DRY AIR 59 35
STALE AIR 69 63
DRAUGHT 28 19
MISSING WINDOW VENTILATION 86 32
42: 6
15
30
15
5
13
29
25
24
6
9
54
0 50 100
AIR-CONDITION (N=1129)
NO AIR-CONDITION (N= 733)

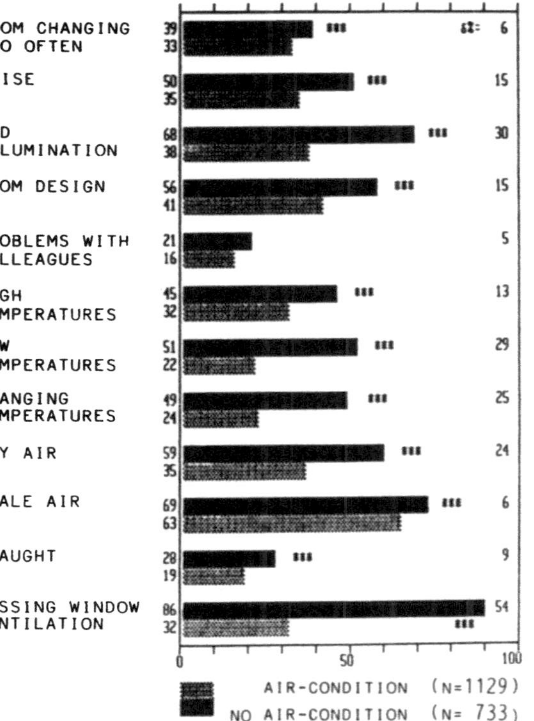

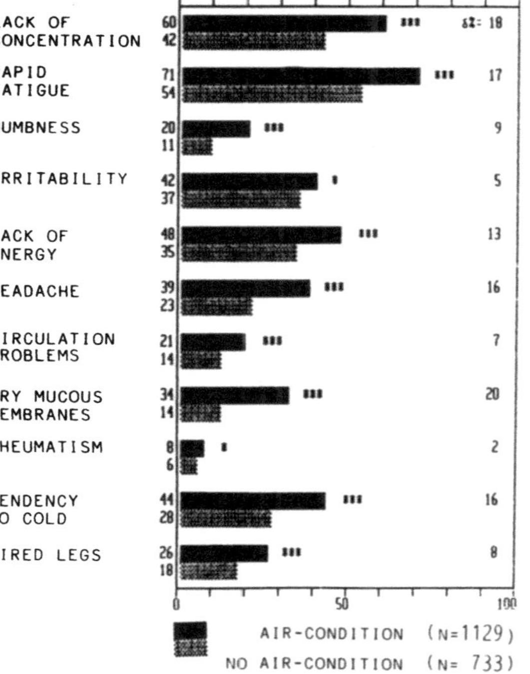

Fig. 6

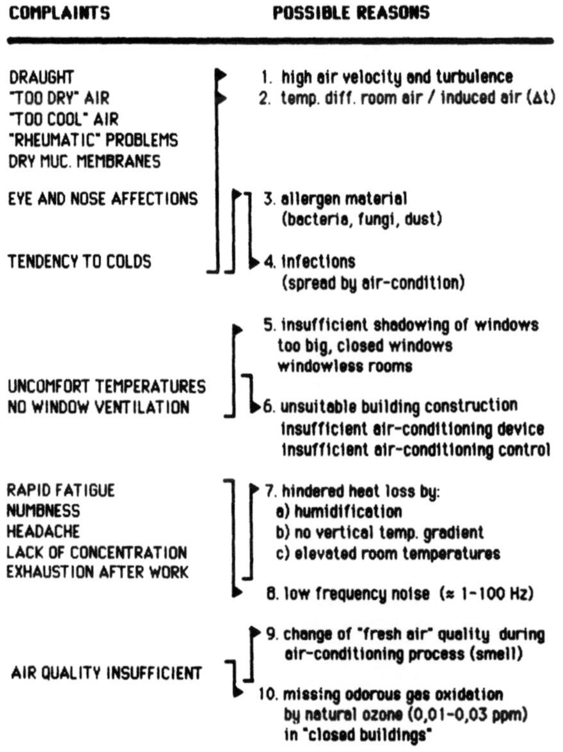
COMPLAINTS
POSSIBLE REASONS
DRAUGHT
"TOO DRY" AIR
"TOO COOL" AIR
"RHEUMATIC" PROBLEMS
DRY MUC. MEMBRANES
EYE AND NOSE AFFECTIONS
TENDENCY TO COLDS
1. high air velocity and turbulence
2. temp. diff. room air / induced air (Δt)
3. allergen material
(bacteria, fungi, dust)
4. infections
(spread by air-condition)
UNCOMFORT TEMPERATURES
NO WINDOW VENTILATION
5. insufficient shadowing of windows
too big, closed windows
windowless rooms
6. unsuitable building construction
insufficient air-conditioning device
insufficient air-conditioning control
RAPID FATIGUE
NUMBNESS
HEADACHE
LACK OF CONCENTRATION
EXHAUSTION AFTER WORK
7. hindered heat loss by:
a) humidification
b) no vertical temp. gradient
c) elevated room temperatures
8. low frequency noise (≈ 1-100 Hz)
AIR QUALITY INSUFFICIENT
9. change of "fresh air" quality during
air-conditioning process (smell)
10. missing odorous gas oxidation
by natural ozone (0,01-0,03 ppm)
in "closed buildings"

Fig. 7

5.LITERATURE

1)Kroeling, P.:
Gesundheits- und Befindensstörungen · in klimatisierten Gebäuden.
Vergleichende Untersuchungen zum "building illness" Syndrom.
Zuckschwerdt-Verlag, Muenchen, Bern Wien (1985)
(The research program has been supported by the
Bundesministerium f. Forschung und Technologie; FKZ Nr.: 01 VD 132)

6. FIGURES

Fig.1: Representative study (A), part1; wellbeing disorders

Fig.2: Representative study (A) part 2; environmental problems

Fig.3: Summary in administration buildings (study B)
      part 1: wellbeing disorders

Fig4: Summary in administration buildings (study B)
      part 2: environmental problems

Fig.5: Summary in school buildings (study B)
      part 1: wellbeing disorders

Fig.6: Summary in school buildings (study B)
      part 2: environmental problems

Fig.7: The main complaints and their possible reasons

Dr. med. P. Kröling
Institut f. Med. Balneologie u. Klimatologie, L.M.-Univ.
Marchioninistr. 17
D-8000 München 70
FRG

BIOAEROSOLS AND OFFICE BUILDING VENTILATION SYSTEMS

E. Pitkänen[1], M. Pellikka[1], P. Kalliokoski[1] and M.J. Jantunen[2]
[1]University of Kuopio, Dept. of Environmental Hygiene, SF-70211
Kuopio, Finland
[2]National Public Health Institute, Dept. of Environmental Hygiene
and Toxicology, SF-70701 Kuopio, Finland

Abstract

Becterial and fungal spore samples were collected from twelve
office building ventilation systems. Measurements were done both
with and without humidification. Ventilation or humidification
systems were not found to act as bioaerosol sources in any case.
No difference was observed between bioaerosol counts in offices
with and without humidification. The microbial levels decreased
in all ventilation systems.

Introduction

The sick building syndrome is a new poorly defined disease entity.
The affected people mainly work in modern, air-conditioned office
buildings and complain such vague syptoms as headache, sore throat,
eye and mucous membrane irritation and fatigue (1.). The causative agent(s) is not known, but the list of suspects is extensive.
It includes bioaerosols, thermal conditions, draught, ions or low
frequency noise of mechanical ventilation as well as organic
compounds released from building materials.

Another disease connected with mechanically ventilated buildings
is the humidifier fever. In this disease the symptoms resemble
those of the sick building syndrome. However, in the humidifier
fever the causative agents are somewhat better characterized.
Clinical studies tend to suggest that symptoms are in many cases
either due to allergic reactions to fungal spores or to pyrogenic
reactions to bacterial endotoxins (2, 3, 4, 5). The common vector

in humidifier fever are the bioaerosols emitted by cooling or
humidifying devices.

In this study we investigated whether the humidification systems
commonly used in Finland or the ventilation systems themselfs
could act as bioaerosol sources. The research is a part of a
larger "Indoor Air Quality and Ventilation Requirements" project
of several Finnish institutes aimed at defining limit values and
recommendations for the quality of indoor climate.

Materials and Methods

Measurements were done in twelve office buildings in some of
which there were workers having symptoms of sick building syndrome. All the buildings had two-way mechanical ventilation and
humidification system. In four offices the humidifiers were hot
vaporizers where the water is evaporated to air flow by heating
it to 80 $^{\circ}$C. In the other eight buildings the humidifiers were
evaporative. The desired moisture of the ventilating air is
achieved by blowing the air through a thick filter wettened by
water from the recirculation tank. The water temperature in
recirculation tank is 10 - 15 $^{\circ}$C.

The bioaerosol samples were taken from the ventilation systems
either when the humidification was on (n=12) or off (n=17). Air
samples for bacterial and fungal spore determinations were
collected by six stage cascade impactors (Andersen 10 - 800)
from outdoors, duct after the fan, room supply inlet, rooms and
recirculation duct in cases recirculation air was used. For
bacterial samples tryptone-glucose agar was used and for fungal
spores Hagen agar. All the samples were incubated for 7 - 10 days
at 20 $^{\circ}$C. The water of the humidifiers was also investigated.
From evaporative humidifiers the samples were taken from the
recirculation tank and from the vaporizers from the water con-
densed in the nozzle. The water samples were cultivated on the

Table I. The geometric mean bacterial counts and the geometric standard deviations in various parts of ventilation systems in office buildings.

measure point	no humidification				humidification				
		< 2 µm		> 2 µm			< 2 µm		> 2 µm
	n	[1]GM	[2]GSD	GM	GSD	n	GM	GSD	GM	GSD
outdoor air	17	20	6	114	4	12	5	5	25	7
air after the fan	11	5	3	12	5	8	5	3	5	3
room inlet air	17	5	3	5	4	12	5	4	5	4
indoor air	11	28	2	87	2	8	23	3	35	3
recirculation air	16	28	4	34	2	11	9	4	13	2

[1] GM=geometric mean-=median (cfu/m^3)

[2] GSD=geometric standard deviation (dimensionless ratio)

Table II. The geometric mean fungal spore counts and the geometric standard deviations in various parts of ventilation systems in office buildings.

measure point	no humidification				humidification				
		< 2 µm		> 2 µ			< 2 µm		> 2 µm
	n	[1]GM	[2]GSD	GM	GSD	n	GM	GSD	GM	GSD
outdoor air	17	72	6	256	4	12	10	4	41	4
air after the fan	11	12	4	38	5	8	6	6	12	5
room inlet air	17	6	2	22	5	12	6	7	6	7
indoor air	11	7	5	21	5	8	6	9	7	7
recirculation air	16	13	4	34	3	11	3	5	9	4

same media and in the same conditions as the air samples.

From two office buildings endotoxin samples were taken at a distance of two meters from the humidifying device when the humidification was both on and off. Samples were collected by a standard dust dampling method (SFS 3860) on Millipore filters (MF-Millipore filter AAWP 3700). The filters were extracted to pyrogen free Srensen buffer (0.067 M, pH 7.0= in a sonicating bath (USF, Finnsonic, W 181) and the endotoxin levels were determined according to Levin and Bang (1964) with a commercial Limulus Amebocyte Lysate (LAL) -test (Sigma Chemical Company, St. Louis, MO, USA).

Results and discussion

Bacteria and fungal spore counts in various parts of the ventilation systems are presented in tables I and II. In the tables the levels of bioaerosols are divided to two separate groups according to their aerodynamic diameter ($< 2\,\mu m$, $> 2\,\mu m$). The total bioaerosol counts varied between $1\text{-}10^2$ cfu*/m^3. In all buildings the bioaerosol levels were lower in ducts than in outdoor air. The filters effectively removed the bacteria-carrying particles and fungal spores bigger than 2 µm in diameter. In that size group the bioaerosol levels were several times lower in air ducts than outdoors. In rooms the levels again increased due to indoor sources (bacteria from humans, fungal spores from flower pots). Filters did not remove the smaller particles so effectively as the bigger ones, however the counts were lower in ducts than outdoors.

*cfu = colory forming unit

The waters in the recirculation tanks were heavily contaminated.
They contained 10^4-10^5 cfu/ml bacteria and 10-10^2 cfu/ml fungal
spores. The microbial counts of the water of vaporizers were
under detection limit (5 cfu/ml). The microbes were however not
emitted to inlet air because the bioaerosol counts in ducts with
and without humidification were on the same level indicating
that no micro-organism-carrying aerosols were formed. In one
office building the endotoxin level reached the detection limit,
0.8 ng/m^3. All the other samples proved to be under that limit.
The detected level is far below the values (130-390 ng/m^3) in
which humidifier fever has been reported (5).

References

1. Berglund, B, Linval, P. and Sundell, J.(Eds.), Indoor Air, Vol.1:
   Recent Advances in the Health Science and Technology, Swedish
   Council for Building Research, Stockholm, 1984, 23-30

2. Banaszak, E.F.: Hypersensitivity penumonitis due to contamination
   of an air conditioner. New Eng. J. Medic. 1970:283-271

3. Fink, J.N., Banaszak, E.F., Thiede, W.H., Barboriak, J.J.:
   Interstitial pneumonitis due to hypersensitivity to an organism
   contaminating a heating system. Ann. Int. Med. 1971:74:80-83.

4. Rylander, R., Haglind, M., Lundholm, M., Mattsby, I., Stengvist,
   K.: Humidifier fever and endotoxin exposure. Clin. Allergy 1978:
   8:511-516.

5. Rylander, R., Haglind, P.: Airborne endotoxins and humidifier
   disease. Clin. Allergy 1984:14:109-122.

6. Levin, J., Mertala Bang, F.B. 1964 Clottable Protein in Limulus
   its clogalization and Kinetics of its coagulation by endotoxin.
   Thromb. Diath. Haemmorrh. 19:186-197.

Pitkänen Eeva, Dept. of Environmental Hygiene, University of
Kuopio, SF-70211 Kuopio, Finland

RADON AND ITS DECAY PRODUCTS IN THE INDOOR ENVIRONMENT: RADIATION EXPOSURE AND RISK ESTIMATION

Werner Burkart

Swiss Federal Institute for Reactor Research, Health Physics Division, Biology & Environment, Würenlingen, CH

Radon, a member of the uranium decay chains, enters the indoor environment from the building subsoil, walls and the drinking water. In areas with high radium content of rocks and soil, considerable indoor levels of radon are measured. A sample of 32 conventional homes in Southeastern Switzerland, an area with generally high uranium concentrations in the subsoil, showed an arithmetic mean radon concentration in living quarters and cellars of 307 Bq/m3 (8.3 pCi/l) and 1410 Bq/m3 (38.1 pCi/l), respectively.
Using UNSCEAR82 conversion factors, the annual effective dose equivalent in this area amounts to 9 mSv (900 mrem). The corresponding dose to the stem cells of the tracheobronchial tissue is about 100 mSv/a (10 rem/a). Life time exposure in such dwellings is in the range of occupational doses collected in modern uranium mines and leading to a significant increase in the lung cancer risk.

1. Introduction

Radon and its short-lived decay products in the indoor environment are the most important single source of exposure to ionizing radiation in many countries with moderate to cold climate. Radon, a noble gas, is transported from the subsoil which always contains traces of uranium into dwellings by air movements driven by temperature differences (stack effect) and pressure differences caused by wind (Burkart, 1986). The short-lived radon decay products polonium-218, lead-214, bismuth-214 and polonium-214 attach to aerosol particles. Upon inhalation, these radon daughters cause considerable radiation doses to the linings of the tracheobronchial tree (UNSCEAR, 1982). The lung cancer incidence in miners populations exposed to moderate to high levels of radon and its decay products is clearly elevated (Archer et al., 1976; Thomas et al., 1985).

Modifications in the construction of dwellings implemented to reduce energy consumption may change indoor pollutant levels and cause substantial long term hazards to the general public (Burkart and Chakraborty, 1984). The following estimate of additional exposure of humans to the ionizing radiation from radon and its short-lived progeny in the indoor air of energy efficient dwellings, i.e. homes showing low air exchange rates, is based on radon measurements in Swiss alpine regions with high terrestrial radiation. Differences in the indoor radon levels between new or retrofitted dwellings are derived from a matched pair study of single family homes in Switzerland (Burkart et al., 1984).

Risk factors for lung cancer induction from radon exposure have to rely mostly on occupational exposures in mines. However, the integral dose over the human life span from radon and its progeny in the indoor air in high background areas may approach or even exceed the values of miner populations showing significant increases in the incidence of malignant lung diseases. Although confounding factors such as poor control of smoking habits and additional chemical toxicants in mine air do not permit a direct comparison of environmental and occupational exposure risks, both radiobiological considerations and the results from human epidemiology point to a linear dose effect relationship.

Table I: Mean values and upper 10 percentiles for indoor radon concentrations and resulting annual effective dose equivalents for a sample of 32 dwellings in Southeastern Graubünden.

	geometric mean	upper 10 percentile
RADON CONCENTRATIONS in Bq/m3 (pCi/l)		
cellar	770 (20.7)	3'560
ground floor	260 (6.9)	830
first floor	180 (4.8)	470
ANNUAL DOSE in mSv (mrem)		
living room (a)	7.5 (750)	24
bedroom (b)	5.2 (520)	14
living area (a + b)/2	6.4 (640)	19

2. Experimental Methods

Time averaged radon levels were measured with passive track etch dosemeters. In most cases three detectors per dwelling were used: one each in the cellar for source strength, living room (generally on the ground floor) and a bedroom (mostly on the first floor), respectively.

3. Results

3.1 Radon Concentrations in a Region of Southeastern Switzerland

The dwellings were situated in the area around Sankt Moritz where cristalline basement rock with relatively high uranium and thorium content reaches the surface. Figure 1 shows the radon concentrations (cellar, ground floor and first floor) in 32 single family homes in this area on a log/probability graph. The marked decrease from cellar to first floor indicates the importance of soil gas radon as compared to emanation from building materials. The geometric mean of the radon concentrations in the living quarters i.e. living room and bedroom amounts to 255 Bq/m3 (6.9 pCi/l) and 176

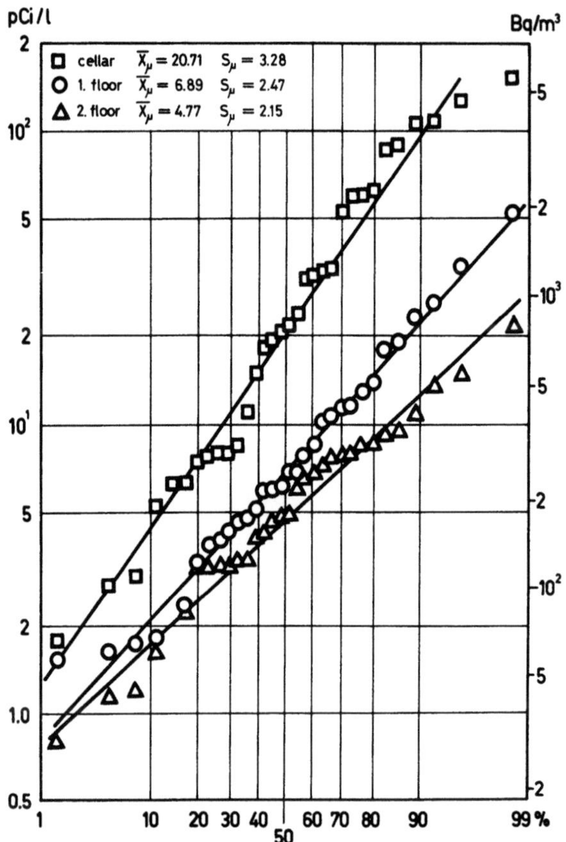

Figure 1: Distribution of the time averaged indoor radonconcentra-
tions during the winter of 1982/83 in a sample of 32
single family houses situated in three alpine valleys of
Southeastern Graubünden.

Bq/m3 (4.8 pC/l), respectively. This is about 4 times higher than
in the average Swiss home. Table I shows the numerical values of
concentrations and annual effective dose equivalents for the means
as well as for the upper 10 percentiles. The dose commitments are
calculated using UNSCEAR82 conversion factors and a radon daughter
equilibrium factor of 0.5 (UNSCEAR, 1982).

3.2 Implications of Climate, Weatherstripping and New Building Technologies on the Radon Source Term

With the rare exception of direct radon emanation from walls in the living area, the radon source term, i.e. the amount of radon entering the indoor environment, is dependent on many and, in some cases, poorly understood parameters. Radon rich soil gas being the main source, radium concentration and permeability of the building grounds as well as cracks and conduits in the cellar floor and walls are recognized as important factors. In addition, the source strength may fluctuate considerably due to partially season dependent parameters such as water content of the subsoil, temperature differences indoor/outdoor (stack effect), wind speed and barometric pressure changes. The reduction of the air exchange rate and the tendency to use part of the cellar and the compulsory bomb shelter for indoor activities are bound to have a decisive influence on the amount of radon entering the living area.

3.3 Assessment of Additional Exposure due to Weatherstripping

For existing building stock, airtightening is the most cost-effective measure to achieve energy savings at unchanged temperature settings. This can be achieved by means of caulking and weatherstripping to seal off airways along windows, doors and blinds. Although air exchange rates in occupied buildings are difficult to assess due to their dependence on a multitude of climatic and behavioural parameters, it is generally accepted that the introduction of central heating systems led to a strong reduction of air exchange rates over the last decades. If a constant source term for indoor radon is assumed, the indoor radon concentration will be inversely proportional to the air exchange rate. Therefore, weatherstripping may increase the risk from radon and its daughters considerably. The calculation of the additional exposures due to an increase in radon levels resulting from airtightening is depicted in table II for the alpine sample previously described. The difference in radon concentration between new or retrofitted energy-efficient buildings and conventional homes is based on a matched pair analysis in the same area (Burkart et al., 1984). In this study, an increase by a factor of 1.5 of the radon levels in the airtight homes was found.

Table II: Parameters assumed for the calculation of radon exposure
due to airtightening.

	conventional house	airtight house
Rn indoor conc. Bq/m3	307	461
equilibrium factor	0.5	
time spent indoors	80 %	
Annual dose, He mSv/a	9.05	13.58
ADDITIONAL DOSE mSv/a (mrem/a)	<u>4.53</u> (453)	

3.4 Risk assessment

To assess the lung cancer risk from exposure to environmental
levels of radon, several assumptions have to be made. Using
UNSCEAR82 (1982) conversion factors which are corrected for lower
breathing rates and volumes in the indoor environment as compared
to mining, the total exposures over a life time in single family
dwellings in parts of the Alps approach or even exceed the
exposures of groups of miners showing statistically significant
increases in the lung cancer incidence. Figure 2 depicts exposure
and annual lung cancer risk due to radon exposure for the many
groups of miners studied. The arrow shows the 30 year exposure at
an average indoor radon gas concentration of 370 Bq/m3 (10 pCi/l).

4. Discussion

In Switzerland, many inhabited areas in the cristalline Alps show
high radon emanation into buildings from the uranium rich subsoil.
In some cases, the lowering of the air exchange rates for the
purpose of energy conservation and new building technologies may
lead to annual effective dose equivalents in the range of the
limits for occupational exposure or even surpass them. The
concomittant risk for lung cancer from a lifetime exposure is
estimated for the energy efficient home in table II as follows: a

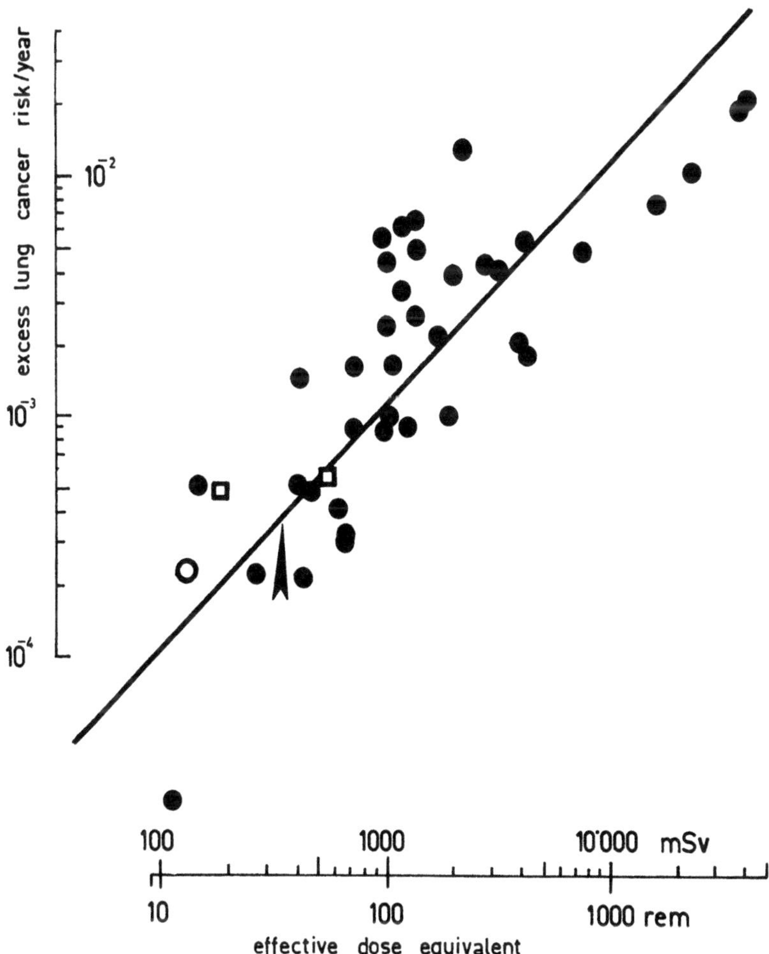

Figure 2: Human epidemiological data on radon induced lung cancer. Each point denotes a population of miners. Risk factors from A-bomb survivors (○) and from Ankylosing spondylitis patients (□) are similar. Details and sources in Burkart and Chakraborty (1984).

fifty year exposure at 16 mSv/a with a lung cancer risk factor of 2.4E-5 per mSv (total risk over 20 year expression period) yields a 2% chance of dying from lung cancer in a high risk area. However, in cases where weatherstripping will also influence the radon source strength, the above assessment may under- or overestimate the risk from energy conservation. Even in geological problem areas of Switzerland, the lung cancer risk to the population from indoor radon is a fraction of the risk from smoking. Therefore, an epidemiological approach to quantify effects from exposure to indoor radon will be difficult.

By the standard used in the radiological protection at the working place, remedial action in dwellings with radon levels above 370 Bq/m3 (10 pCi/l) is warranted. Several methods to reduce the infiltration of radon into the living areas are described in the literature. Forced ventilation to enhance the air exchange will both decrease the radon level as well as those of other indoor pollutants.

References

Archer, V.E., Gillam, J.D. and Wagoner, J.K. (1976) Respiratory disease mortality among uranium miners. Ann NY Acad Sci. 271, 280-295

Burkart, W. and Chakraborty, S. (1984) Possible Health Effects of Energy Conservation: Im-pairment of Indoor Air Quality due to Reduction of Ventilation Rate. Env. International 10, 455-461

Burkart, W., Wernli, C., and Brunner, H.H. (1984) Matched pair analysis of the influence of weatherstripping on radon concentration in Swiss dwellings. Rad. Prot. Dos. 7, 299-302

Burkart, W. (1986) An estimation of radiation exposure and risk from airtightening of homes in an alpine area with elevated radon source strength. Env. International, in press

Thomas, D. C., McNeill, K. G. and Dougherty, C. (1985) Estimates of lifetime lung cancer risks resulting from Rn progeny exposure. Health Phys. 49, 825-846

UNSCEAR 82 (1982) Ionizing Radiation: Sources and Biological Effects. Report of the UN scientific committee on the effects of atomic radiation, United Nations, New York

Dr. Werner Burkart, Biologie & Umwelt, Abt. 81, EIR, CH-5303 Würenlingen, Switzerland

V Phytopathology

EXS 51:
Advances in Aerobiology
©1987 Birkhäuser Verlag Basel

UNIFORMITY OF AIRSPORA CONCENTRATIONS

Charles L. Kramer and Merle G. Eversmeyer

Division of Biology and USDA-ARS, Dept. of Plant Pathology,
Kansas State University, Manhattan, Kansas 66506 USA

Concentrations of airspora in the atmosphere are influenced by biometerological factors that affect either production, release or deposition of fungal spores, and pollen. Temperature and precipitation are especially critical in the growth of plants and fungi, and affect the amount and timing of spore and pollen production. Wind is probably the most important factor in effecting the release and deposition of aerobiological particles.

Most airspora data used in plant disease epidemiological studies and pollen and fungal spore allergy alerts are obtained from single samplers located at one sampling station in an area. Our intent is to discuss several studies that help to elucidate the question of uniformity in the distribution of airspora within an air mass. A better understanding of the variation in the concentration and composition of airpsora that may occur within the atmosphere may help in interpreting results obtained from single samplers for areas surrounding the sampling location.

The escape of spores into the atmosphere from their source, produces a localized concentration or spore cloud. The size of the cloud will vary with the size of the source area. However, as the spore cloud is carried away from the source area, it becomes larger and less concentrated due to both horizontal and vertical dispersal. As dispersal progresses, greater uniformity occurs in the distribution of spores within the air mass.

To gain better understanding of the concentration of airspora within a localized air mass, we (2) used 15 Rotorod samplers placed on five towers to form a 2 x 2 x 2-m grid. There were three samplers on each tower, 2 m apart with the lowest 2 m above ground. A single on/off switch operated all samplers from a single power source. This 5-tower unit was set up 3 m downwind from a 50 x 50-m spore source plot of wheat.

No significant differences were found in spore concentrations horizontally between the five samplers at either of the three heights for <u>Puccinia</u> <u>recondita</u>, <u>Cladosporium</u>, or <u>Alternaria</u>. However, there were significant differences vertically between the collections at the three heights. The differences vertically indicate that the wheat plot was the source of those spores and that vertical dispersion had not yet produced uniformity within the spore cloud (2).

In contrast to that situation, ascospores, as well as basidiospores, did not show a significant difference between any of the sampler locations, either vertically or horizontally. The uniformity of those spores in the air mass suggests that they have been transported from some remote source rather than originating from the wheat plot (2).

We completed a similar study (2) using the same 15-sampler unit, but this time the samplers were located near the downwind edge of a 50 x 50-m tilled area away from any spore source. At this distance of at least 50 m from the nearest spore source, there were no significant differences between the catches of the 15 samplers. This indicated that the distance those spores had traveled was sufficient to allow them to become uniformly distributed in the air mass.

In order to further elucidate the patterns of uredinio-spore movement away from a source area such as a wheat field, we undertook two additional studies. In one of these studies (unpublished), five 3-m towers with Rotorod samplers located at 1, 2, and 3 m above ground were placed downwind from a 16 x 24-m plot. One of the towers was placed at the downwind edge of the plot, one was 10 m downwind and the remaining three were in a row 40 m downwind from the plot. The center tower of the three at 40 m was in line with the one at 10 m and the one at the edge of the plot. There also were three towers, each with a single sampler located 3 m above ground on the windward side of the plot to monitor incoming spores. All samplers were operated with a single on/off switch from a single power source. Leaf rust infection in the wheat plot was approximnately 10%. The area around the plot was tilled.

Very few urediniospores of P. recondita were collected on the windward side of the plot indicating that the plot was the source of nearly all of the spores trapped downwind by the 15 samplers on the five towers (unpublished).

A significantly greater number of urediniospores were collected at the lowest sampler position on the tower at the downwind edge of the plot than at any other sampler position. The next largest number of spores was trapped by the sampler at the lowest position on the tower 10 m downwind. Concentrations at that position on the tower at 10-m distance were significantly greater than at the same position on the towers at 40-m distance. It also was true that concentrations trapped at the lowest position on each of the five towers was significantly greater than those at either of the higher positions on the same tower. In contrast, there were no significant differences in numbers of spores trapped on any of the five samplers at the 3-m position (unpublished).

In the case of Cladosporium, the wheat plot was not the source of most of the spores. Spore counts upwind from the plot were not significantly different from those trapped on the downwind side with the exception of the lowest sampler position on the tower closest to the downwind edge of the plot. The significantly higher numbers at the lowest position, indicate that the wheat plot was the source for some, but comparatively few of the Cladosporium spores entering the atmosphere from the source plot (unpublished).

In another study of the dispersal of P. recondita and P. graminis, we (1) made collections of urediniospores 6 meters

above ground at 60, 120, and 180 m downwind from a source plot
of wheat. Average numbers of urediniospores of both species
were significantly lower at each more distant location when the
wind was from the direction of the source plot. However, when
the wind was from another direction, spore numbers were not
significantly different at any of the three locations.

We also undertook studies that were concerned with
uniformity or variation in spore concentrations vertically. In
one particular study (4), we had volumetric samplers at 1.5, 9,
and 30-m elevations on a 30-m tower.

When adequate moisture was available to support spore
production locally, spore counts usually were significantly
higher at 1.5 m than at either of the other heights. For
example, during a 7-day period of wet weather, _Cladosporium_
spore numbers at 1.5 m exceeded those at 30 m by 3.5 times.
However, numbers at 9 and 30 m were not significantly different
from each other indicating that spores collected at 9 m or above
during any kind of conditions, probably were from remote rather
than local sources (4).

In contrast, when moisture was insufficient to sustain
spore production locally, all spore forms were uniformly distributed vertically in the atmosphere indicating that they were
from remote sources (4).

In an attempt to gain a better understanding of the
uniformity of airspora over a wide area, we compared simultaneous collections from as many as eight sites (3). Six of
those were located within or just outside the city limits of
Manhattan, Kansas, while the other two were located 10 km to the
north and 10 km to the south of the city.

At any given time, differences in the number of airspora at two or more locations were comparatively great. When
we averaged the number of spores trapped per m^3 of air over 3-h
periods, differences between sites were often in the thousands.
When we made similar comparisons based on 24-h mean concentrations, there was still considerble variation between sites. In
the case of _Cladosporium_, 26% of the between-site comparisons

were significantly different. In ascospores it was 20% and in basidiospores, 40% (3).

As average numbers were compiled for longer periods of time, between-site differences became progressively less. For example, when mean spore concentrations were compiled for three 10-day periods during 1976, between-site differences occurred in 11% of the comparisons with _Cladosporium_ and 17% with ascospores and basidiospores (3).

When we averaged the concentration of _Cladosporium_ spores for ten 10-day periods during 1977, there were no significant differences between sites during any of the ten periods (3). When we averaged the total spore concentrations for each of those same ten 10-day periods, between-site differences occurred in only 5% of the comparisons (3).

Finally, when spore concentrations were averaged over 30 and 100-day periods, there were no significant differences between any of the sites for any of the spore groups (3).

<u>References</u>

1. Eversmeyer, M. G. and C. L. Kramer (1980) Horizontal dispersal of urediospores of _Puccinia_ _recondita_ f. sp. _tritici_ and _P._ _graminis_ f. sp. _tritici_ from a source plot of wheat. Phytopathology _70_:683-685.

2. Eversmeyer, M. G. and C. L. Kramer (1986) Single versus multiple sampler comparisons. Grana _25_:___-___ .

3. Kramer, C. L. and M. G. Eversmeyer (1984) Comparisons of airspora concentrations at various sites within a ten kilometer radius of Manhattan, Kansas, USA. Grana _23_:117-122.

4. Lyon, F. L., C. L. Kramer and M. G. Eversmeyer (1984) Vertical variation of airspora concentrations in the atmosphere. Grana _23_:123-125.

Prof. Charles L. Kramer, Division of Biology, Ackert Hall, Kansas State University, Manhattan, Kansas 66506 U.S.A.

EXS 51:
Advances in Aerobiology
©1987 Birkhäuser Verlag Basel

OZONE AND ACIDIC INPUTS INTO FORESTS IN NORTHWEST USA

R. L. Edmonds and F. A. Basabe

University of Washington, College of Forest Resources, Seattle, WA, USA

1. Introduction

Forest ecosystems near urban and industrialized areas are increasingly becoming under stress as a result of atmospheric inputs (ULRICH AND PANKRATH 1983; ACID RAIN FOUNDATION 1985; LEGGE and KRUPA 1986). Forest decline has been noted in Europe in recent years, particularly in Germany (BLANK 1985). A number of hypotheses have been proposed to explain it including; acidic inputs (both dry and wet - including fog), ozone, combinations of pollutants (sulfur dioxide, nitrogen oxides, ozone, heavy metals, etc.), excess ammonium, changing climate and pathogens (BLANK 1985; NIHLGARD 1985). None of these, however, has been proven to be the cause of forest decline in Europe.

Spruce/fir forests in the eastern U.S. are also exhibiting symptoms of decreased health and vigor, most noticeable at higher elevations (ADAMS ET AL. 1985). However, research to date has not clearly linked atmospheric deposition to the health of these forests. On the other hand there is a fairly strong link between ozone concentrations in the San Bernadino Mountains of southern California and damage to ponderosa pine trees (MILLER 1973). Furthermore, SKELLY (1980) indicates that photochemical oxidants (primarily ozone) pose a more serious problem to forests than any other single air pollutant.

Forests in the Pacific Northwest U.S. are considered to be growing in a relatively pollution free environment. However, recent population expansions in this area, particularly in the vicinity of Seattle, Washington, will no doubt mean that atmospheric pollution will increase and we now have an opportunity to monitor acid precipitation and ozone levels relative to changes in forest health.

The objectives of this study are to present data on precipitation pH and ozone levels in the Puget Sound region of Washington, which includes Seattle and several smaller cities and to relate these levels to potential problems to forests in this area.

2. Materials and Methods

2.1 Precipitation Monitoring

A network of rain samplers was operated in the Puget Sound area by Dr. TIM LARSON of the Department of Civil Engineering, of the University of Washington, in February and March 1985. In addition, rainfall pH was determined in 1985 from a site in the Hoh River River Valley of the Olympic National Park, approximately 40 km from the Pacific Ocean (Fig. 1). This site represents one of the cleanest in the U.S. and should give background levels for rainfall pH in this area.

2.2 Ozone Concentrations and Meteorological Data

Ozone data were obtained from a network of ozone monitors operated by the Washington State Department of Ecology. The Dasibi monitors were continuously operated from June to August 1985 at a number of locations near and in forested areas (Port Angeles, Arlington, Graham, Sumner and Pack Forest). The locations of the stations are shown in Fig. 1. Additional ozone data were obtained from the Thompson Research Center of the College of

Forest Resources, University of Washington which is located in the Cedar River Watershed southeast of Seattle (Fig. 1). The Dasibi model 1008 AH ozone monitor at this site is monitoring air directly above a 30 m high Douglas-fir (_Pseudotsuga_ _menziesii_) canopy. This study is part of an overall acid precipitation/ forest effects study being funded by the Electric Power Research Institute (EPRI). All stations are located below an elevation of 300 m.

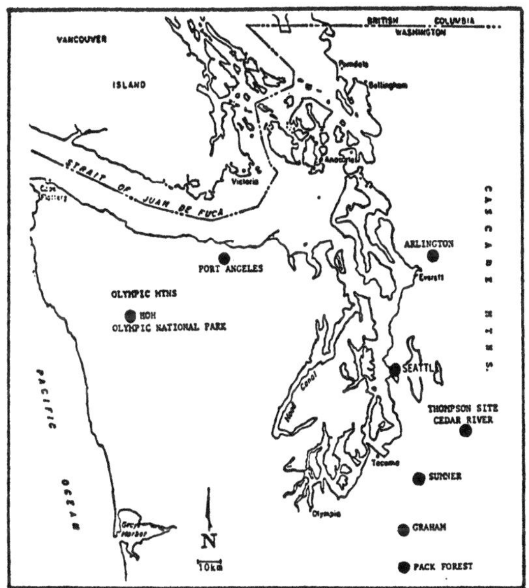

Fig. 1. Location of study sites in western Washington.

3. Results and Discussion

3.1 Precipitation pH

Rainfall pH ranged from 4.5 to 4.9 in the Puget Sound area in February and March 1985 (TIM LARSON, pers. commun.) with lowest values near Tacoma (Fig. 1). This probably represents the influence of a copper smelter in Tacoma which has since been closed. At the pristine Olympic National Park site in the Hoh River Valley rainfall pH in 1985 averaged 5.2 which represents "clean air". Thus rainfall in the Puget Sound area is slightly acidified. Urban fog is more acid than rain (as low as pH 2.9) but fog on the Olympic Peninsula generally has a pH slightly higher than rain (near 6) (TIM LARSON, pers. commun.)

In the northeastern U.S. average values of precipitation pH of less than 4.0 are common. A pH less than 3 occurs in some events (NORTON ET AL. 1980). Wet deposition pH in southern German forests ranged from 4 to 5.2 from 1976 to 1983 but there was no increase in this period (BLANK 1985). The threshold for occurrence of visible injury from direct acidity is between pH 3.0 and 3.6 (SHRINER 1986). Thus it appears that direct acidity is unlikely to be a problem to Pacific Northwest forests.

3.2 Ozone

Table I shows mean hourly and maximum and minimum hourly ozone concentrations for five sites in or near forested areas in western Washington in June, July and August 1985. Highest concentrations occurred in the foothills of the Cascade Mountains east of Puget Sound while the lowest occurred at Port Angeles on the Olympic Peninsula west of Seattle. The highest ozone concentrations in the summer of 1985 occurred at Pack Forest south of

Table I. Ozone concentrations in June, July and August 1985 at 5 stations in western Washington

Station	Ozone concentrations (ppb)		
	Mean hourly	Min. hourly	Max. hourly
Port Angeles	33	20	50
Arlington	49	20	110
Graham	42	20	110
Sumner	53	0	110
Pack Forest	57	10	130

Seattle where the maximum hourly concentration was 130 ppb. Interestingly, maximum ozone concentrations in British Columbia, Canada, 8 km north of the U.S. border are nearly twice those reported in Washington.[1]

Typically diurnal changes in ozone concentrations occur. For example, on May 31, 1986 above the forest canopy at the Thompson Research Center, highest average hourly values occurred at 1500-1600 hours (122 ppb) with a minimum value of 28 ppb at 0600 hours.

Oxidant (ozone) damage to vegetation has been observed in southern California (MILLER 1973) and ozone has been suggested to be involved with forest decline in the eastern U.S. (SKELLY 1980) and in Germany (BLANK 1985). REICH AND AMUNDSON (1985) also found ambient levels of ozone (10-130 ppb) to reduce net photosynthesis in tree and crop species in the eastern U.S. Fluctuating ozone levels typical of diurnal patterns may also result in different effects than exposure to constant ozone levels (HOGSETT ET AL. 1985). Forest tree species have variable responses to ozone (MILLER 1973). The exact sensitivity of Douglas-fir, the main commercial species of northwest forests, to ozone is not known. In fact, the sensitivity of most species under field conditions is not known. We observed foliage flecking in 1985 in Douglas-fir and other species at Pack Forest but we do not know whether this was attributable to ozone or other causes. We need to do more physiological research to determine how sensitive Douglas-fir and other northwest species are to ozone.

4. References

Acid Rain Foundation (1985) Air pollutants effects on forest ecosystems. (Acid Rain Foundation, St. Paul, Minn.).

---

1. Brit. Columbia Min. Environ., Waste Mgt. Branch

Adams, H.S., S. L. Stephenson, T. J. Blasing and D. N. Duvick (1985) Growth trend declines of spruce and fir in mid Appalachian subalpine forests. Environ. Exp. Bot. 25, 315-325.

Blank, L. W. (1985) A new type of forest decline in Germany. Nature 314, 311-314.

Hogsett, W. E., M. Plocher, V. Wildman, D. T. Tingey and J. P. Bennett (1985) Growth response of two varieties of slash pine seedlings to chronic ozone exposures. Can. J. Bot. 63, 2369-2376.

Legge, A. H. and S. V. Krupa (1986) Air pollutants and their effects on the terrestrial ecosystem. (John Wiley, N. Y.).

Miller, P. R. (1973) Oxidant-induced community change in a mixed conifer forest. p. 101-117, In J. A. Naegele (ed.) Air Pollution Damage to Vegetation. Adv. Chem. Series 122. (Am. Chem. Soc. Washington, D.C.).

Nihlgard, B. (1985) The ammonium hypothesis - an additional explanation to the forest dieback in Europe. Ambio 14, 1-8.

Norton, S. A., D. W. Hanson and R. J. Campana (1980) The impact of acidic precipitation and heavy metals on soils in relation to forest ecosystems. p. 152-157, In P. R. Miller (ed.) Proc. Symp. on Effects of Air Pollutants on Mediterranean and Temperate Forest Ecosystems. USDA For. Serv. Pac. S.W. For. Range Exp. Sta. Gen. Tech. Rep. PSW-43.

Reich, P. B. and R. G. Amundson (1985) Ambient levels of ozone reduce net photosynthesis in tree and crop species. Science 230, 566-570.

Shriner, D. S. (1986) Terrestrial ecosystems: wet deposition. p. 365-388, In A. H. Legge and S.V. Krupa (ed.) Air Pollutants and Their Effects on Terrestrial Ecosystems. (John Wiley, N.Y.).

Skelly, J. M. (1980) Photochemical oxidant impact on Mediterranean and temperate forest ecosystems: real and potential impacts. p.38-50, In P.R. Miller (ed.) Proc. Symp. on Effects of Air Pollutants on Mediterreanean and Temperate Forest Ecosystems. USDA For. Serv. Pac. S.W. For. Range Expt. Sta. Gen. Tech. Rep. PSW-43.

Ulrich, B. and J. Pankrath (1983) Effects of accumulation of air pollutants in forest ecosystems. (D. Reidel Publ. Co. Boston).

Prof. Robert L. Edmonds, College of Forest Resources, University of Washington, Seattle, Washington 98195, USA

EXS 51:
Advances in Aerobiology
©1987 Birkhäuser Verlag Basel

EFFECTS OF ACIDIC FOG CONTAINING H_2O_2 ON THE SENSITIVITY OF AGRICULTURAL CROPS TO IMPORTANT FUNGAL DISEASES

G. Masuch, V.H. Paul and R.K.A.M. Mallant

Fachbereich 13 - Chemie, Universität Paderborn, 4790 Paderborn, FRG Fachbereich 9 - Landbau, Universität Paderborn, 4770 Soest, FRG, and Netherlands Energy Research Foundation ECN, 1755 ZG Petten, NL

Introduction

During the recent years acidic fog and its effects on agricultural plants and forest trees have been more and more the subject of investigations, especially because fog is much more acidic than acid rain. Analyses of fog and cloud water have shown that it may contain a considerable amount of H_2O_2 with a concentration of 100 to 5000 ppb (1,2). H_2O_2 is an oxidant well soluble in water. After it has been proved that H_2O_2 plays an important role in forest decline, we now want to present first results about its influence on certain agricultural plants especially under the aspect of host-parasite interactions.

Experimental

Pinto bean plants *(Phaseolus vulgaris L.)* and cucumber plants *(Cucumis sativus L.)* were cultivated in the greenhouse. The artificial infection of pinto bean by uredospores of bean rust *(Uromyces phaseoli (Pers.) Wint.)* and of cucumber by conidia of powdery mildew *(Erysiphe cichoracearum DC. ex Mer.)* was carried out in the two-leaf-stage of the plants. Then, the plants were transferred to exposure chambers. The experiments with acidic fog (pH 4) containing H_2O_2 were performed at the Energy Research Foundation in Petten, The Netherlands. The plants were nebulized every day from 6 to 9 a.m. over a period of six weeks, beginning with the 30th of July 1985. Three concentrations of H_2O_2 have been chosen for experimental work: 700 ppb, 2000 ppb and 5000ppb.

The fog chambers used in these experiments were 1 m³ glass
boxes, placed in the laboratory (3). Fog generation, analytical
and botanical methodology have already been described (4).

Results

Effects of bean rust and H_2O_2 on leaves of pinto bean
With increasing concentration of H_2O_2 in acidic fog the leaves
of pinto bean showed progressive chlorotic effects. Under the
influence of bean rust the plants remained smaller, but the
green colour of the leaves became darker green when treated
with H_2O_2. *Uromyces phaseoli* infects the pinto bean leaves through
the stomates. The germination tubes of the uredospores penetrate
the leaf preferably through the lower epidermis and develop
an intercellular dicaryotic mycelium. With increasing concentration of H_2O_2 in acidic fog   the amount of hyphae decreased in
the intercellular space. The amount of haustoria and generative
mycelium in the palisade and spongy mesophyll cells was reduced
as well. The percentage of palisade and spongy mesophyll cells,
infected by the rust haustoria and generative mycelium was
lowered, depending on the concentration of H_2O_2. The tissues
of the mesophyll were altered by the influence of acidic fog
   containing H_2O_2 as well as by rust infection. The intercellular
volumes of the bean leaves increased under the influence of
high concentrations of H_2O_2, but there was no increase of the
intercellular spaces under the combined influence of H_2O_2 plus
rust infection. A more detailed picture was obtained by comparing the spongy mesophyll area with the intercellular space
area. Compared with the spongy mesophyll area the intercellular
space area is a bit larger in the reference series and in the
experimental series, treated with 700 ppb H_2O_2, but this difference is not significant. This effect is more serious with
increasing concentration of H_2O_2. This enlargement of the intercellular space is being lost under the additional influence
of the fungal infection. The intercellular space area decreased

in relation to the spongy mesophyll in the case of the untreated
reference and at the lowest concentration of 700 ppb H_2O_2,
where the disease is rather strong. A concentration of 2000 ppb
H_2O_2 inhibited the rust disease heavily. In this case the
areas of spongy mesophyll and intercellular space were distributed to equal parts. At the highest concentration of 5000 ppb
H_2O_2,where the strongest rust inhibition was to be found,
the intercellular space area was better developed than the
spongy mesophyll tissue area.

Effects of powdery mildew and H_2O_2 on leaves of cucumber

The cucumber plants, infected by the powdery mildew, showed
a higher sensitivity to the disease with increasing concentration of H_2O_2. Some of the marked characters of the powdery
mildews as ectoparasites are their superficial hyaline
mycelium, their haustoria in the epidermal cells of their hosts
and the high water-content of their large turgid airborne conidia. Typically the hyphae radiate singly with branching from
an infection point and grow closely appressed to the host.

The density of powdery mildew (hyphae and conidia) increased
under the influence of H_2O_2 in acidic fog. The strongest
effect occured at 5000 ppb H_2O_2, where the cucumber plant
was already dead. The upper epidermis was extremely more
attacked by powdery mildew than the lower cucumber epidermis.

The histological comparison between infected plants without
H_2O_2 and with 700 ppb H_2O_2 showed specific alterations in the
host tissue. The cells of the upper and lower epidermis of
the polluted plants were hypertrophied and the intercellular
spaces were enlarged. Under the influence of 700 ppb H_2O_2
the amount of fungal parts (mycelium, conidiophores and conidia)
as seen in semithin sections increased significantly.

Discussion

Bean rust and powdery mildew are obligately parasitic fungi.

While bean rust is an endoparasit, developing in the intercellular space of the host leaf and feeding itself through
intracellular haustoria, powdery mildews live as ectoparasites
superficially on the epidermis of plant leaves, sending their
haustoria into the epidermal cells of their hosts. Both groups
of fungi inhibit the vegetative development of the host plants
and reduce their yields. The reaction of the bean rust to
H_2O_2 is a gradual going-down, dependent on the concentration
of H_2O_2 in the acidic fog. The bean plant grows less under
the influence of H_2O_2, but it recovers from the effects of the
bean rust. The favorable effect of dew on powdery mildews has
already been reported by UBRIZSY in 1946 (5), fog is said to
have the same effect (6). Until the cucumber plants died in
our experiment under the influence of infection by powdery
mildew combined with the effects of high concentrations of
H_2O_2, the parasitic fungi developed better even in the foggy
atmosphere containing H_2O_2. This oxidant increased the fungal
development as long as the host lived. The susceptibility of
bean to rust attack was decreased under the influence of H_2O_2
and that of cucumber to powdery mildew was increased. The
causes of the effects induced by H_2O_2 are unknown. It seems
that other photooxidants also tend to cause extension of the
intercellular space in pinto bean leaves as it was shown recently
by VOGELS and MASUCH (7) for ozone.

References

(1) KADLECK, J., MCLAREN, S., CAMAROTA,N., MOHNEN,V., WILSON,J.
    1982/83): Cloudwater chemistry at Whiteface Mountain.
    In "Precipitation Scavenging, Dry Deposition and Resuspen-
    sion", Proc. 4th Int. Conf., Santa Monica, Cal.,29.Nov.-
    3. Dec.,1982
    (Elsevier Sci. Publ. 1983) 103-113

(2) KOK, G.L. (1980): Measurements of hydrogen peroxide in
    rainwater. Atmos. Environ. **4:** 653-656

(3) MASUCH, G., MALLANT, R.K.A.M., (1986) Ambio, in press

(4) MASUCH, G., KETTRUP, A., MALLANT, R.K.A.M., SLANINA, J.
    (1986):
    Int. J. Analyt. Environ. Chem., in press

(5) UBRIZSY, G. (1946) (Contributions to the knowledge of the
    Erysiphaceae of Nyirseg Acta)Mycol. Hung. $\underline{3}$, 28-33

(6) CARTER, C.N. (1915) A powdery mildew on citrus. Phytopatho-
    logy $\underline{5}$, 193-196

(7) Vogels, K., Masuch, G. (1984) Metabolic and substructural
    changes of bean leaves caused by seperate and simultaneous
    ozone sulfur dioxide exposures. Proc.Int. Workshop on
    Eval. and Assessment of the effects of photochem. oxidants
    on human health, agricult. crops, forestry, materials
    and visibility.
    Göteborg Feb. 29 - March 2, 1984

Prof. Dr. Georg Masuch, Fachbereich 13, Warburger Straße 100,
Universität, D-4790 Paderborn

EXS 51:
Advances in Aerobiology
©1987 Birkhäuser Verlag Basel

SPREAD OF BARLEY MILDEW BY WIND AND ITS SIGNIFICANCE FOR PHYTO-
PATHOLOGY, AEROBIOLOGY AND FOR BARLEY CULTIVATION IN EUROPE*

E. Limpert

Technical University of Munich, Chair of Agronomy and Plant Breeding, Freising-Weihenstephan, Germany (FRG)

Background and objectives. Barley mildew, _Erysiphe graminis_ f.sp.
hordei, is the most important pathogen of barley in Europe, causing losses estimated to surpass one thousand million DM on an
average per year in the countries of the European Community.
     Although a variety of host resistance genes and syste-
mic fungicides have been used for mildew control, previously
successful methods of control have regularly been overcome after
extensive use. In order to get a better understanding of the
distribution of the pathogen by wind, as well as of the selective
forces acting on the pathogen population, the mildew situation
has been investigated during the recent past in several countries
of northwestern Europe.
     Such an investigation has become feasible through the
development of a new method of sampling and analysing spores
(for detailed description see LIMPERT, 1985). By means of an
apparatus mounted on the roof of a car and containing a jet spore
sampler** (SCHWARZBACH, 1979), representative random samples
of the spores were taken from the air along transects through
regions of interest for barley cultivation and for the epidemiology of barley mildew. Single spore progenies were subsequently
analysed in the laboratory for their ability to attack and overcome defined host resistance genes. Certain examples of the fre-
quency-distribution of these race-specific virulences will be
given.

1. Spread of spores enhanced in main wind direction (MWD)

New, unusual virulence characters provide a good tool for aerobio-
logic research. In regions where the varieties with corresponding
resistance genes are cultivated, they are selected and occur
for the first time in measurable amounts. The drift of spores

---

* This work has been supported by the German Research Foundation
** Product of Burkard Manufacturing Co. Ltd., Rickmansworth,GB

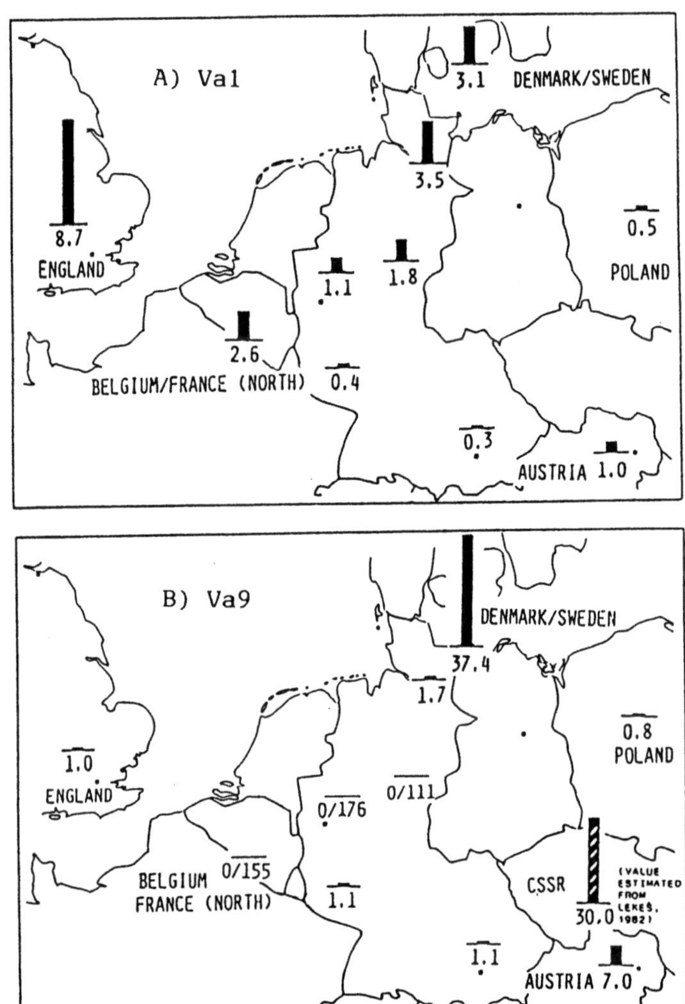

Fig. 1. Distribution pattern of virulence characters in north-western Europe in 1981. The frequency (%) of Va1 (A) was highest in England and decreased steadily to the east whereas Va9-frequency (B) sharply decreased from its foci to the west. These patterns are probably due to the spread of mildew spores with prevailing winds from the west.

from such regions can be easily traced, as they are labelled by their unusual character, which is not needed for survival in other regions. The occurrence of unusual and unneccessary characters in the mildew population of a given area thus provides direct evidence for the introgression of mildew from another region.

Examples 1 and 2 (Fig. 1) indicate that the distribution of spores is enhanced in and reduced against MWD. Va1 was selected in England, where it occurred most frequently. In other European regions (besides Denmark) the virulence was unnecessary. The observed distribution of the virulence (Fig. 1A) indicates a considerable introgression from England, the region of origin of these pathotypes, to the continent, where the frequency steadily decreased eastward.

On the other hand Va9 (Fig. 1B) was found most often in Denmark, Czechoslovakia and Austria. From these centres of origin the frequency of this virulence decreased sharply to the west, thus indicating that the spread of mildew was reduced in this direction.

A third example (table I) is given for the spread of spores within Southern Germany. The region of selection of Va7+Vk was Rheinland-Pfalz (in Southwest Germany). In the neighbouring countries south, west and north of Rheinland-Pfalz the frequency of Va7+Vk was low, and in Bavaria selection for this virulence was absent. Therefore the considerable amount of Va7+Vk found in Bavaria (table I) is supposed to have originated in Rheinland-Pfalz. This assumption is confirmed by the result of a comparison of frequencies of certain virulence combinations in the pathogen populations from both regions (LIMPERT, 1985).

Table 1. Barley and barley mildew in Southwest Germany (Rheinland-Pfalz) and in Bavaria

---

1. Frequencies (%) of varieties with Lyallpur-resistance

	Southwest Germany	Bavaria
∅ 1974-1978	35,9	0,5
∅ 1979-1981	19,5	1,0

2. Frequency of mildew spores with the specific matching virulence

---

∅ 1979-1981	58,2	44,0

---

2. Velocity of drift of mildew populations

In the region of Weihenstephan a peculiar development of the frequency of virulence Va7+Vk was recognized, which was rather similar to that in Rheinland-Pfalz, the region of origin (Fig. 2). As there was an interval of time of three years between these developments, it is concluded that the mildew population from Rheinland-Pfalz needed this lapse of time to drift to Weihenstephan (for details see LIMPERT, 1985). This corresponds to an

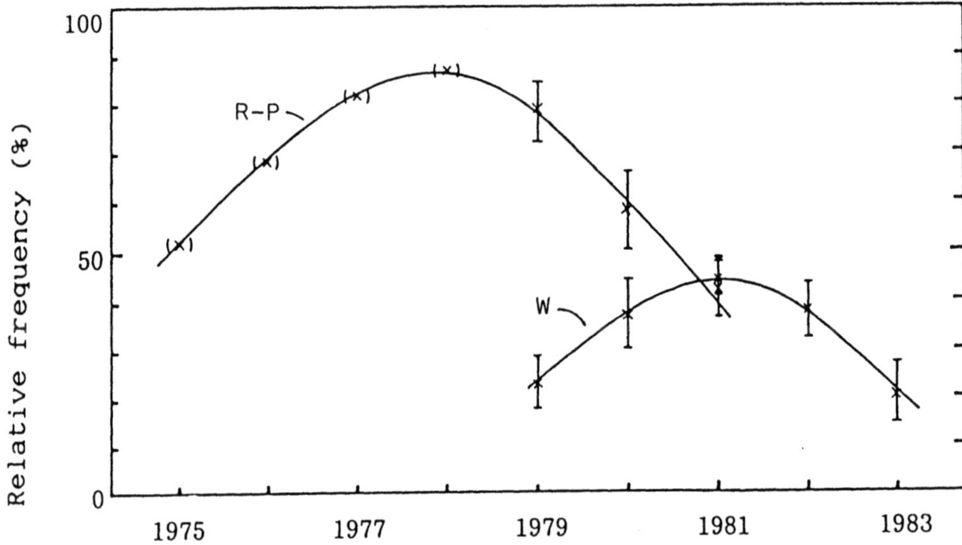

Fig. 2. Development of the frequency of Lyallpur-virulence (Va7 +Vk) in Southwest Germany (Rheinland-Pfalz, RP) and in Bavaria (Weihenstephan, W). The difference in time between the maximum and the subsequent decrease of the virulence frequencies in the two regions indicate that the mildew population from Rheinland-Pfalz needed three years drift for the 330-km-distance to the Weihenstephan area.

---

annual drift of approximately 110 km on an average.

These examples of drift show, that vast regions of Europe have to be regarded as an epidemiologic unit. That means that even in regions where winter barley is cultivated and mildew is regarded to be generally endemic, a considerable influence from neighbouring regions is present.

3. Further remarks on wind and drift

Factors influencing drift. In Europe wind mostly comes from westerly directions; in large regions approximately 50% of the winds are from a 90° sector around due west. The preferred eastward drift of windspread pathogens can, however, be even more favoured by the fact that the velocity of these winds is around 50% higher than that of winds from the other directions. In addition, during the movement of these mostly maritime and humid air masses the stratification of the atmosphere is less stable. This creates turbulence with the aid of which spores are liberated and may be lifted into the air. For these reasons, it is suggested that in wet years populations will be farther displaced than in dry years, when their is the influence of high pressure in the east.

<u>Comparison of velocities of spread.</u> With the normal movement of air masses spores may travel 1000 km a day. That, however, will be the case for only a tiny part of the population. Newly emerging diseases in Europe, like tobacco blue mold (<u>Peronospora tabacina</u>) or grapevine powdery mildew (<u>Uncinula necator</u>) were spread on their susceptible hosts at around 500-1000 km per year (for a survey see GREGORY, 1973). In comparison with these values 110 km/year is not much. In the latter case, however, the drift of whole populations across a region in which the pathogen is already present everywhere, is being considered.

<u>Space, time and altitude of spore dispersal.</u> Although a distance of 100 km/year is quite long, the drift of spores need not necessarily occur at great altitudes. Given the conditions of barley cultivation it seems reasonable to suppose that there may be 4-5 stages, of 20-25 km each, to cover this distance. Assuming a terminal velocity of fall of mildew spores of 1.2 cm/sec (YARWOOD & HAZEN, 1942, cited by GREGORY, 1973) and normal wind velocities (6m/sec, data Deutscher Wetterdienst, unpublished), a rather low altitude of 40 m would be sufficient for the spores to bridge the distance of one stage within one hour. As spores are liberated more easily with higher wind velocities, the duration of drift under such conditions would be even shorter. Under these conditions viability of conidia would hardly be impaired. For a survey on these questions of spore dispersal see also FRINKING, this volume.

4. Conclusions

This knowledge about the drift of mildew throughout Europe seems to be comparable with the tip of an iceberg, the dimensions of which need to be plumbed by further investigations to make the best use of the underlying principles. It may be assumed that these basic principles of drift would - in suitably modified forms - apply to other kinds of economically important pathogens spread by wind throughout Europe, e.g. for powdery mildew of wheat and for a variety of rusts and smuts.

The present knowledge could be used to improve strategies for the use of host resistance and of fungicides for mildew control in Europe. HAMILTON and STAKMAN (1967) have shown that spores of rust diseases in Northern America drifted northward with prevailing winds during spring from their regions of overwintering in the south, and during autumn in the opposite direction. Therefore FREY et al. (1977) proposed to use different sources of resistance in three successive regions across the main wind direction, to make the best use of resistance genes under these peculiar conditions of climate and spore migration.

Caused by different epidemiological conditions of barley mildew in Europe (drift mainly in one direction, from west to east; overwintering of the pathogen everywhere possible) a different strategy which is described in Fig. 3 would be better suited for the use on this continent. With growing awareness of the obvious principles of winddrift of mildew spores in Europe, host plant resistance as well as fungicides could be used with greater success.

Period	Z o n e s			
	A	B	C	D
1	Ly	Ar	Sp	Ru,MC
2	Ar	Sp	Ru,MC	Ly
3	Sp,NR1	Ru,MC	Ly	Ar
4	Ru,MC	Ly	Ar,NR2	Sp,NR1
5 ($\hat{=}$ 1')	Ly	Ar,NR2	Sp,NR1	Ru,MC

Fig. 3. System for a Europe-wide use of race-specific resistance against barley mildew. The whole area of cultivation is devided into several zones (A-D) across the main wind direction (MWD), in each of which certain resistance genes will be used during a certain period. Every new period cultivation moves for one stage against MWD or from A to D, respectively. Varieties with new resistance genes can be integrated at any position. During the fifth period positions of departure are reached again; diversity has increased, however, in the meantime by the addition of new resistance genes (NR1, NR2). Symbols of resistance genes: Ly = Lyallpur (Mla7+Mlk); Ar = 'arabische' (Mla12); Sp = Spontaneum (Mla6); Ru = Rupee (Mla13); MC = Monte Cristo (Mla9).

Authors address:
Dr. Eckhard Limpert, Technische Universität München, Lehrstuhl für Pflanzenbau und Pflanzenzüchtung, D-8050 Freising-Weihenstephan

References:
Frey, K., J.A. Browning & M.D. Simmons, 1977. Management systems for host genes to control disease loss. In: The genetic basis of epidemics in agriculture, P.R. Day (ed.), Annals of the New York Academy of Sciences Vol. 287, 255-274.

Gregory, P.H., 1973. Microbiology of the atmosphere. Leonhard Hill, 377 pp.
Hamilton, L.M. & E.C. Stakman, 1967. Time of stem rust appearance on wheat in the Western Mississipi Basin in relation to the development of epidemics from 1921 to 1962. Phytopathology 57, 609-614.
Limpert, E., 1985. Ursachen unterschiedlicher Zusammensetzung des Gerstenmehltaus, Erysiphe graminis DC.f.sp. hordei Marchal, und deren Bedeutung für Züchtung und Anbau von Gerste in Europa. Diss. München 183 S.
Schwarzbach, E., 1979. A high throughput jet trap for collecting mildew spores on living leaves. Phytopath. Z. 94, 165-171.

FALLING BEHAVIOUR OF STERILIZED AND SLEPT
FRUITFLY AND MELONFLY

by

Akira AZUMA, Yoshio ONDA
Institute of Interdisciplinary Research,
Faculty of Engineering, The University of Tokyo.

and

Ryo-hei ICHIKAWA
Japan Agricultural Aviation Association

SUMMARY

Oriental fruitfly, Dacus dorsalis Hendel, and melonfly, Dacus cucurbitae Conquillett, are species of noxious insects for fruits and vegetables.
An effective method to stamp out noxious insects is to send sterile males into
breeding area.   The sterile insects treated by applying the cobalt rays are
made sleep by freezing and dropped from the air.   The terminal rate of fall
and the height loss or the time to get the terminal rate, both of which are
important parameters for determining the flight speed and altitude of a dispersal aircraft, are functions of the drag-area-to-mass ratio f/m and the
opening angle of the wings, Υ of the insets.   From the free-falling test and
the vertical-wind-tunnel test, it was found that the rate of fall of slept
flies were appreciably reduced by increasing either the drag-area-to-mass
ratio f/m or the opening angle Υ, and the reduced rate of fall were resulted
from both the increased drag area by taking the shallow attitude (close to the
horizontal attitude) and the lift of the wing generated by the spinning motion
of the body.

INTRODUCTION

Aircraft such as airplane and helicopter have been used to disperse
either the seeds of crops or the insecticidal sprays in the agricultural
aviation.   The dispersion of sterile males in the breeding area is an effective way to exterminate the noxious insects without giving any severe damage
on other innoxious species in the living creatures.
An actual aerial application to the extermination of fruitfly with a
partial dispersion of the insecticidal sprays at the Southwestern Islands
(Ryukyu Retto) in 1968 through 1985.   The embargo of the fruits at the islands were removed perfectly in 1985.   The project of extermination for the
melonfly is still going on the same islands.
Shown in Fig.1 are Oriental fruitfly, Dacus dorsalis Hendel and
Melonfly, Dacus cucurbitae Coquillett.   The mass of sampled bodies are given
in Table 1.   Since the mass of Oriental fruitfly was different between male
and female the sex was also described.

FREE FALLING TEST

As shown in Fig.2a,b, free falling test of the melonfly was performed at the front of a black curtain. The falling flies were photographed with a rule under a series of flashed lights. The rate of fall was determined from the distance of two adjoining images of the photographs and the period between two success flashes.

Fig.3 shows the rate of fall versus the falling distance of normal flies and of dried flies which were forcedly dried by a desiccant in order to find the effect of the drag-area-to-mass ratio f/m on the rate of fall. As clearly seen from the figure, the rate of fall are separated into two-groups, normal and dried which are represented by two branches (solid lines) estimated theoretically by $f/m = 0.07$ m^2/kg and 3.94 m^2/kg respectively.

The theoretical value of the rate of fall can be given by solving the following equation :

$$m(dU_Z/dt) + \frac{1}{2}\rho U_Z^2 f - mg = 0 \qquad (1)$$

$$dZ/dt = U_Z \qquad (2)$$

where Z and U_Z are the vertical distance and the rate of fall (downward positive), m and ρ are the mass of fly and air density, and f and g are the drag and the gravity acceleration respectively. Then the solution of the above equation can be given by

$$U_Z = \sqrt{U_{Z,0}^2 + \{2g/\rho(f/m)\}[\exp\{\rho(f/m)Z\} - 1]} \cdot \exp\{-\frac{1}{2}\rho(f/m)Z\} \qquad (3)$$

where $U_{Z,0}$ is the initial falling speed. Then the terminal speed $U_{Z,\infty}$, can be given by

$$U_{Z,\infty} = \sqrt{2g/\rho(f/m)} . \qquad (4)$$

It is also found from Fig.3 that the rate of fall for dried flies reaches nearly the terminal speed within the falling distance of 0.5m whereas the normal flies are still accelerated within the 0.5m and reaches the falling distance of approximately 2~5m to get the terminal speed. Careful observation further revealed that the flies attained the terminal speed in early stage of the falling have the large opening angle Υ, which is given from the dihedral angle Γ by $\Upsilon = \pi - 2\Gamma$, and make spin automatically (autorotation) during the fall. It is clear that the spinning motion contributes to reduce the rate of fall by the following effects : (i) change of the flight attitude from the deep angle (close to the vertical) to the shallow angle (close to the horizontal), which brings the increase of the projected area or the increase of the drag area, and (ii) generation of the lift on the rotary wings, which increases the equivalent drag area.

WIND TUNNEL TEST

In addition to the free falling test, a vertical wind tunnel was used to measure the rate of fall and the rate of spin.

Fig.4a,b show an arrangement of a vertical wind tunnel and test apparatus. The free flight of fruitfly and melonfly was performed in a trimmed up-current of the tunnel. The up-current was adjusted to nullify the rate of fall of a sampled fly and to keep the position of the fly in a fixed point in the test section. Then the fly was photographed by 16 mm movie camera and 35 mm still camera with the assistance of stroboscopes. The terminal speed and the rate of spin were determined by knowing respectively the wind speed and the frequency of flash of the stroboscope which is synchronized with the spin rate of the sampled fly.

About 90 % of the tested samples in both species, fruitfly and melonfly, made the autorotational flight. Their flight attitudes are shown in Fig.5a ~ c. As seen from Fig.5b and c, the melonfly took two typical body attitude in which the deeply inclined body had low rate of spin and thus high rate of fall and the shallow one had high rate of spin and low rate of fall.

The rate of fall for both species are, as shown by the probability density in Fig.6, mostly concentrated at the speed of $U_{Z,\infty}$ = 3.5 ~ 4.5 m/s. These values are, from equation (1), corresponding to have the drag-area-to-mass ratio of f/m = 1.3~0.8 m^2/kg. The rates of fall for the respective species are also shown in Fig.7 and 8 as a function of the rate of spin Ω (in rpm) and of the opening angle Υ respectively.

In Fig.7 the data are plotted to make clear the difference of the date when the test was conducted by considering the dried condition of the insects after the treatment. It can be seen that within a couple days after the treatment no appreciable difference can be recognized.

In Fig.8 the data are distinguished by the difference of opening angle which is either larger or smaller than 90°.

It is observed that the rate of fall decreases as the opening angle increases (or the dihedral angle Γ decreases) and the rate of spin increases. That is to say, the rate of spin increases with the opening angle of the wings.

APPLICATION TO GENERAL PURPOSE

In order to apply the present study to more general purpose such as for determining the terminal speed of other insects and for finding the horizontal distance during the fall from the air, it is necessary to nondimensionalize the equations of motion. By dividing equation (3) with the above terminal speed, the following nondimensional equation can be obtained:

$$u_Z = U_Z/U_{Z,\infty} = \sqrt{(U_{Z,0}/U_{Z,\infty})^2 + [\exp\{(\rho/\rho_m)C_D z\} - 1\}]} \cdot \exp\{-\tfrac{1}{2}(\rho/\rho_m)C_D z\} \quad (5)$$

where

$$z = Z/\ell \ , \quad \ell = V/S \ , \quad \rho_m = m/V \ , \quad C_D = f/S \quad (6)$$

and where V, S and ℓ are volume, reference area (or horizontally projected area in trimmed flight) and reference length (or body length) respectively.

Then the terminal speed given by equation (4) can be rewritten as

$$U_{Z,\infty} = \sqrt{2(\rho_m/\rho)(\ell g)/C_D} \ . \quad (7)$$

It can be said from these equations that (i) effects of the drag force is strongly dependent on the density ratio between the falling body and the air, ρ_m/ρ, and (ii) the terminal speed $U_{Z,\infty}$, increases in proportion to the square root of the density ratio times the reference length, $\sqrt{(\rho_m/\rho)\ell}$, and in inversely proportion to the square root of the drag coefficient $\sqrt{C_D}$.

If the insect is dropped from a flying aircraft, the horizontal speed, $U_X = \dot{X}$ is decelerated by the drag as follows :

$$m(dU_X dt) + \tfrac{1}{2}\rho f U_X^2 = 0. \quad (8)$$

This is equivalent to make g = 0 in equation (1). Then, by referring to equation (3), the velocity U_X can be given by

$$U_X = U_{X,0} \exp\{-\tfrac{1}{2}\rho(f/m)X\} = 1/\{\tfrac{1}{2}\rho(f/m)t + 1/U_{X,0}\} \qquad (9)$$

or in nondimensional form

$$u_X = \exp\{-\tfrac{1}{2}(\rho/\rho_m)C_D x\} = 1/\{\tfrac{1}{2}(\rho/\rho_m)C_D \tau + 1\} , \qquad (10)$$

and the nondimensional distance is given by

$$x = \ln\{\tfrac{1}{2}(\rho/\rho_m)C_D\tau + 1\}/\tfrac{1}{2}(\rho/\rho_m)C_D \qquad (11)$$

where

$$u_X = U_X/U_{X,0} , \quad x = X/\ell , \quad \tau = (U_{X,0}/\ell)t. \qquad (12)$$

In the above equations the elapsed time t is considered to be the time until the falling body approaches to the ground after the falling has been initiated. Since, usually the insects reach their terminal speed very soon after the fall, the time can be approximated by

$$t \cong Z/U_{Z,\infty} \qquad (13)$$

Table 2 gives the related parameters and resulted quantities obtained by the present experiments for the melonfly. It can be seen that the equivalent drag coefficient takes very large value in comparison with, for example, that of the flat plate, $C_D \cong 1.0$. This fact suggests that the lift is generated by the spinning motion of the body.

CONCLUSION

The falling behaviour of sterile insects, Oriental fruitfly and melonfly, treated by applying the cobalt rays and slept by freezing are studied by the free falling test and the vertical-wind-tunnel test. During the fall, many flies make spin or autorotation and attain the reduced terminal speed by the increase of the drag area and the resulted lift of the wing. The spin rate is related to the opening angle or dihedral angle of the wing and the rate of fall decreases as the rate of spin increases. In the tested flies, the rate of fall is mostly ranged from $U_Z = 3.5\sim4.5$ m/s if the dispersion is performed within a couple of days after the treatment. The equivalent drag coefficient is about $C_D = 1.2$

SYMBOLS

C_D drag coefficient $= f/S$, where S is the reference area (m^2)
f drag area (m^2)
g gravity acceleration (m/s^2)
m mass of insect (kg)
U_X horizontal speed, forward positive, (m/s)
U_Z vertical speed, downward positive, (m/s)
$u_X = U_X/U_{X,0}$
$u_Z = U_Z/U_{Z,\infty}$
X horizontal distance, forward positive, (m)
$x = X/\ell$, where ℓ is the reference length (m)
Z vertical distance, downward positive, (m)
$z = Z/\ell$

γ opening angle $= \pi - 2\Gamma$ (deg)
Γ dihedral angle (deg)
ρ air density (kg/m^3)
ρ_m density of insect $= m/V$ (kg/m^3) , where V is the volume (m^3)
τ reference time $(U_{Z,0}/Z)t$

Subscripts

    $(\ )_0$ initial value of $(\ )$

    $(\ )_\infty$ terminal value of $(\ )$

Table 1. Mass of Species

	species	mean mass $\overline{m}$,(mg)	standard deviation Δm,(mg)	number of sample	conditioned*
For free falling test	Melonfly ♀ (Dacus cucurbitae)	18.5	——	100	normal
For free falling test	Melonfly ♀ (Dried Dacus cucurbitae)	3.3	——	50	dried
For wind tunnel test	Oriental fruitfly ♂ (Dacus dorsalis)	15.7	1.72	68	normal
For wind tunnel test	Melonfly ♂ (Dacus cucurbitae)	12.9	2.57	50	normal

Table 2. Parameters and quantities determined by experiments

Items	Symbol and unit	Melonfly
Reference length	l (m)	6×10^{-3}
Reference area	S (m^2)	$(1.4 \sim 0.8) \times 10^{-5}$
Volume	V (m^3)	9.2×10^{-9}
Density	ρ_m (kg/m^3)	1.4×10^3
Drag-area-to-mass ratio	f/m (m^2/kg)	$1.3 \sim 0.8$
Equivalent drag coefficient	C_D	$1.20 \sim 1.24$

* Normal means that the flies were measured and tested in the next day after the sterilized whereas dried means that the flies were measured and tested in long time later after the flies were sterilized and forcedly dried by a desiccant.

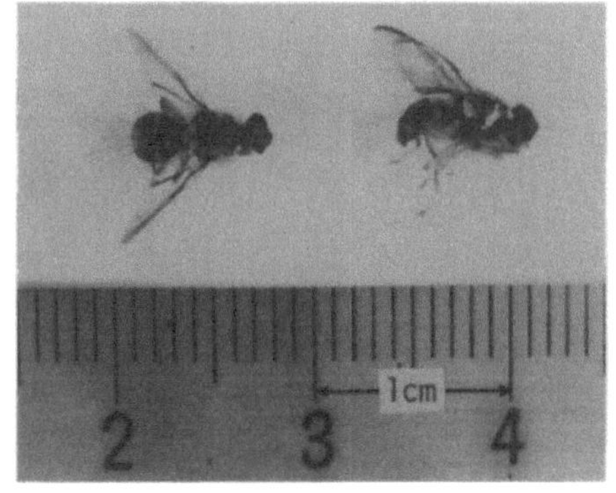

(a) Oriental fruitfly

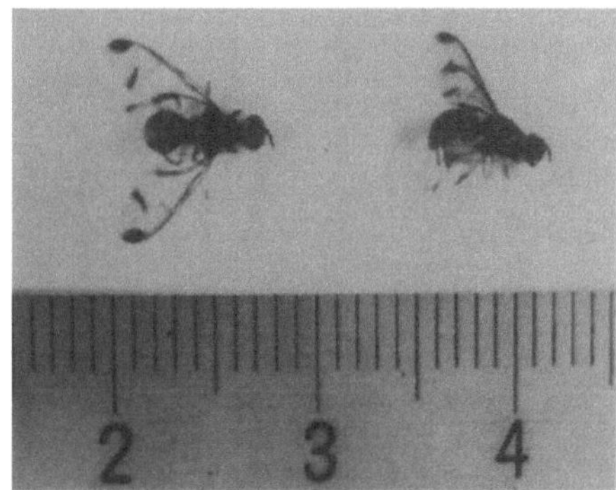

(b) Melonfly

Figure 1. General of Oriental fruitfly and Melonfly

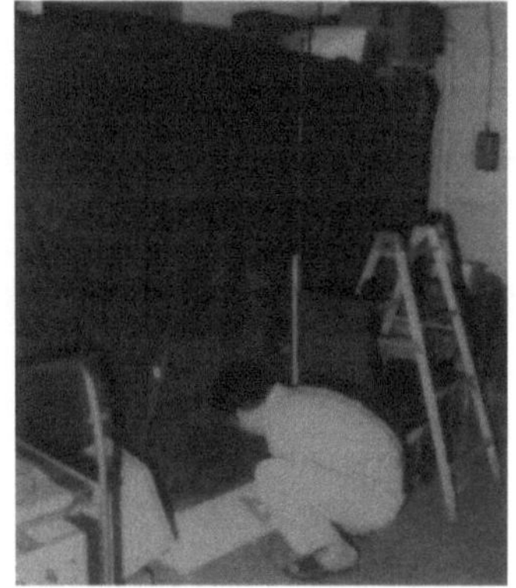

(a) Arrangement of apparatus   (b) A series of flashed picture
Figure 2. Free falling test

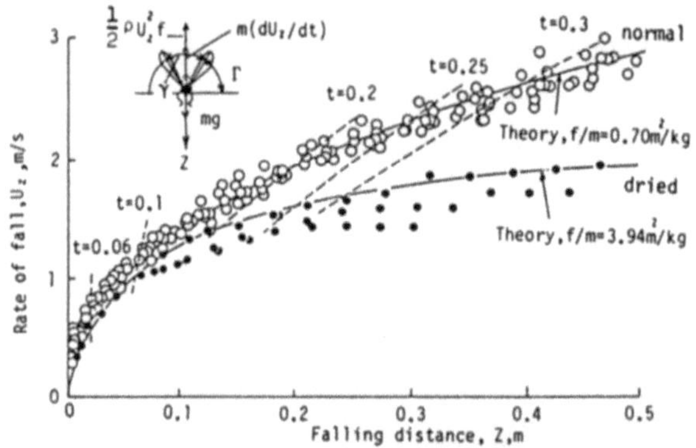

Figure 3. Rate of fall versus falling distance (Melonfly)

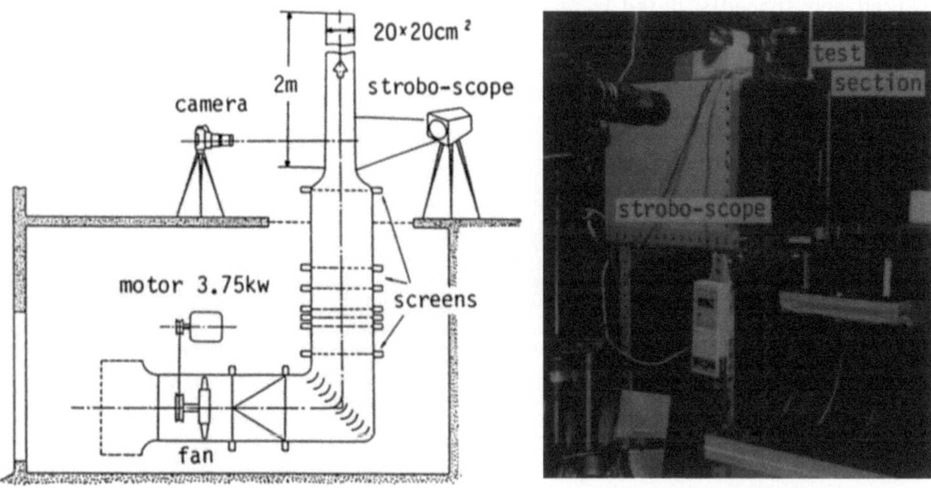

(a) Arrangement of the vertical wind tunnel   (b) Test apparatus
Figure 4. Arrangement and apparatus for the wind tunnel test

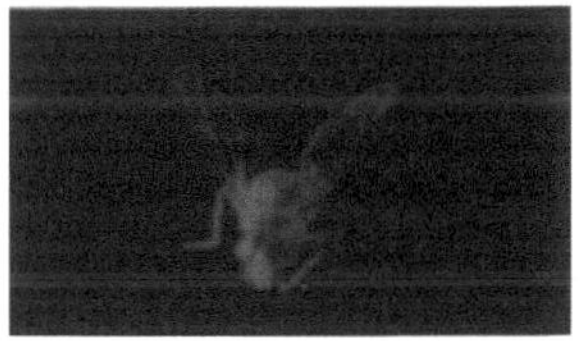
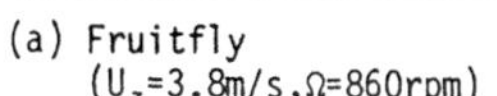

(a) Fruitfly
$(U_z=3.8m/s, \Omega=860rpm)$

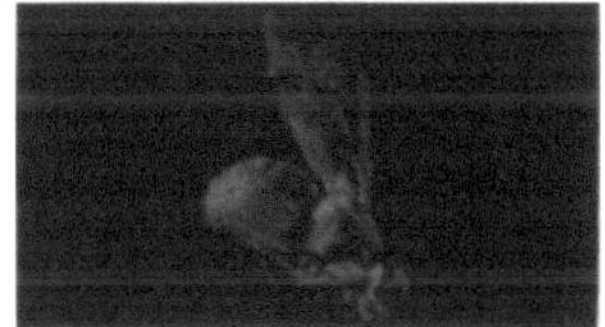

(b) Melonfly in low spin rate
$(U_z=4.0m/s, \Omega=551rpm)$

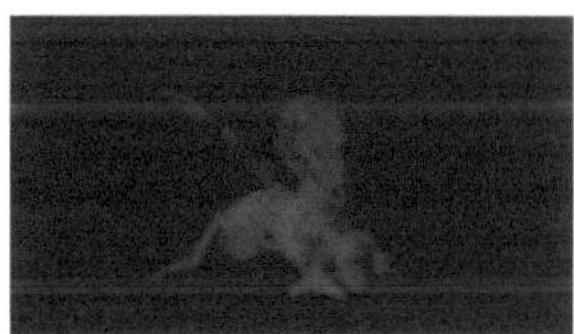

(c) Melonfly in high spin rate
$(U_z=2.3m/s, \Omega=1,795rpm)$

Figure 5. Flight attitude of Melonfly and Oriental fruitfly

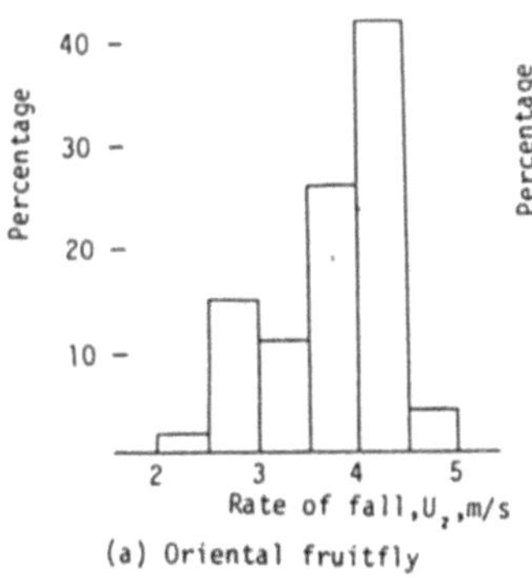
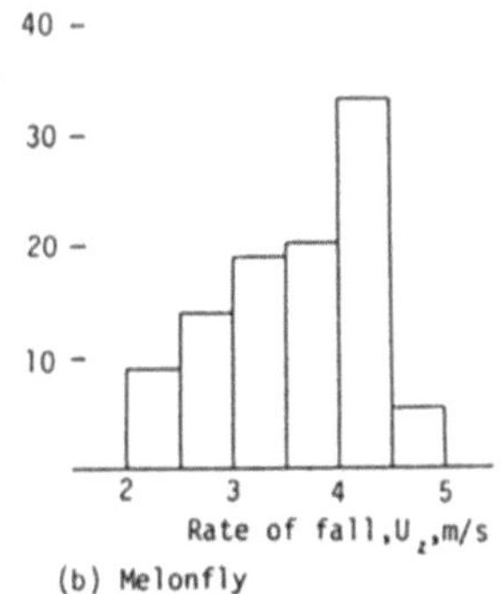

(a) Oriental fruitfly   (b) Melonfly

Figure 6. Probability density of the rate of fall

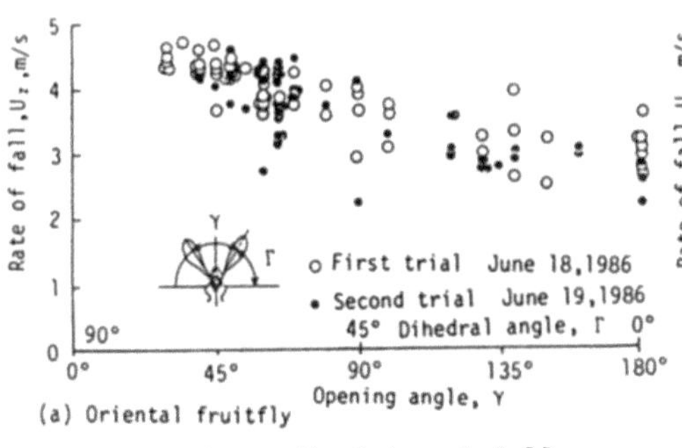

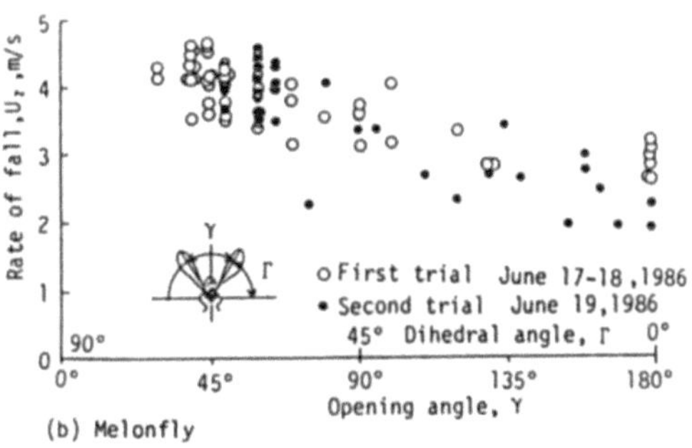

(a) Oriental fruitfly   (b) Melonfly

Figure 7. Rate of fall versus opening angle

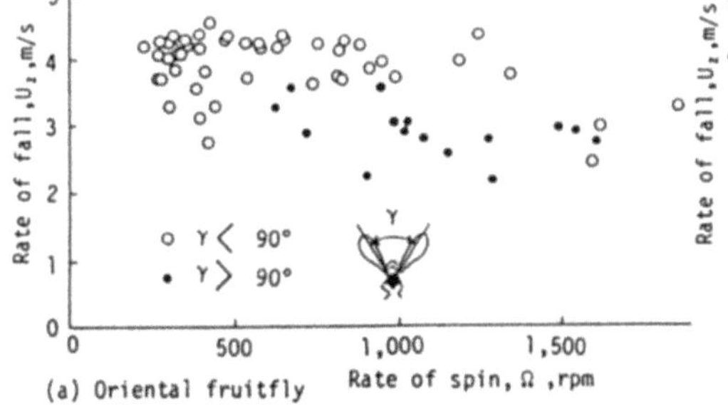

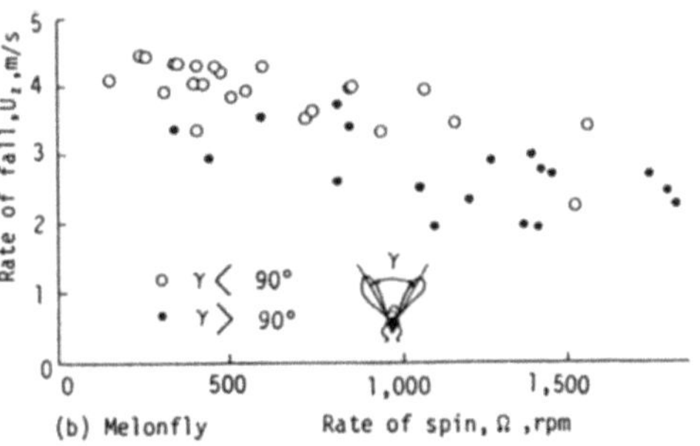

(a) Oriental fruitfly   (b) Melonfly

Figure 8. Rate of fall versus rate of spin

VI Microbiology

EXS 51:
Advances in Aerobiology

AIRBORNE SPREAD OF MICROORGANISMS AFFECTING ANIMALS

R.F. Sellers

Animal Disease Research Institute, Nepean, Ont., Canada.

Introduction

Microorganisms affecting animals can be transported by the airborne route in particles (3 - 20 um) and by insects. In each instance an infectious dose has to be delivered to the animal downwind. Most airborne particles will enter by the respiratory route, but some will be deposited on the ground and transferred by splash or contact. Insects transmit infection by biting, but the site on the animal's body varies. One infectious dose should be sufficient, but some routes require more e.g. 10^5 infectious units of foot-and-mouth disease virus (FMDV) are required for alimentary infection compared to 25 for respiratory.

Particle spread

Various factors affect the final delivery of the infecting dose to animals at risk. There must be sufficient microorganisms at source. Differences occur between microorganisms (e.g. pigs infected with FMDV excrete more than pigs infected with Aujeszky's disease), between animal species (e.g. pigs produce more airborne FMDV than cattle or sheep) and distribution of particle size (larger particles deposit rapidly, smaller particles may be inactivated). During downwind carriage RH affects survival, FMDV survives better at RH values greater than 60%. Natural secretions surrounding the microorganisms may protect.

Once airborne the plume of infective particles is dispersed.
The less the vertical and horizontal dispersion, the higher the
remaining concentration. Dispersion is greater over land than
sea owing to roughness of the surface, and more during the day
due to insolation than at night. The concentration downwind
can be estimated by Pasquill's equation, which applies to
downwind dispersion of all types of particles.

The greater the numbers of susceptible animals downwind and the
longer the duration of the plume, the more chance there is of a
successful infection.

Spread by insects

Insects may transmit infection over long distances biologically
or mechanically. In biological transmission multiplication of
the microorganism takes place in the insect and the insect may
not be able to transmit infection until 7 - 10 days after
biting an infected animal. Although the insect may need to
take a blood meal every 3 - 4 days, actual transmission may
only occur at the second blood meal or later and this requires
survival of sufficient infected insects until this time. In
mechanical transmission the microorganism survives in the
mouthparts of the insect. Unless there is intermittent
feeding this means survival for 3 - 4 days until the insect
takes the next blood meal. Whereas for microorganisms that
undergo biological transmission this is their main method,
organisms that are mechanically transmitted can be divided into
those for which it is the main method e.g. myxoma virus, and
into those for which other methods are available e.g. sheep and
goat pox.

Insects fly unaided below the boundary layer, but above it may
be carried for considerable distances on the wind. The insect
species carried long distances appear to be those occupying
temporary habitats, and their flights are adaptive migrations

to take advantage of temporary or seasonal conditions i.e. food sources such as migrating animals or breeding sites.

Insects are active between certain temperatures e.g. in the tropics and subtropics between 15° and 30°C. Warm winds are important for their dispersal and are found north and south of the equator associated with movements of the Intertropical Convergence Zone (ITCZ). Outside the limits of the ITCZ, maritime tropical air flowing northwards from the Gulf of Mexico, warm southeasterlies in the Tigris and Euphrates valleys and warm northeasterlies flowing into southern Africa carry insects.

The insects are carried in association with fronts. High concentrations of _Culicoides_ have been caught in association with the ITCZ over Sudan. Rift Valley fever could have resulted from the carriage of infected insects due to the northward movement of the ITCZ to Aswan, Egypt. Where fronts meet, convergence occurs. Insects are unable to endure low temperatures and thus deposition occurs over a period leading to concentration of insects. Recent investigations indicate that some of the epidemics of equine encephalitis in North America may have resulted from the concentration of infected insects at points of convergence, where fronts meet. Movement of insects can also be associated with stable airstreams resulting from inversions. Such may have occurred with the movements of insects carrying myxoma virus from France to England and of _Culicoides_ carrying bluetongue virus from the Eastern Mediterranean coast to Cyprus.

These investigations have been made through retrospective analysis of past outbreaks. Sutiable winds, sources of infection and similarity of isolates of organisms have been demonstrated and insects have been captured in association with fronts. Isolation of organisms from aerial sampling of insects has not yet been achieved. Clearly there is the opportunity for more detailed investigations in order to obtain methods for disease prediction.

General comments

With both particle and insect transmitted infections coincidence of occurrence of a number of factors is important in transmission.   Examples are the spread of FMDV across the English Channel from Brittany in March 1981 and the possible spread of bluetongue to Portugal in 1956 when, in both instances sources of infection, suitable meteorological conditions and susceptible animals downwind coincided.   At the fringes of disease distribution such coincidences may only happen every 5 to 20 years, but in endemic centres, they may occur annually or more frequently.

R.F. Sellers, Agriculture Canada, Animal Diseases Research Institute, P.O. Box 11300, Station "H", Nepean, Ontario, K2H 8P9, Canada.

General references

Donaldson, A.I. (1978) The factors influencing the dispersal, survival and deposition of airborne pathogens of farm animals. Vet. Bull. 48, 83-94

Donaldson, A.I. (1983) Quantitative data on airborne foot-and-mouth disease virus; its production, carriage and deposition. Phil. Trans. R. Soc. Lond. B. 302, 529-534

Gloster, J. (1983) Forecasting the airborne spread of foot-and-mouth disease and Newcastle disease. Phil. Trans. R. Soc. Lond. B. 302, 535-541

Pedgley, D.E. (1983) Windborne spread of insect-transmitted disease of animals and man. Phil. Trans. R. Soc. Lond. B. 302, 463-470

Sellers, R.F. (1983) Seasonal variations in spread of arthopod-borne disease agents of man and animals; implications for control. Phil. Trans. R. Soc. Lond. B. 302, 485-495

EXS 51:
Advances in Aerobiology
©1987 Birkhäuser Verlag Basel

IMPROVEMENT OF MATHEMATICAL MODELS FOR PREDICTING THE AIRBORNE SPREAD OF FOOT-AND-MOUTH DISEASE

A.I. Donaldson, M. Lee* and C.F. Gibson,

Animal Virus Research Institute, Ash Road, Pirbright,
Woking, Surrey, GU24 ONF, England

* Meteorological Office, London Road, Bracknell,
Berkshire, RG12 2SZ, England

Introduction

Throughout the last eighty or so years numerous reports have been compiled, mainly following epidemics of the disease in north-western Europe, indicating that in certain circumstances the spread of foot-and-mouth disease (FMD) has occurred over long distances by the agency of the wind (1). Spread over a maximum distance of 60 Km over land (2) and over 250 Km across the sea has been claimed (3).

Initial computer prediction model

A computer-based mathematical model has been developed jointly by the Animal Virus Research Institute and the Meteorological Office for analysing the risk of airborne FMD spread over land during outbreaks of the disease (4).

The model was developed by amalgamating data relating to the physical parameters influencing the dispersion of particles

in the atmosphere with data of the aerobiological properties
of FMD virus.   The model can be used to analyse airborne
virus dispersion up to a distance of 10 Km from a source.

To operate the model the following information is required:
(i) an estimation of the duration and total quantity each
day of airborne FMD virus dispersed from the infected
premises; (ii) hourly or 3 hourly observations of
windspeed, wind direction, relative humidity, cloud cover and
precipitation (or a qualitative indication of rain such as
that given in the synoptic observation work) in the region of
the outbreak; (iii) latitude and, if possible, (iv)
topographical features of the area.

In the United Kingdom this information is provided by an
Epidemiological Team (3) sent to the infected premises and,
in addition, by a nearby weather station.   The information
is passed to the Meteorological Office at Bracknell for
processing and computer plots of airborne virus dispersion
can be made available within 1 hour after receipt.

The quantities of airborne infectivity which may be released
at source have been determined experimentally and are
dependent on the species of animal, the strain of virus, the
stage of disease and the number of animals affected (5, 6).
At the time of peak airborne virus excretion a pig, for
example, can liberate 277,000 $TCID_{50}$ (bovine thyroid tissue
culture infectious units) of airborne virus per minute (7).
The peak amount of airborne virus emitted by an infected
steer or sheep is, by contrast, about 170 $TCID_{50}$ per minute
(8).

In the absence of experimental information about the number
of $TCID_{50}$ which equate to the minimum dose to infect an adult
bovine or ovine - the species of animals most likely to be
infected by airborne FMD virus - the dose for both species

had hitherto been arbitrarily defined as 1 $TCID_{50}$ (2).
Consequently, the accuracy of the computer model to predict
the precise extent of airborne spread may have been sub-
optimal.

<u>Improved computer model</u>

In order to rectify this deficiency, experimental studies
have recently been carried out to quantify the minimum doses
of airborne FMD virus which can initiate infection in sheep
and cattle.   Using an O_1 strain of FMD virus it has been
found that a dose of 10 $TCID_{50}$ is sufficient to infect a
sheep (9) and a dose of 25 $TCID_{50}$ can infect a calf
(Donaldson and others, unpublished findings).   The model has
been altered to take account of these findings.

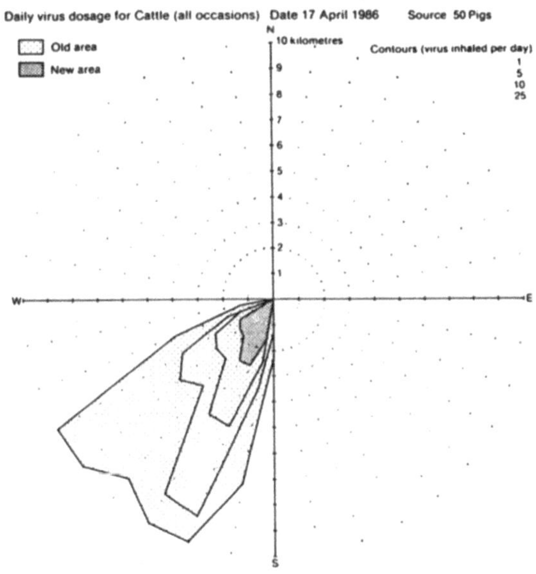

Fig. 1. An hypothetical example of airborne virus
dispersion on 17 April 1986 from a farm containing 50 pigs
affected by foot-and-mouth disease.

Figure 1 shows an example of the areas of likely infection of cattle, under the old (1 $TCID_{50}$) and under the new (25 $TCID_{50}$) inhaled dose criteria. The reduction in the area now considered to be at risk is considerable. A similar presentation can be produced for sheep and in a more speculative way for pigs, for which research is still required to establish a minimum infective dose.

All quantities quoted by the model are given in terms of the dose per day. Since laboratory experiments on infective doses were conducted over a period of minutes, there must be some uncertainty as to compatability of the two sets of figures. It is not known what influence, if any, an animal's normal respiratory clearance mechanisms will have on the same dose of virus accumulated over a prolonged period of exposure and so the level of uncertainty introduced by extending the dose period from minutes to hours cannot be quantified. It is probable that a higher dose will be required to infect an animal over a period of a day than over a few minutes, but since the model has worked well in the past (4) the difference should not be large, perhaps a factor of two.

A relationship between sensitivity to infection and the period over which the dose is administered, if indeed a relationship exists, could be added to the model with ease, and further enhance its usefulness and reliability in the event of a foot-and-mouth disease outbreak.

<u>References</u>

1. Donaldson, A.I. (1979) Airborne foot-and-mouth disease. Vet. Bull. <u>49</u>, 653-659

2. Gloster, J., Sellers. R.F. and Donaldson, A.I. (1982) Long distance transport of foot-and-mouth disease virus over the sea. Vet. Rec. <u>110</u>, 47-52

3. Donaldson, A.I., Gloster, J., Harvey, L.D.J. and
    Deans, D.H. (1982) Use of prediction models to forecast
    and analyse airborne spread during the foot-and-mouth
    disease outbreaks in Brittany, Jersey and the Isle of
    Wight in 1981. Vet. Rec. 110, 53-57

4. Gloster, J., Blackall, R.M., Sellers, R.F. and
    Donaldson, A.I. (1981) Forecasting the airborne spread
    of foot-and-mouth disease. Vet. Rec. 108, 370-374

5. Sellers, R.F. and Parker, J. (1969) Airborne excretion
    of foot-and-mouth disease virus. J. Hyg., Camb. 67,
    671-677

6. Donaldson, A.I., Herniman, K.A.J., Parker, J. and
    Sellers, R.F. (1970) Further investigations on the air-
    borne excretion of foot-and-mouth disease virus. J.
    Hyg., Camb. 68, 557-564

7. Donaldson, A.I., Ferris, N.P. and Gloster, J. (1982) Air
    sampling of pigs infected with foot-and-mouth disease
    virus: comparison of Litton and cyclone samplers. Res.
    vet. Sci. 33, 384-385

8. Sellers, R.F. (1971) Quantitative aspects of the spread
    of foot-and-mouth disease. Vet. Bull. 41, 431-439

9. Gibson, C.F. and Donaldson, A.I. (1986) Exposure of
    sheep to natural aerosols of foot-and-mouth disease
    virus. Res. vet. Sci. In press.

EXS 51:
Advances in Aerobiology
© 1987 Birkhäuser Verlag Basel

FINE PARTICLE AEROSOLS IN EXPERIMENTAL LEGIONNAIRES' DISEASE: THEIR ROLE IN INFECTION AND TREATMENT

R.B. Fitzgeorge, A. Baskerville and A.S.R. Featherstone

Experimental Pathology Laboratory, Public Health Laboratory Service Centre for Applied Microbiology and Research, Porton Down, Wiltshire, England SP4 OJG

Introduction

In Legionnaires' disease (LD) lung infection is initiated in the respiratory bronchioles and alveoli (WINN et al. 1978), intracellular growth of L. pneumophila (the predominant Legionella species) being supported by alveolar macrophages (JEPRAS, FITZGEORGE and BASKERVILLE, 1985). To reach this site infective particles must be of a respirable size, $\leq 5\mu m$ in diameter.

Alveolar inflammatory cells are situated on the airways side of the lung (de NUCCI and MONCADA, 1985) and in LD there is relatively little plugging of the airways by exudate (WINN et al. 1978). Treatment of LD by inhalation of fine particle aerosols is therefore likely to provide more direct and immediate access of antibiotic to L. pneumophila and alveolar macrophages containing them than is possible by diffusion from the circulation after parenteral administration. The evaluation of this hypothesis is the subject of this paper.

Procedures and Results

Guinea-pigs were infected with 10 LD_{50} L. pneumophila (Corby strain serogroup 1) given via a Collison spray as a fine particle aerosol (≤ 5 µm diameter). After infection guinea-pigs were exposed to a similar aerosol of antibiotics at a 1% concentration.

As a preliminary, aerosols of 3 antibiotics were administered for 7 hours to groups of 10 guinea-pigs infected 24 hours previously. Rifampicin and ciprofloxacin prevented all deaths and the former also prevented pyrexia. Erythromycin was less effective under these conditions giving only 20% protection.

The most effective antibiotic, rifampicin, was investigated further in respect of inhalation time required to prevent death and its effect when administered late in LD.

Table I shows that rifampicin inhalation times may be reduced to as little as 3 hours and still be effective. Additionally, inhalation for as little as 1 hour gave 50% survival and extended average survival time (AST) by approximately 2 days.

Table I.   Effect of rifampicin aerosol for periods up to 7 hours on the survival of guinea-pigs infected with L. pneumophila 24 hours previously

Duration of treatment (hours)	% survival (10 guinea-pigs/group)	AST[*]
1	50	5.1
2	80	7
3 to 7	100	>21
No treatment	0	3.3

[*]AST: average survival time, calculated after exclusion of survivors to 14 days or more, by dividing the total number of days survived before dying by the total number of animals which died.

The effect of antibiotic inhalation therapy administered late in LD was investigated by the administration of rifampicin aerosols for 2, 4 and 7 hours to groups of guinea-pigs in which 20% mortalities had already occurred. Table II shows late inhalation therapy to be of value.   Infected animals not

Table II. Effect of rifampicin inhalation antibiotic therapy on guinea-pigs in the later stages of LD (exhibiting 20% mortality at time of treatment)

Duration of treatment (hours)	% survival (10 guinea-pigs/group)	AST[*] (days)
2	10	3.8
4	30	4.1
7	40	5.2
No treatment	0	3.5

[*]AST: average survival time.

given antibiotic therapy all died with an AST of 3.5 days. Similar guinea-pigs given 2, 4 and 7 hours rifampicin inhalation therapy showed 10%, 30% and 40% survival rates with AST of 3.8, 4.1 and 5.2 days respectively.

Discussion and Conclusions

The 1976 outbreak of LD in Philadelphia, which gave the disease its name, had a mortality rate of 19%. A recent outbreak in 1985 of a similar size occurred in Stafford, England, and had a mortality rate of 28%. This apparent lack of progress in the antibiotic treatment of LD is in part due to the time required for diagnosis of the disease, thus delaying suitable therapy. Diagnosis is most frequently made by serological means and may incur a delay of 7 to 10 days or longer whilst the antibody response develops. It is therefore important upon diagnosis that antibiotic therapy produces a fast and effective response.

This study has shown that antibiotic therapy for a period of 3 hours is sufficient to prevent mortality in guinea-pigs lethally infected with L. pneumophila. Even in guinea-pigs where 20% mortality has already occurred, it is possible to reduce mortality in comparison with control animals.

Antibiotics administered parenterally enter the bloodstream and are diluted and dispersed to sites away from the focus of infection. Dilution is minimised if antibiotics are inhaled giving high concentrations in the lung, aiding diffusion into alveolar macrophages.

In summary, antibiotic inhalation therapy of LD offers the possibility of treatment specifically directed at the site of infection and may be effective even if administered late in the disease.

References

De Nucci, G., Moncada, S. (1985). Inhaled corticosteroids for respiratory distress? Lancet ii, 1061.

Jepras, R.I., Fitzgeorge, R.B. and Baskerville, A. (1985). A comparison of virulence of two strains of Legionella pneumophila based on experimental aerosol infection of guinea-pigs. J. Hyg., Camb. 95, 29-38.

Winn, W.C., Glavin, F.L., Perl, D.P., Keller, J.L., Audes, T.L., Brown, T.M., Coffin, C.M., Sensecqua, J.E., Roman, L.N. and Craighead, J.E. (1978). The pathology of Legionnaires' disease, fourteen fatal cases from the 1977 outbreak in Vermont. Arch. Path. Lab. Med. <u>102</u>, 344-350.

R.B. Fitzgeorge, Experimental Pathology Laboratory, Public Health Laboratory Service, Centre for Applied Microbiology and Research, Porton Down, Salisbury, Wiltshire, SP4 OJG, England.

EXS 51:
Advances in Aerobiology
© 1987 Birkhäuser Verlag Basel

THE SIZE DISTRIBUTION OF WHIRLPOOL-GENERATED DROPLETS,
THEIR ABILITY TO CONTAIN BACTERIA AND THEIR DEPOSITION
POTENTIAL IN THE HUMAN RESPIRATORY TRACT

Klaus Willeke[1] and Paul A. Baron[2]

[1]University of Cincinnati, Cincinnati, OH 45267, USA
[2]National Institute for Occupational Safety and Health
Cincinnati, OH 45226, USA

Bacterial respiratory infections have been associated with a variety of water sources. The transmission of the bacteria from the water to
the air and then to the human respiratory system requires the generation of
water droplets small enough to reach and be deposited in the respiratory
tract, yet large enough to contain the bacteria and keep them in a viable
condition. A number of outbreaks of disease have been studied in which a
transmission mechanism of this type has been hypothesized. In recent years,
the most noted disease of this type has been Legionnaires' disease and its
variant, Pontiac fever. It appears that little attempt has been made so far
to measure the sizes of liquid aerosols that have disease potential. Measurements of the sizes and concentrations of liquid droplets equal in size or
larger than the suspected bacteria would confirm that a source-receptor
relationship exists for the disease-carrying bacteria in these outbreaks.

The generation of droplets from bubbles bursting at the water surface is a very common occurrence. Bubbles are present in liquids as trapped
air due to wave motion in natural bodies of water, or they are injected into
the liquid, e.g., by aeration in health club whirlpools and in sewage treatment plants. The bubbles move upward and burst when breaking through the
surface layer of the liquid (see Fig. 1). When the surface film of the air
bubble bursts, "film droplets" are created. In addition, when the top of the
bubble bursts, part of the film is pulled back into the surface of the water
depression by water tension and a pencil-like water jet is formed at the
center of the depression. It shoots up at high velocity and breaks up into
several droplets. These "jet droplets" tend to be larger than the film droplets. As the bubble rises through the water, it collects particles such as

bacteria to its surface by interception. This can result in an enrichment of bacterial concentration in the bubble surface relative to the bulk concentration in the liquid. Both the film droplets and the jet droplets that are formed from the bubble surface can have an enrichment of bacterial concentration up to several hundred. The bacterial enrichment depends on the distance the bubble rises through the water and on water contaminants such as surfactants and oils.

The size distribution of airborne liquid droplets is critical for delivering viable bacteria to the respiratory system. Legionella bacteria have been measured by electron microscopy to be 0.3 to 0.9 μm in diameter and more than 2 μm long. We assume, for the present discussion, that the minimum droplet diameter to contain such bacteria is 2 μm. The investigation of a Pontiac fever outbreak at a whirlpool in a health spa in Rochester, Michigan, provided an opportunity to use a newly developed instrument, the Aerodynamic Particle Sizer (APS), to make in situ measurements of the water droplet size distribution. Measurements were made at this whirlpool and at another one of a different type in an attempt to elucidate the conditions under which droplets are emitted and are likely to transmit bacteria. During the measurement of the water droplet size distribution, the APS was suspended over the pool by a scaffolding constructed of light structural steel beams. The APS was inverted with the inlet nozzle pointing down so that the instrument body did not interfere with the upward motion of the aerosols emitted from the water surface.

Typical size distribution (1) of the number of water droplets above the whirlpool surface are shown in Fig. 2. This whirlpool (Pool A) was located in an open atrium in the center of a health club in Cincinnati. During use, an air injection system caused increased jet action in the pool. The jets were aimed up toward the surface and this increased the turbulence and bubbling at the surface. The measurements were taken in the region above the whirlpool where people might inhale the droplets.

The process of droplet generation, growth and evaporation is dependent on the water surface conditions and the relative humidity in the air surrounding the water surface. One variable that was available for our con-

trol was the whirlpool temperature. By raising the water temperature, both
the rate of water evaporation and the difference in water and air temperatures were increased. These increases contributed to higher saturation of
water vapor in the air above the surface. Fig. 2 shows (a) the droplet size
distributions for the initial condition in Pool A with no massage jets turned
on; (b) the distribution in the same pool at the same temperature, with the
jets and aeration on; and (c) the distribution in the same pool, with jets
and aeration on, at a higher temperature. Since the jets in Pool A angled
up, the surface of the whirlpool was quite turbulent. With no surface turbulence, it can be seen that the droplet concentration above 1 μm is relatively
low. The particle concentration indicated between 0.8 and 1 μm is representative of number concentrations found in ambient air environments. The high
number concentration in this size range corresponds to very little mass.
Fig. 2 shows that when water jets and aeration are turned on, water droplets
are generated. For instance, at 1 μm the increase is 5-fold, at 2 μm 36-
fold, and at 5 μm over 100-fold. At the higher temperature of 40.6° C, there
is a dramatic jump in the concentration of droplets in the 1 to 7 μm range.

The low concentration of droplets larger than 7 μm is largely due
to their increased settling rate. It should be noted that the normal operating range of the whirlpool was indicated to be about 39° to 43°. Thus, the
latter measurement condition is within the normal operating range of the pool
and indicates a typical aerosol distribution. Assuming that Legionella bacteria are present in the whirlpool and that the droplets need to be 2 μm in
diameter or larger to potentially contain these bacteria, this typical aerosol size distribution shows that bacteria may be present in the air above the
whirlpool. In order to determine the amount of deposition in the human respiratory system of droplets potentially containing bacteria, the droplet
number distribution, as seen in Fig. 2, or the corresponding droplet mass
distribution can be multiplied by the size-dependent deposition curves. For
instance, for the top curve in Fig. 2., i.e., normal operation of the whirlpool, the greatest mass of droplets is available to the tracheobronchial
region, with significant amounts also available to the alveolar and extra-
thoracic surfaces (1).

Reference

1. Baron, P.A. and Willeke, K. (1986) Respirable droplets from whirlpools: Measurements of size distribution and estimation of disease potential. Environmental Research <u>39</u>, 8-18.

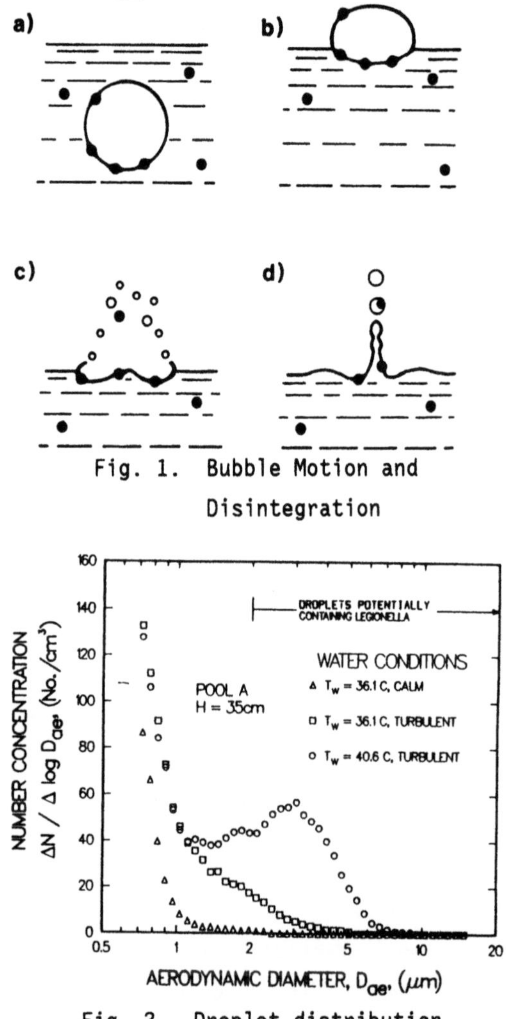

Fig. 1. Bubble Motion and Disintegration

Fig. 2. Droplet distribution above a whirlpool.

Prof. Klaus Willeke, Department of Environmental Health, University of Cincinnati, Cincinnati, Ohio 45267, USA.

EXS 51:
Advances in Aerobiology
© 1987 Birkhäuser Verlag Basel

BACTERIAL ATMOSPHERIC CONTAMINATION IN WASTEWATER TREATMENT PLANTS

P. Boutin[*], M. Torre[*], J. Moline[**], E. Boissinot[**]
* CEMAGREF, B.P.3, F33610 CESTAS, ** C.H.R. TOURS, F37044 TOURS CEDEX, FRANCE

In aerocontamination studies, wastewater treatment plants were usually consi-dered as single emitting systems, and aerosol sources were not distinguished. A more accurate approach is necessary to identify the main emission points and to specify the sanitary risks according to work stations. Consequently, tests were carried out on 2 plants, with emphasis on aerosol granulometry and pulmonary deposition of viable particles.

1. Methodology

1.1 Plants and investigation program

Two activated-sludge plants with deep air injection were studied, Tours (La Riche: 21,000 kgBOD/d, 47,000m^3/d) and Bordeaux (L. Fargues: 18,000 kgBOD/d, 90,000 m^3/d). At each plant, uncontaminated comparison sites (Tt1 and Tt2, Bt) were put upwind. In Tours (06/29/1985), 8 sampling sites were chosen:
    -- T1, pretreatment and nightsoil delivery platform,
    -- T2, T3, pretreatment and primary settlers, downwind,
    -- T4, tow screening,   -- T5,T6,T7, aeration tanks, downwind,
    -- T8, Archimedean screw (treated effluent pumping).
In Bordeaux (10/16/1985), 5 sites only were considered:
    -- B1, raw water pumping screw,
    -- B2,B3, aeration basins, with antifoam spraying (treated wastewater),
    -- B4, same site as B3, without spraying,   -- B5, nightsoil delivery.
3 Andersen impactors (ANDERSEN, 1958) were used for the numbering of viable particles (VP) and the aerosol granulometry (1.8m above the ground; air flow: 28.4 1/mn, sampling time: 20-30 mn). They were completed with Petri dishes (TSA medium - incubation: 72h at 30°C). In Bordeaux 3 selective media were used (TTC agar, 44°C, for thermotolerant coliforms, Hektoen for enterobacteria, Slanetz & Bartley for "faecal" streptococci). Exposed dishes were stored in an icebox and then incubated. In Tours, the airborne bacterial content was determined by means of 4 continuously rinsed out aerocyclones (ERRINGTON, 1969). The analysis of the rinsing liquid stored in the icebox was initiated the next day.
Stage by stage impact numbers ("positive holes") on Andersen impactor dishes were statistically corrected (ANDERSEN, 1958) before calculating the particle amount per m^3 of air. Bacteria numbers from aerocyclones were converted into CFU/m^3 of air. Temperature, wind direction and speed were continuously monito-red, whereas relative humidity was measured every 2 hours.

1.2 Andersen impactor use

Diameter ranges correspond to "equivalent aerodynamical diameters" (EAD), *i.e.* the size of spherical particles of density 1, which behave like the aerosol particles. The impactor divides VP in 6 EAD classes (stage 1: > 9.2 um, stage 2: 5.6-9.2 um, stage 3: 3.2-5.6 um, stage 4: 2.0-3.2 um, stage 5: 1.0-2.0 um, stage 6: <1.0 um), whose limits are in a quasi-geometric progression. After having checked the log-normality of the distributions, the mean VP diameter (Dm) and the standard deviation (s) were computed. Dispersion was appreciated through the ratio: u = s/Dm. The pair (Dm, u) defines the aerosol granulometry.

1.3 Respiratory model

The marginal product of the two distributions: aerosol granulometry and a quantified respiratory model, makes it possible to compute the bronchopulmonary deposition sites of VP. The model (Table I) was interpolated with one of CHRETIEN's tables (1976) to fit impactor EAD classes. Considering aerodynamical modifications with inhaled flow, 2 standard volumes were selected (450 and 900 ml), for 2 activity levels (.45 and 1 m^3/h inhaled respectively).

Table I: Respiratory model

Impactor stage n°	Aer. eq. diameter	Vt = 450 ml				Vt = 900 ml			
		RP	TB	DL	EX	RP	TB	DL	EX
1	> 9.2	15	63	16	6	13	67	14	6
2	5.6-9.2	2	48	37	13	3	54	37	6
3	3.2-5.6	0	30	39	31	2	35	45	20
4	2.0-3.2	0	16	32	52	1	17	54	28
5	1.0-2.0	0	8	25	67	0	9	49	42
6	< 1.0	0	6	12	82	0	6	29	65

RP: rhinopharynx - TB: trachea and bronchi - DL: deep lung - EX: exhalation.

2. Experimental results

Meteorological conditions were the following:
    -- Tours - T: 22.5-25°C; air humidity: 55-73%; wind (NE): 2.2-3.9 m/s,
    -- Bordeaux - T: 20°C; air humidity: 65%; wind (NNE): 3.5-4.5 m/s.
Experimental results are displayed in tables II, III and IV.

3. Interpretation and discussion of results

3.1 Problems raised by the interpretation of numbering

With the Andersen device, stage by stage colony numbering makes it possible to determine VP granulometry. The process overlooks unviable particles; large VP are poorly captured (LUNDHOLM, 1983): only the .5-20 um range is representatively sampled. High speed screening probably breaks up some aggregates. These phenomena tend to distort the actual granulometry. Observed EAD differs from the initial diameter due to the evaporation of droplets.

Table II: Tours - Viable particle aerosol (Andersen impactor, VP/m^3)

Stage	Tt1	T1	T2	T3	T4	T5	T6	T7	T8	Tt2
1	25	364	124	65	760	101	37	74	178	18
2	18	583	81	65	1131	67	41	76	311	21
3	14	1016	237	49	4643	85	25	19	827	4
4	9	369	87	37	4643	4	76	23	422	5
5	7	79	26	7	1830	37	23	9	4643	4
6	5	23	12	16	592	16	9	7	479	5
Total	**78**	**2434**	**567**	**239**	**13599**	**355**	**211**	**208**	**6860**	**57**

Table III: Bordeaux - Viable particle aerosol (Andersen impactor, VP/m^3)

Medium	Stage	Bt	B1	B2	B3	B4	B5
T.S.A.	1	36	333	307	1228	84	250
	2	52	388	466	1645	100	379
	3	29	350	1810	786	59	1183
	4	22	122	1093	495	47	814
	5	9	38	410	474	7	253
	6	4	57	59	43	16	81
	Total	**162**	**1288**	**4145**	**4671**	**313**	**2960**
T.T.C. (44°C)	1	0	3	2	29	5	0
	2	0	5	17	69	0	2
	3	0	3	33	72	0	10
	4	0	5	41	95	0	5
	5	0	3	22	7	0	0
	6	0	7	3	2	0	2
	Total	**0**	**26**	**118**	**274**	**5**	**19**
HEKTOEN	1	0	21	5	7	0	0
	2	0	12	7	17	0	2
	3	0	5	14	28	5	17
	4	0	3	9	33	0	7
	5	0	0	7	12	2	2
	6	0	2	0	0	0	3
	Total	**0**	**43**	**42**	**97**	**7**	**34**
SLANETZ & BARTLEY	1	0	2	0	3	2	0
	2	0	12	5	10	0	9
	3	0	5	9	36	0	2
	4	0	3	7	21	0	2
	5	0	0	5	10	0	2
	6	0	0	0	2	0	5
	Total	**0**	**22**	**26**	**82**	**2**	**20**

368

Table IV: Tours - Air bacteriology (Aerocyclones, CFU/m^3)

Site	"Total" flora /TSA	Enterobact. /ENDO	/EMB	"Faec." coli. /TTC	Presum. E. coli /EMB	Presum. Klebs /ENDO	Pseudo-monas	Staphy-lococci /Chapm.
T1	$7.7.10^8$	$9.8.10^4$	$1.3.10^5$	$1.7.10^4$	$7.7.10^4$	$8.4.10^3$	$1.5.10^5$	$6.8.10^3$
T2	$6.9.10^5$	$1.5.10^4$	$2.9.10^4$	$3.6.10^3$	$8.9.10^3$	$6.9.10^3$	$1.3.10^4$	$1.2.10^3$
T3	$7.1.10^5$	$1.6.10^4$	$2.0.10^4$	$2.8.10^2$	$1.9.10^3$	---	$1.7.10^4$	$1.6.10^3$
T5	$1.2.10^3$	$1.1.10^2$	$2.2.10^2$	$7.2.10^1$	$3.6.10^1$	$1.8.10^1$	$1.4.10^2$	$1.8.10^1$

Yeasts, moulds (/Sabouraud) < 30 CFU/m^3 everywhere.

EAD dispersion results in an even higher dispersion of particle volumes. A few particles bear the main part of bacterial numbers; their diameter is probably larger than the range correctly covered by impactors. Even though aerocyclones make it possible to assess the bacteria number per m^3 of air, our technique is thus inconsistent for evaluation of VP germ density, which would be a preliminary approach to calculate the actual number of inhaled germs.

3.2 Aerosol granulometry - Contamination levels
With sufficient VP densities (N > 75 VP/m^3), EAD distributions adequately fit a log-normal model, apart from cases of overrepresentation of finest particles (stage 6), attributable to a partial splitting of larger elements. Granulometric distributions (/TSA) appear quite homogeneous in the range Dm = 3-6 um and u = 1.0-1.5. Exceptions are T7 (large granulometry) and T8, which depart from the model. Taking granulometry as a basis (pair Dm, u), it seems impossible to individualize aerosol types and to reveal particularities according to sources, or an evolution during VP air carrying. The only discriminating parameter seems to be the VP number/m^3, N, from which 5 classes can be distinguished:
     -- IV - very heavily contaminated (N > 8,000): site T4
     -- III - heavily cntaminated (N > 2500): sites T8, B2, B3, B5
     -- II - fairly contaminated (N > 800): sites T1, B1
     -- I - slightly contaminated (N > 200): sites T2, T3, T5, T6, T7, B4
     -- 0 - uncontaminated (N < 200).
Surprizingly, deep air injection (fine bubbles) in aeration tanks produces relatively few particles, less than 500VP/m^3. Whith antifoam devices (treated water spraying), the density comes up to 5,000 VP/m^3, which is not far from what was observed nearby tanks with surface aeration units. Bubble bursting produces a smaller number of droplets than water dispersion by rotating mechanical devices or droplet ejection through spraying. WANNER (1975) obtained similar results. Figures without spraying are of the same order of magnitude as LUE-HING (1982) results. They are lower than the observations made near tanks equipped with turbines (BOUTIN, 1984). Contamination is restricted to the surroundings of the works (at first rank Archimedean screws and some particular operations). It hardly reaches the management buildings, even when they are downwind to the treatment plant (Tours).

3.3 Sanitary interpretation

The main factors which determine pulmonary deposition sites are aerosol density and granulometry, and the activity level of the subject (Table V).

Table V: Inhaled PV numbers/h according to activity level

Plant	Site	Growth medium	At rest (Vt = 450 ml)					Activity (Vt = 900 ml)				
			RP	TB	PP	**Total**	EX	RP	TB	PP	**Total**	EX
TOURS	Tt(1+2)	TSA	2	12	2	**8**	8	3	31	21	**55**	12
	T1		22	394	372	**788**	307	97	974	974	**2045**	389
	T2		8	92	80	**178**	77	23	232	210	**465**	102
	T4		61	1469	1958	**3488**	2632	272	3536	6120	**9928**	3672
	T5		6	59	42	**107**	53	18	142	131	**291**	64
	T7		5	34	25	**74**	20	12	102	67	**181**	27
	T8		0	432	833	**1265**	3087	0	1029	3156	**5185**	2675
BOR-DEAUX	Bt	TSA	3	27	21	**51**	22	6	66	57	**129**	33
	B1	TSA	29	232	174	**435**	145	64	567	451	**1082**	1288
	B2	TSA	19	504	616	**1139**	726	83	1244	1865	**3192**	953
		TTC	0	13	17	**30**	53	0	33	54	**87**	31
	B3	TSA	105	862	631	**1598**	504	187	2102	1682	**3971**	700
		TTC	4	41	39	**123**	84	8	99	118	**225**	49
		HEKTOEN	0	13	14	**27**	16	3	31	43	**77**	20
		SLANETZ	0	10	12	**22**	15	1	23	38	**61**	21
	B4	TSA	7	56	41	**103**	37	16	134	113	**263**	50
	B5	TSA	13	373	440	**826**	506	30	974	1302	**2279**	681

Major differences result from density uneveness; however point T8, where large particles are uncommon, clearly shows the granulometry effect. Exhalation reaches 60% at rest, 40% when fully active, that partially compensates for locally high densities. The rhinopharyngeal stop of impactor sampled particles is low (some % in average).

The maximum number of <u>retained</u> TSA growing VP (RP+TB+DL) per hour, observed in Tours (T1), reaches 2,000-3,000 at rest, 10,000 for an active subject. In Bordeaux, figures drop to 800-1,600 and 3,000-4,000 VP/h. With regard to "faecal" bacteria only, numbers hardly exceed 250 VP/h.

It is presumed that risks for workers stem from the presence of pathogenic bacteria and viruses in wastewater. Intensive biological treatments, as considered here, divide germs inequally between treated effluent and sludges, which concentrate about 90% of them. Germ mortality is important during the air routing of spread droplets, supposedly reaching 90-95% after desiccation. Faecal bacteria are present in the aerosol (Table IV), but no salmonella nor shigella was identified. It is to be noticed that presumed klebsiellae are frequently identified; some klebsiella strains are admitted pathogens.

Epidemiological surveys on Tours plant workers did not reveal any noteworthy pathology surely attributable to bacteria from effluents and sludges. The observed symptoms were synthetized by scandinavian authors under the name of "sewage worker's syndrome" (eye inflammation, sinusitis, fever and shivering, diarrhoea). Fleeting and harmless, they mainly strike the newly hired employees or workers resuming their job. They are attributed to bacterial allergens, probably endotoxins of gram-negative rods, liberated by cell destruction and then disseminated through aerosol. The hazard is proportional to the inhaled mass (RYLANDER, 1983). This syndrome is a minor risk for people in good health. Caution is recommended for either immuno-depressed subjects or people with a broncho-pulmonary past history.

4. Conclusion

High airborne contamination is located nearby specific works: Archimedean screws, pretreatments, tow screening (Tours), nightsoil delivery (Bordeaux). Farther off, contamination falls to a level slightly higher than upwind. Contrary to expectation, air injection tanks are not major emitters, except when the antifoam spraying device is operating. Epidemiological surveys only state a few mild cases of "sewage worker's syndrome" and remain unsufficient. As assessed here, risk seems limited. Considering penetration of microbial aerosols into the respiratory tract, the potential risk must not be neglected, especially in the event of previous respiratory illness or immunodepression.

References

Andersen, A.A. (1958) New sampler for the collection, sizing and enumeration of viable airborne particles, J. Bact., 76 , 471-484

Boutin, P. et al. (1984) Contamination atmosphérique et respiratoire par les aérosols d'eaux résiduaires, Rev. Mal. Resp., 1 , 125-131

Errington, F.P., Powell, E.O. (1969) A cyclone separator for aerosol sampling in the field, J. Hyg. Camb., 67 , 387-399

Lue-Hing, C. et al. (1982) Environmental impact of the microbial aerosol emissions from wastewater treatment plants, Wat. Sci. Techn., 14 , 289-309

Lundholm, M. (1982) Comparison of methods for quantitative determination of airborne bacteria and evaluation of total viable counts, Appl. Environ. Microbiol., 44 , 179-183

Rylander, R., Snella, M.C. (1983) Endotoxins and the lung-cellular reactions and risk for disease, Prog. Allergy, 33 , 332-344

Wanner, H.U. (1975) Air pollution from treatment plants, Progr. Wat. Techn., 11 217-222.

Pierre Boutin, CEMAGREF, Groupement de Bordeaux, 50 Avenue de Verdun, Gazinet, B.P. 3, F 33610 CESTAS PRINCIPAL (France).

EXS 51:
Advances in Aerobiology
©1987 Birkhäuser Verlag Basel

AIRBORNE GRAM NEGATIVE BACTERIA ASSOCIATED WITH THE HANDLING OF DOMESTIC WASTE

B. Crook, S. Higgins and J. Lacey

Rothamsted Experimental Station, Plant Pathology Department, Harpenden, Herts, England.

Introduction

In the United Kingdom, domestic refuse is collected from houses and taken to transfer stations where it is put in bulk containers, sometimes after storage in bunkers. It is then transported by road, rail or river to landfill disposal sites where the bulk containers are emptied, the refuse is levelled by bulldozers and then buried. Alternatively, the waste may be sorted, separated and recycled or collected in storage bunkers and incinerated to provide heat or electricity. All of these processes generate dust and aerosols which may be laden with micro-organisms from decomposing food and other waste. Employees handling the refuse may thus inhale large numbers of airborne bacteria and fungi. As part of a nationwide survey of airborne micro-organisms associated with domestic refuse disposal, gram negative bacteria were isolated, identified and counted.

Materials and Methods

Air was sampled at various places within 7 refuse transfer stations, 5 landfill sites, 2 incineration plants and one recycling plant in different parts of England. Samples were taken from refuse reception/tipping halls, refuse compactors, conveyor belts carrying whole or pulverised refuse, grab cranes removing refuse from storage bunkers, cabs of bulldozers levelling refuse on landfill sites and outside to give background counts. Samples have been collected regularly since February 1984.

Multistage all glass liquid impingers (Hants R & D Glassware, Southampton) sampling at 55 l/min were operated for 20 min periods. Airborne particles drawn into these samplers were separated by aerodynamic size into three fractions and impinged into collection fluid (May, 1966). Airborne particles were also collected on polycarbonate membrane filters (Nuclepore; Sterilin Ltd., Feltham, Middlesex) in aerosol monitors (Millipore UK Ltd., Harrow, Middlesex), then resuspended in 0.05% Tween 80 solution. The aerosol monitors were connected to portable pumps running at 4 1/min for 1h.

Cell suspensions collected from the multistage impingers and aerosol monitors were serially diluted, spread on violet red bile glucose agar plates (Oxoid) and incubated at 37°C for 48h. Representative colonies were isolated and identified using standard biochemical tests AP120E *Enterobacteriaceae* identification kits (API Ltd., Basingstoke, Hants.). For *Salmonella* isolation, aliquots of cell suspension were incubated in enrichment broth and selective isolation medium (van Schothorst and Renaud, 1983).

Results

Concentrations of airborne gram negative bacteria were usually greatest in enclosed environments such as tipping halls, near to grab cranes and especially next to conveyor belts used to move refuse through the transfer stations (table I). In these environments, gram negative bacteria were isolated from more than 90% of air samples, frequently at concentrations exceeding 1,000 colony-forming units (CFU)/m^3 and with a maximum of 208,000 CFU/m^3 air sampled. The mean concentration of airborne gram negative bacteria near to conveyor belts was 29,000 CFU/m^3, when they formed 25.9% of the total bacteria isolated at 37°C. Numbers were about 1000 times greater than the average number of gram negative bacteria isolated outside the transfer stations.

Pseudomonas spp. were the most common group of gram negative bacteria, comprising 59.3% of all gram negative bacteria isolated. *Klebsiella/Enterobacter* spp. accounted for a further 21.4% of the total, *Serratia* spp. for 10.5% and *Escherichia coli* for 0.3%. Of the species identified (table II), the most commonly isolated were *Pseudomonas fluorescens*, *Ps. putida*, *Enterobacter cloacae*, *Ent. agglomerans* and *Serratia liquefaciens*.

Table I. Total gram negative bacteria isolated at 37°C from different areas of transfer stations and landfill sites

Sample Area	Mean concentration (CFU/m³)	% samples containing GNB	>1000 GNB	% GNB in total bacteria
Tipping Hall	11,900	93.4	56.0	22.2
Refuse compactors	725	88.6	28.6	3.8
Conveyor belts transporting refuse	29,000	94.1	61.8	25.9
Grab cranes/ storage bunkers	20,000	92.0	60.0	21.1
Landfill site bulldozer cabs	2,400	69.2	20.5	1.9
Outdoor upwind of sites	26	62.5	0	0.6

CFU = colony-forming units; GNB = gram negative bacteria

Table II. Frequency of occurrence of identified gram negative bacteria in air samples from refuse handling facilities

Species	Frequency of isolation from samples (%)
Acinetobacter spp.	15.2
Alcaligenes spp.	4.0
Citrobacter freundii	5.1
Enterobacter aerogenes	17.2
Ent. agglomerans	35.4
Ent. cloacae	45.5
Ent. sakazakii	21.2
Escherichia coli	5.1
Klebsiella oxytoca	25.3
Kleb. pneumoniae	16.2
Pseudomonas cepacia	26.3
Ps. fluorescens	56.6
Ps. maltophilia	25.3
Ps. putida	42.4
Ps. stutzeri	29.3
Serratia liquefaciens	46.5
Ser. odorifera	4.0

Frequency of isolation= No. of occasions isolated/No. of samples x 100.

Enrichment and selective isolation techniques, although used successfully to isolate *Salmonella typhimurium* from mixed culture control suspensions, were unsuccessful in recovering *Salmonella* and *Shigella* spp. from any of the 345 air samples collected over the two year survey period.

Discussion

Gram negative bacteria were frequently present in the air when domestic refuse was being handled in transfer stations and landfill disposal sites and were often found in large concentrations. Workers handling domestic refuse and sewage sludge have been reported to exhibit symptoms resembling influenza, with fever, chills and irritation of the upper respiratory airways, that could be caused by inhalation of endotoxin (lipopolysaccharide from gram negative bacterial cell walls), or of abdominal pains and diarrhoea, which could be caused by enteric gram negative bacteria (Lundholm and Rylander, 1980). Gram negative bacteria, such as *Enterobacter cloacae*, *Ent. agglomerans*, *Klebsiella oxytoca* and *Pseudomonas putida*, all of which have previously been found to possess strong endotoxin activity (Salkinoja-Salonen *et al.*, 1982), were frequently isolated from the air in the refuse disposal facilities surveyed, but enteric *E. coli* were few and potentially hazardous bacteria such as *Salmonella* and *Shigella* spp. were not isolated from any samples.

There are, at present, no standards regulating the number of airborne bacteria permissible in general working environments. However, Rylander *et al.* (1983) suggested a safe level of less than 1000 gram negative bacteria per m^3 of air. In our survey, this concentration was exceeded in more than half of the indoor samples. Consequently, precautions should be taken by workers to avoid unnecessary exposure to aerosols generated during the disposal of domestic refuse.

Acknowledgements

This work was supported by Health and Safety Executive Grant No. 1/MS/126/643/82.

References

Blomquist, G., Palmgren, U. and Ström G. (1984) Improved techniques for sampling airborne fungal particles in highly contaminated environments. Scand. J. Work Environ. Health 10, 253-258.

Lundholm, M. and Rylander, R. (1980) Occupational symptoms among compost workers. J. Occup. Med. 22(4), 256-257.

May, K.R. (1966) Multistage liquid impinger. Bact. Rev. 30(3), 559-570.

Rylander, R., Lundholm, M. and Clark, C.S. (1983) Exposure to aerosols of micro-organisms and toxins during handling of sewage sludge. In: Biological Health Risk of Sludge Disposal to Land in Cold Climates. P.M. Wallis and D.L. Lohmann eds (Univ. Calgary Press, Canada), 69-78.

Salkinoja-Salonen, M.S., Helander, I. and Rylander, R. (1982) Toxic bacterial dusts associated with plants. In: Bacteria and Plants. M.E. Rhodes-Brown and F.A.Skinner eds., Soc. Appl. Bact. Technical Series 10, (Academic Press, London), 219-233.

Schothorst, M. van and Renaud, A.M. (1983) Dynamics of salmonella isolation with modified Rappaports medium (R10). J. Appl. Bacteriol. 54, 209-215.

Dr. Brian Crook, Plant Pathology Department, Rothamsted Experimental Station, Harpenden, Herts. AL5 2JQ, England.

VII Methods in aerobiology

EXS 51:
Advances in Aerobiology
©1987 Birkhäuser Verlag Basel

SPECIAL APPLICATIONS OF THE BURKARD-SAMPLER FOR COLLECTING AIRBORNE POLLEN AND SPORES

I. Semi-quantitative continuous determination of dust-particles, soot, flue ash etc. which may activate diseases of the respiratory tract.
(Paper given at the satellite symposium of the 3rd. International Conference on Aerobiology)

G. Boehm, Basel, Switzerland

The air outdoors contains suspended dust, which includes inanimate particles from various sources which are present in low concentrations, flue ash, soot and certain other particles resulting from industrial activity. These suspended particles, like the well-known injurious gases, may aggravate or perhaps simply 'prepare the way for' a number of respiratory ailments. Measurements of the dust in the outside air, in order to assess the level of pollution with the particles mentioned above, are made using a wide variety of methods. These include simple sedimentation on adhesive foils, methods in which air is sucked up and blown onto prepared surfaces for investigation, and finally a variety of physical methods. All these installations, however, require special apparatus, which may be very expensive and need constant maintenance.

The well-known Burkard Apparatus, for the collection of pollen grains and fungal spores from the air, is in continuous use in many places in Europe. In this apparatus air is sucked up, and the particles suspended in it are trapped on a sheet of transparent 'Melinex'[(R)] adhesive foil, which is moved slowly along at a constant speed. Besides pollen-grains and spores, all kinds of dust particles can be collected - including, incidentally, dust from the Sahara (1., 2., 3.).

The apparatus produces '24-hour samples' in the form of Melinex$^{(R)}$ strips 48mm long (2mm = 1 hour). The strips are placed on ordinary microscope slides, covered with 'Gelvatol' (a polyvinyl alcohol), and examined under the microscope for the presence of pollen and fungal spores. On these preparations it is possible even with the naked eye to see 'shadows', or more sharply demarcated 'dark stripes', which are due to the presence of dust particles. It is possible to determine precisely at what time of day these stripes were produced. A semi-quantitative assessment of the amount of material present in the shadows and stripes can be obtained from densitometric measurements. These can be made using equipment for the evaluation of serum electrophoresis on foils or glass plates, which is available in all modern hospitals.

The information obtained from such curves may be particularly valuable when it is possible to establish that besides a small load of dust particles in the air, remaining constant over long periods, there has been a surge of dust, soot, flue-ash or particles from an industrial process at a definite time. It is often possible, when the direction of the wind is known, to give some indication of the source of the particles. Such observations can be even more important if it can be shown that the increase in the load of particles frequently occurs at the same time of day (see Fig. 1). The preparations can be stored in an archive for comparative purposes. It would be possible to use them to find out whether measures taken to improve the quality of the air have been successful - for example - in health resorts.

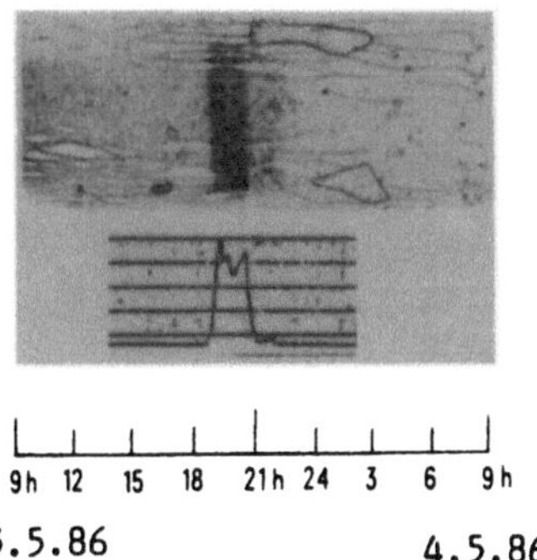

Fig. 1. 24-hour preparation from 3.5.1986 - 4.5.86 with two dark stripes due to dust etc., together with a part of the corresponding densitometric curve. The autoradiographic findings from this preparation - some days after the disaster of Chernobyl - will be described elsewhere.

<u>References:</u>

1. G. BOEHM: 'Eine einfache wenig aufwendige Methode zur halbquantitativen Bestimmung von Staub in der Luft'. Annalen der Meteorologie, N.F. <u>22</u>, 92 - 94, 1985. Internat.Tagung f.Human-Biometeorologie v. 2.- 4. Okt. 1985; Offenbach 1985;Selbstverl.Deutsch. Wetterdienst.

2. G.BOEHM und Ruth LEUSCHNER: 'Registrierung von Saharastaubfällen mit Hilfe der Burkard Pollen- und Sporenfalle'. Experientia <u>30</u>, 1974 pp 574 - 576.

<u>Authors's address:</u>

Prof.Dr.med. G. Boehm, Herrengrabenweg 51, CH-4054 Basel, Switzerland

EXS 51:
Advances in Aerobiology
© 1987 Birkhäuser Verlag Basel

SPECIAL APPLICATIONS OF THE BURKARD-SAMPLER FOR COLLECTING
AIRBORNE POLLEN AND SPORES

II.   Record of radioactive dust in the air

G. Boehm, Basel, Switzerland
(Paper given at the satellite symposium of the 3rd.
International Conference on Aerobiology)

         For special analyses of the attached particles the
Melinex$^{(R)}$ strips, together with their adhesive layer, can be
carefully detached from the microscope slide with almost no
loss of material. That was of particular relevance after the
Chernobyl disaster: it was possible by the use of autoradio-
graphy to determine whether, and at what time, radioactive
particles were present in the air above Basel. For example,
the autoradiogram from the 24-hour sample from 1.5.1986 -
2.5.1986 (Fig. 1) quite clearly shows 'hot spots' with radioactive particles, probably ^{131}I and ^{137}Cs.This was not simply
a case of radioactive fallout resulting from sedimentation,
like that observed at that time in Sweden (1) or, for example,
on the leaves of meadow plants near München (2). It was possible to show with our simple procedure at what times the air over
Basel - the air that we were breathing - was bringing radioactive material with it.

         It would be interesting to compare our results with
those from preparations made over the same time-period in other
places in Europe with 'Burkard' equipment. This could still be
done, because it will be possible to detect the radiation from
^{137}Cs for a long time.

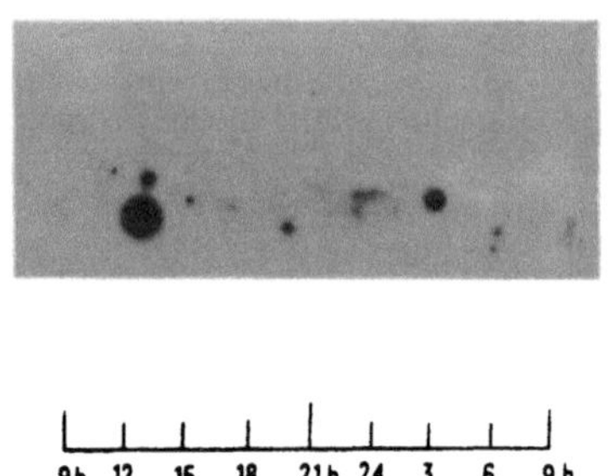

Fig. 1. 24-hour preparation from 1.5.1986 - 2.5.86. The 'hot' radioactive spots can be seen as dark spots and patches. For technical reasons - which will be discussed elsewhere - the Melinex$^{(R)}$ foil (48 x 20 mm) was cut in half lengthways. The outline of the foil cannot be seen in the picture. The time-scale drawn below the photograph makes it possible to assess the times during the 24-hour period when radioactive dust was collected.

References:

1. PSCHYREMBEL Wörterbuch:'Radioaktivität, Strahlenwirkung, Strahlenschutz'
         Bearb.v.d.Pschyrembel-Redaktion; Christoph ZINK (Hrsg.). Berlin;
         New York; Walter de Gruyter, 1986, p 5.
2. L.DEVELL, H.TODEVALL, U.BERGSTRÖM, A.APPELGREN, J.CHYSSLER, L.ANDERSSON:
         'Initial observations of fallout from the reactor accident at
         Chernobyl. Nature 321, 15. May 1986, 192 - 193.

Author's address:

Prof.Dr.med G. Boehm, Herrengrabenweg 51, CH-4054 Basel, Switzerland

EXS 51:
Advances in Aerobiology

EFFICIENCY OF A NEW BIOAEROSOL SAMPLER IN SAMPLING BETULA POLLEN FOR ANTIGEN ANALYSES

A. Rantio-Lehtimäki[1], E. Kauppinen[2], A. Koivikko[3]

[1] University of Turku, Department of Biology, Turku, Finland
[2] Technical Research Centre of Finland, Laboratory of Heating and Ventilating, Espoo, Finland
[3] Turku University Hospital, Department of Pediatrics, Turku, Finland

Abstract

A new bioaerosol sampler consisting of Liu-type atmospheric aerosol sampling inlet, coarse particle inertial impactor, two-stage high-efficiency virtual impactor (aerodynamic particle sizes respectively in diameter: $\geq$ 8 µm, 8-2.5 µm and 2.5 µm; sampling on filters) and a liquid-cooled condenser was designed, fabricated and field-tested in sampling birch (_Betula_) pollen grains and smaller particles containing _Betula_ antigens. Both microscopical (pollen counts) and immunochemical (enzyme-linked immunosorbent assay) analyses of each stage were carried out. The new sampler was significantly more efficient than Burkard trap e.g. in sampling particles of _Betula_ pollen size (ca. 25 µm in diameter). This was prominent during pollen peak periods (e.g. May 19th, 1985, in the virtual impactor 9482 and in the Burkard trap 2540 _Betula_ p.g. x m^{-3} of air). _Betula_ antigens were detected also in filter stages where no intact pollen grains were found; in the condenser unit the antigen concentrations instead were very low.

Introduction

Instead of counting airborne particles (both living and dead, allergenic and non-allergenic) aerobiologists should concentrate more on developing methods for analysing chemically actual allergen concentrations in air. Allergic symptoms could be expected to correlate with actual allergen concentrations, not with particle counts. Also concentrations of allergens in particles depositing in the different parts of the respiratory tract must be known. E.g. HABENICHT & al. (1984) and SOLOMON & al. (1983) have used sampling methods to size-fractionate aerosols into "small" and "large" particles without having well defined collection characteristics in the samplers used.

Pollen allergies occur also in periods when allergenic plants do not produce any pollen grains or harmful moulds are not sporulating. This could be due to allergens associated with respirable atmospheric aerosols.

Many allergenic pollen and spores have been studied for their contents of allergenic antigens and immunochemical methods exist for analysing several different antigen compositions simultaneously.

Material and methods

No bioaerosol sampler suitable for collecting samples for both microscopical and immunochemical analyses was available. Therefore a new bioaerosol sampler (Fig. 1) with the Liu-type atmospheric aerosol sampling inlet (LIU & PUI 1981), the coarse particle inertial impactor (stage 1), the high-efficiency two-stage virtual impactor (stages 2 and 3) and liquid-cooled condenser (stage 4) was designed by one of us (Kauppinen).

The coarse particle impactor (cut-point ca. 50 μm) removes water droplets and insects from the sample. Estimated cut-point of the first virtual impactor is ca. 8 μm. Fine particles are separated by the second virtual impactor with the

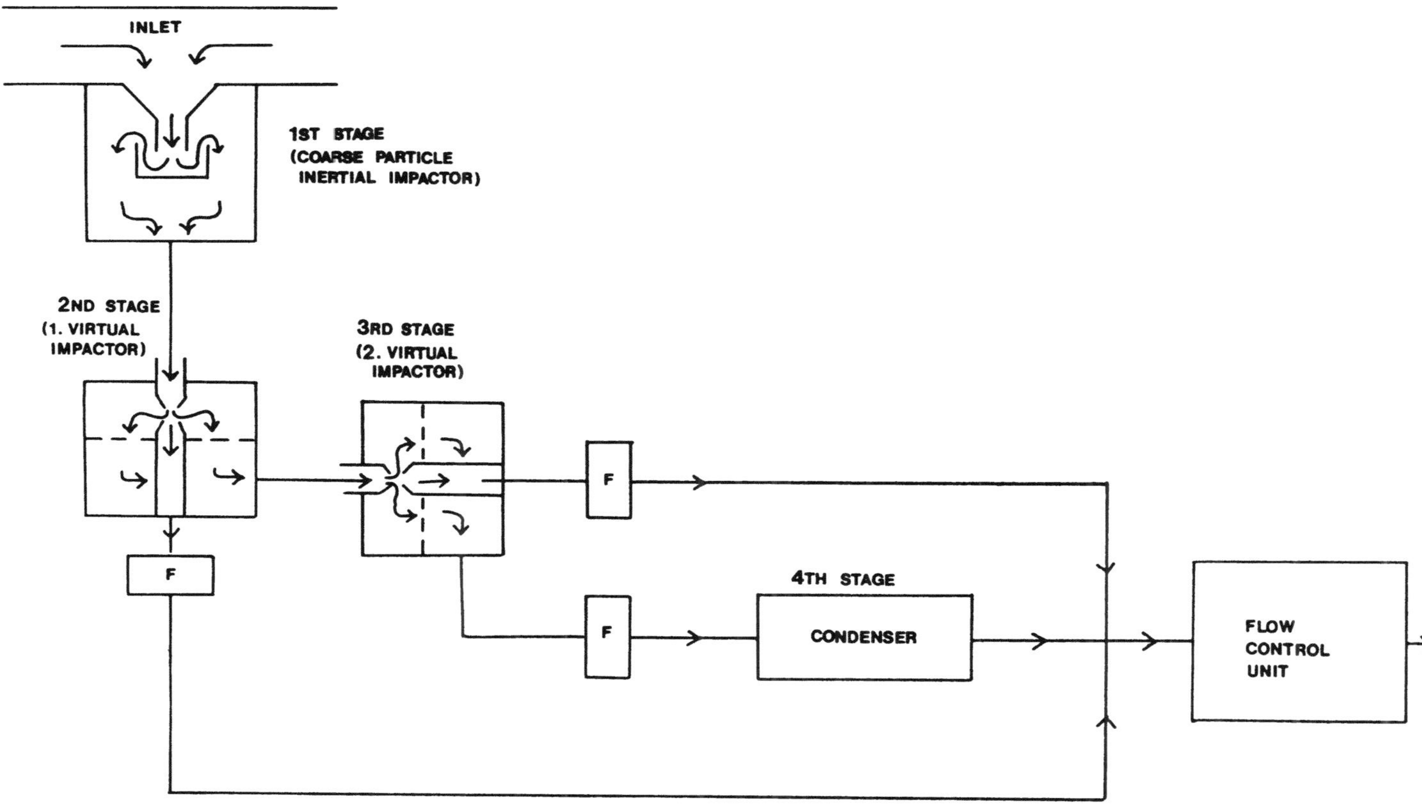

Fig. 1. Sceme of the new bioaerosol sampler.

cut-point about 2.5 µm (LOO & al. 1979).

To detect the peak flowering seasons and spore seasons, also pollen grains and spores should be collected for microscopical analyses.

Virtual impactors were calibrated with monodisperse fluorescein particles in the size range 1-15 µm (HILLAMO & al. 1986). Losses were less than 5 % in the virtual stage 1 and less than 20 % in the virtual stage 2.

The sampler was tested mainly during birch pollen season, because birch antigens are well-known (VIANDER 1979) and most chemicals needed are commercially available. Air samples were taken daily (usually ca. 24 hours´sampling time; volume ca. 26 m^3) during birch flowering, and twice a week after the flowering period, in 1985.

The sampling site was situated on the roof (ca. 15 meters above the ground) of Turku University buildings, close to the place where the Finnish Aerobiology Group has its continuously operating Burkard sampling site for general monitoring of airborne pollen and spores.

Nuclepore Membra-Fil membranes of pore size 0.7 µm were used in the stage 1 (coarse particle inertial impactor) and pore size 0.45 µm in both virtual impactor stages. Particle counts of four squares (ca. 0.4 sq.cm) of each gridded membrane, cleared with immersion oil, were analysed microscopically and the results were compared with Burkard counts.

For immunochemical ELISA analyses filter discs of 5 mm in diameter were cutted. Immunochemical analyses were performed also from the liquid in condenser unit (never exceeding 20 ml per day). The analysing method of IgG with ELISA is shown in Fig. 2.

Results and discussion

The new sampler effectively collects particles of <u>Betula</u> pollen size as shown in Fig. 3, but for very big particles as <u>Pinus</u> pollen (Fig. 4) or small ones as <u>Cladosporium</u> spores (Fig. 5)

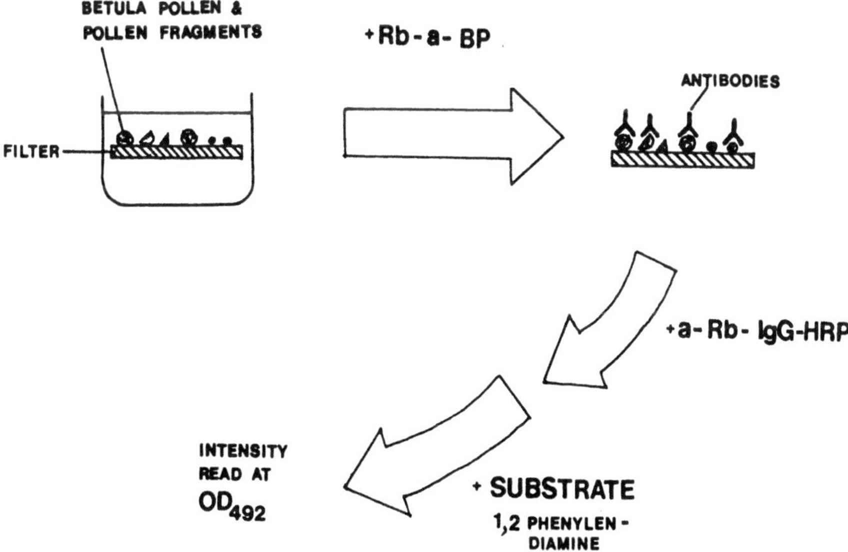

Fig. 2. Scematic diagram of the steps needed to measure birch specific antibody of the IgG class.

higher counts have been obtained from Burkard trap. However, results from the two samplers were very significantly correlated in 1985. Error limits of both samplers have to be investigated in detail.

Betula antigens were present in all four stages. Highest concentrations were in the stages 1 and 2. Ca. 98 % of intact birch pollen grains were catched by the stage 1. In condenser unit only very low concentrations of Betula antigens were detected.

Our intention was to test some other filter types during the flowering season in 1986, but unfortunately Betulas were pollinating very poorly in spring 1986.

Due to numerous error sources, especially in filters used, the method is not standardized so far that we could give also quantitative results of antigen concentrations in each stage.

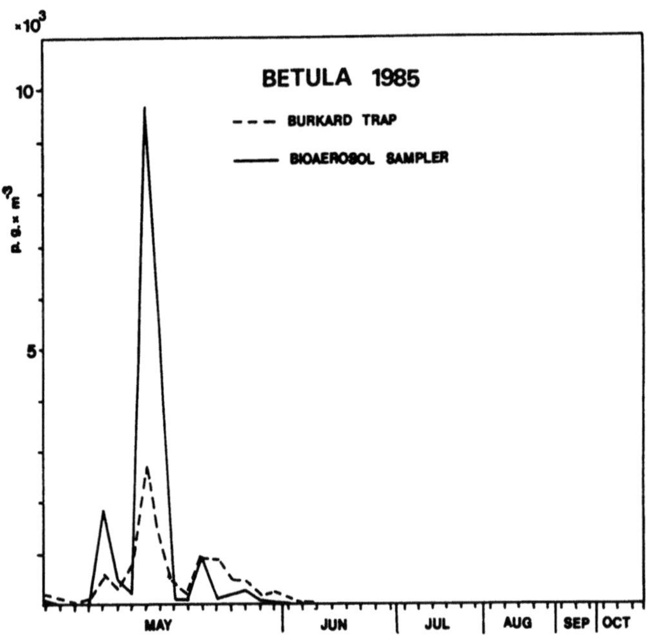

Fig. 3. Frequencies of airborne Betula pollen grains
in Burkard samples and on filters of bioaerosol sampler.

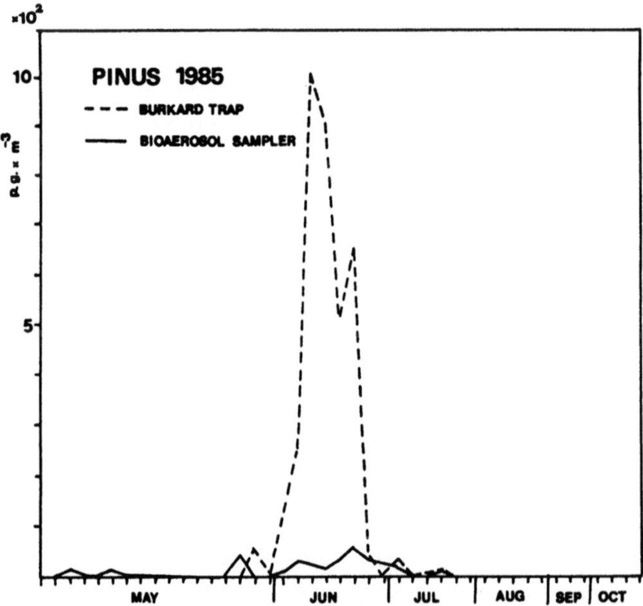

Fig. 4. Frequencies of airborne Pinus pollen grains
in Burkard samples and on filters of bioaerosol sampler.

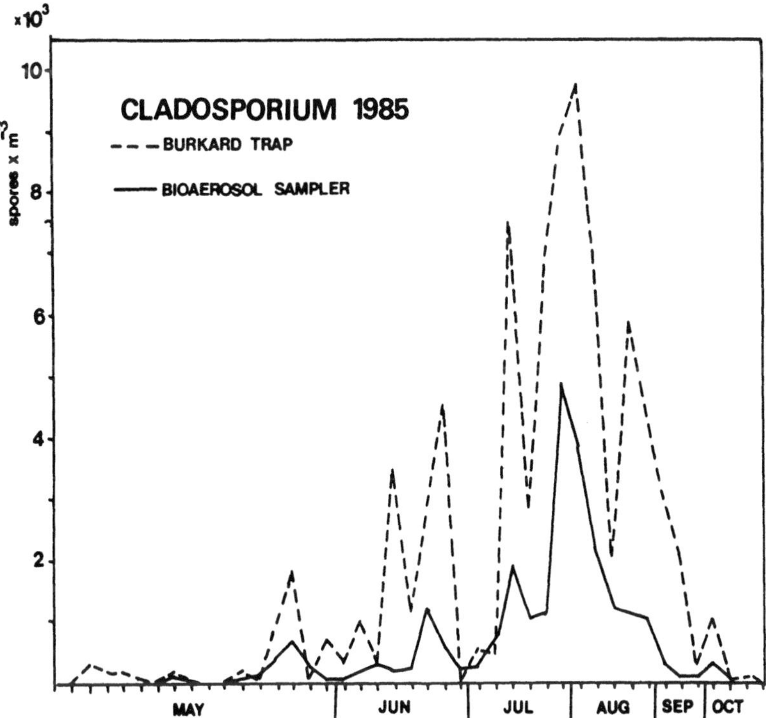

Fig. 5. Frequencies of airborne Cladosporium spores
in Burkard samples and on filters of bioaerosol sampler.

In the future maybe all aerobiological reports are
based on analyses done by using the methods which now seem rather
difficult. Much is to achieve because e.g. early spring allergies
to spores maybe could be explained by using antigens-in-small-
particles theory.

Acknowledgements

The authors would like to thank Dr. Markku Viander for his helpful advice in immunological analyses and Mr. Arto Lankinen for
analysing pollen grains.

References

Habenicht, H.A., Burge, H.A., Muilenberg, M.L. & Solomon, W.R.
(1984) Allergen carriage by atmospheric aerosol. II. Ragweed-
pollen determinants in submicronic atmospheric fractions. J.
Allergy Clin. Immunol. 74, 64-67.

Hillamo, R., Kauppinen, E., Rouhiainen, P., Hakkarainen, T. & Ruuskanen, J. (1986) Calibration of compressible flow low pressure impactor. 2nd Int. Aerosol Conf., Berlin, Sept. 1986.

Liu, B.Y.H. & Pui, D.Y.H. (1981) Aerosol sampling inlets and inhabable particles. Atmos. Env. 15, 589-600.

Loo, B.W., Adachi, R.S., Cork, C.P., Goulding, F.S., Jaklevic, J.M., Landis, D.A. & Searles, W.L. (1979) A 2nd generation dichotomous sampler for large-scale monitoring of airborne particulate matter. Lawrence Berkeley Lab. Rep. LBL-8725.

Solomon, W.R., Burge, H.A. & Muilenberg, M.L. (1983) Allergen carriage by atmospheric aerosol. I. Ragweed pollen determinants in smaller micronic fractions. J. Allergy Clin. Immunol. 72, 443-447.

Viander, M. (1979) Immune response to birch pollen antigens. Academic dissertation. Departments of Medical Microbiology and Pediatrics, University of Turku, Finland. 58 pp.

Dr. Auli Rantio-Lehtimäki, Department of Biology, University of Turku, SF-20500 Turku, Finland.

EXS 51:
Advances in Aerobiology
©1987 Birkhäuser Verlag Basel

COLLECTION EFFICIENCY OF TWO SAMPLERS FOR MICROBIOLOGICAL AEROSOLS

Eva Henningson and Ingrid Fängmark

National Defence Research Institute, Umeå, Sweden

Introduction

Airborne microorganisms can cause respiratory diseases, like infections and allergy, or contaminate sterile products. Evaluation and control of contamination by airborne microorganisms requires samplers that collect the relevant particle sizes and preserve the microorganisms in a viable state. In this study a dynamic test system was used to test collection efficiency, inlet excluded, for a cyclone sampler and a modified personal impinger. Some results on survival of microorganisms during sampling are also presented.

Materials and methods

Samplers tested.

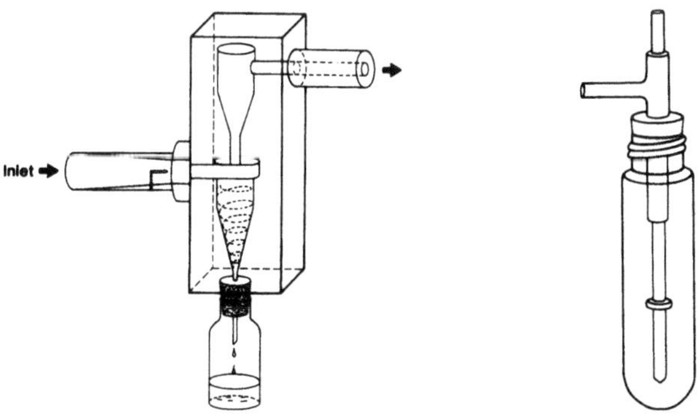

Figure 1. The cyclone sampler (left) and the modified personal impinger (MPI) (right).

The plexiglass cyclone (ERRINGTON et al 1969) is operating at 75 l/min (fig 1). A collection liquid is delivered at approximately 1 ml/min by a syringe pump. The liquid used in the tests was tri(hydroxymetyl)aminomethane buffer with glycerol and polyglycol, (BOVALLIUS et al 1978)

The modified personal impinger (MPI) has a jet diameter of 0.42 mm, sampling rate 1.3 l/min and a liquid capacity of 10 ml (MACHER et al 1984), figure 1. Water was used as collecting fluid in the tests and the jet tip was placed 4-5 mm from the bottom.

<u>Efficiency test method</u>. Monodisperse polystyrene latex (PS) and polyvinyltoluene (PVT) spheres (Dow PS 0.3, 0.5, 0.8, 1.1, 2 and PVT 2.0 µm) in suspensions of 0.025 % Triton X-100 and distilled water were aerosolized from a Collison nebulizer. The aerosol was dried by a diffusion dryer and neutralized by a ^{85}Kr source before sampling. Reference air samples were collected alternately (cyclone) or in parallell (MPI) with the sampler, onto membrane filters (Millipore corp type HA). After sampling the liquids from the cyclone and the MPI were filtered, the filters dryed and sections mounted on a glass slide for particle counting in a light microscope (0.8 - 2 µm) or an electron microscope (smaller particles). Collection efficiency was calculated as the ratio between the number of particles collected in the cyclone and the average number in the air samples collected immediately before and after the sampler. The MPI efficiency was calculated as the ratio between the MPI and the air filter.

<u>Viability test</u>.
Survival of an aerosol of <u>Escherichia coli</u> when sampled by the cyclone was studied. An All Glass Impinger (AGI-30) was used as reference sampler. The viable microorganism in the aerosol sample was cultivated and the total number of bacteria was determined by staining with etidiumbromide and counting in light microscope (fluorescence).

Results

The efficiency, measured and theoretical, for the cyclone is shown in fig 2. The measured cut off (the particle size which is collected to 50 %) is 0.6 µm which is lower than theoretical, 1.0 µm. However the theoretical

calculations assume that the cyclone is operating without liquid. A few
tests without liquid indicate that the efficiency of a dry cyclone is less
than a wet cyclone. The very low efficiency for the 2.0 μm particles might
be explained by particle bounce off or reentrainment.

As often is found for impactors, the efficiency curve for the MPI is not as
steep as the theoretical, calculated from impactor theory (fig 2). The
difference between theoretical cut off, 0.3 μm, and measured, 0.7 μm, could
be an effect of jet Reynolds number or the ratio of separation distance
between jet tip and bottom, to jet diameter.

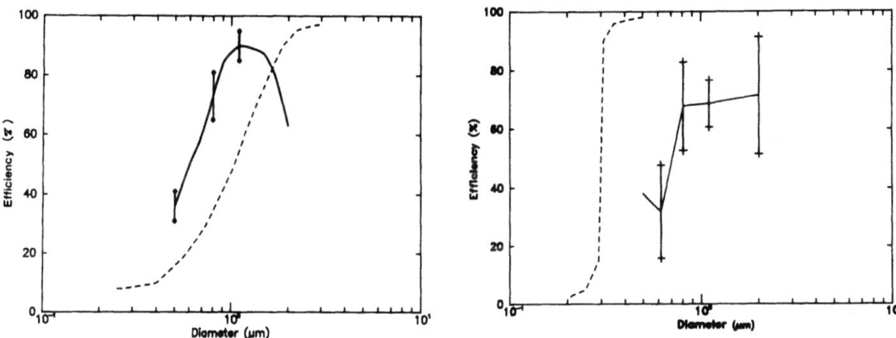

Figure 2. Collection efficiency of the cyclone sampler (left) and
the modified personal impinger (right) as a fuction of particle
aerodynamic diameter. Theoretical efficiency (- - - -) and
measured efficiency (________).

Aerosolized bacteria showed reduced survival compared to bacteria stored in
spray medium (saline). Loss of survival can occur during the aerosol
generation, drying, neutralization and/or sampling moment. A significant
difference in survival of <u>Escherichia coli</u> when sampled with the cyclone
compared to the AGI-30 was found (fig 3). Regression data analysis gave 58 %
survival of the bacteria in the cyclon sampler and 26 % in the AGI-30. There
is also a significant difference in the total number of bacteria collected
in the cyclone compared to the AGI-30 which could be explained by a
difference in collection efficiency.

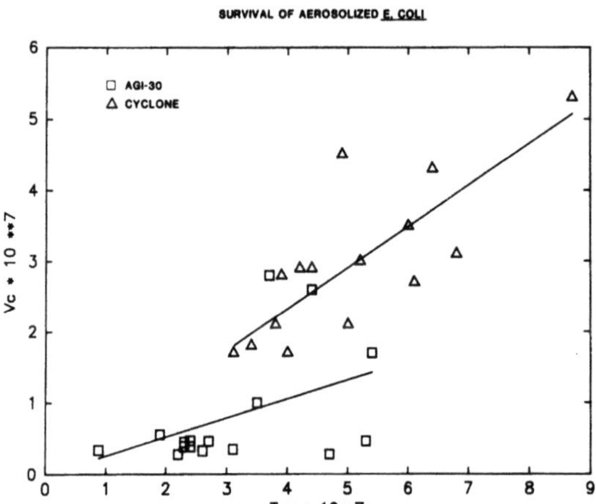

Figure 3. Survival of Escherichia coli when sampled in the cyclone sampler (△) and in the AGI-30 (□). VC = viable count, TCC = total number.

Further studies of the samplers collection efficiency and survival of collected microorganisms and suitable collecting liquids will be made.

References.

Bovallius, Å., B. Bucht, R. Roffey and P. Ånäs. (1978) Three year investigation of the natural airborne bacterial flora at four localities in Sweden. Appl. Env. Microb. 5, 35, 847-852.

Errington, F.P. and Powell, E. O.(1969) A cyclone separator for aerosol sampling in the field. J. Hyg. Camb. 67, 387-3.

Macher, J.M. and M. W. First. (1984) Personal air samplers for measuring occupational exposure to biological hazards. Am. Ind. Hyg. Assoc. J. 45,2, 76-83.

Author's address

Eva Henningson, National Defence Research Institute, Division of cell- and microbiology, 901 82 Umeå, Sweden.

EXS 51:
Advances in Aerobiology

MICROORGANISMS AS BIOLOGICAL INDICATORS OF AIR POLLUTION

S. Waldner-Sander, K. Botzenhart

Hygiene-Institut, Universität Tübingen, Tübingen, FRG

Introduction

Many countries have recently established networks for monitoring air pollution, which generally employ physicochemical methods of analysis. In many places, biological indicator systems are also used to evaluate air quality. Some highly standardized methods employ lichens or tobacco, which react selectively with particulate air pollutants (ARNDT et al., 1985), or cultures of grass which accumulate heavy metals (VDI, German Assoc. of Engineers, 1978). Frequently a combination of different species of green plants and lichens is used to detect the effects of various air pollutants. Such bio-indicators, being living organisms, react to a sum total of toxic effects in the air and thus may yield valuable information more relevant to the health of man than that provided merely by physicochemical measurement of air pollutants. Bacteria, the smallest organisms with a complete metabolism, have long been used to assess the efficacy of antimicrobial agents or equipment and have recently been shown to be very convenient indicators of DNA-damaging substances, e.g., in conjunction with the Ames test (AMES et al., 1975).

In 1968, DRUETT and PACKMAN demonstrated an adverse effect of unknown air constituents on the viability of bacteria. This effect was confirmed by DRUETT and MAY (1968) and was later termed "open air factor" (OAF). DEMIK (1976) showed a correlation between OAF concentration, on the one hand, and the concentration of ozone and motor vehicle pollution, on the other hand; he also documented the damaging effect of ozonized cyclohexane on bacterial DNA. Many other ozonized hydrocarbons have been found to be germicidal and mutagenic in smog chamber experiments; toxic effects on microorganisms under such conditions can be documented by changes in the death rate calculated from numbers of colony-forming units (CFU) detected over a given period of time (NOVER and BOTZENHART, 1983, 1985). The method described in the present paper was developed to quantitate the toxic effects of air pollutants using specially treated microorganisms exposed to open air.

Method and Sampling Sites

Prior to exposure of the test strains to open air at a given site, two sampling chambers were prepared as follows: Membrane filters were loaded with 100-200 cells of a given test strain by adsorption of cells removed from the culture medium during the exponential growth phase. 30 filters of each strain were placed in the incubation chambers. Subsequently, the chambers were purged with synthetic air, adjusted to the desired relative humidity

using saturated salt solution, sealed, and transported to the exposure site.
After a minimum exposure time of 30 min, transfer of the membrane
filters of one chamber to agar plates was initiated and continued at regular
intervals up to a maximum exposure time of 90 min. Immediately thereafter,
filters which served as controls were removed from the second chamber which
remained sealed up to this time. After 1 or 2 days incubation, colonies were
counted and the reduction rate λ was calculated from the difference between
control and exposure diagrams.

The test strains were exposed at 3 field stations in the state of
Baden Württemberg, Federal Republic of Germany (FRG) which are equipped with
extensive instrumentation for monitoring meteorlogical parameters and levels
of: sulfur dioxide, nitrous oxides, ozone, and carbon monoxide. One station
is located in an elevated region of the Black Forest near the
Kälbelescheuer. At this site, immission levels of SO_2 and NO_x are generally
very low, those of ozone high. The other sites are situated in 2 forest
areas near Stuttgart: Welzheimer Wald and Schönbuch. In the Welzheimer Wald,
NO_x and SO_2 are detected sporadically and ozone levels correlate with levels
in other West German regions of similar altitude. In the Schönbuch; the NO_x,
SO_2, CO and O_3 levels are affected by nearby motor vehicle traffic.

Results and Discussion

The feasibility of employment of bacteria, yeasts and molds as
bioindicators of air pollution was tested in preliminary experiments under
field conditions. Spores of fungi and Bacillus subtilis were readily excluded
due to their stability during the given exposure period. Considerably higher
sensitivity to the open air was revealed by the yeast Candida albicans and
the bacteria: Staphylococcus epidermidis, Serratia marcescens and Micrococcus
luteus. Serratia marcescens proved to be unstable, however. A pronounced loss
of viability during transport and under test conditions caused impaired
reproducibility of results and thus the applicability of this bacteria as an
indicator of air pollution was limited.

An example of the results obtained with two of the sensitive microorganisms mentioned above is provided in Fig. 1. The reduction diagrams

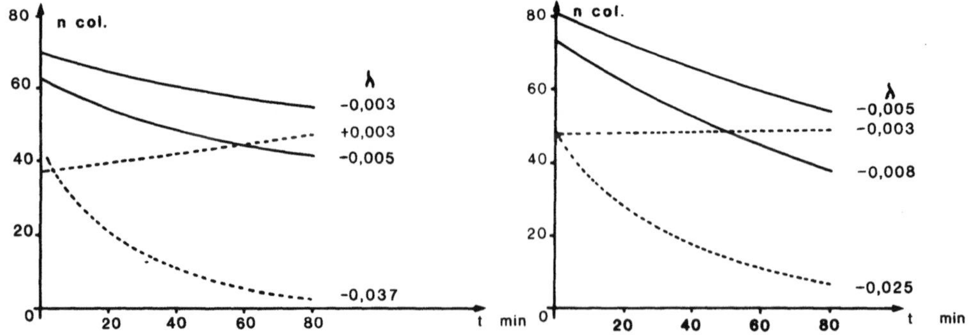

Fig. 1. Response of Candida albicans (left) and Staphylococcus
epidermidis (right) to outdoor air exposure on two different days (20 June
1985 ——, 17 October 1985 ---). The upper curve of each pair of reduction
diagrams represents the control results.

clearly reveal increased death rates of Candida albicans and Staphylococcus epidermidis on a day (17 October 1985) on which nitrous oxides amounted to 110 $\mu g/m^3$ and carbon monoxide 529 $\mu g/m^3$ - compared to merely 11 $\mu g/m^3$ NO_x and 69 $\mu g/m^3$ CO on the other day (20 June 1985). Whereas temperature and relative humidity were virtually the same on both days, the markedly increased immission levels on 17 October 1985 suggest that factors resulting from air pollution may account for the observed loss of viability on this day.

To date, clear correlation has not been observed between a given gasseous pollutant and changes in the death rate of test strains exposed to open air. This problem was examined in the framework of the present study in laboratory tests initially focussed on the toxic effects of ozone on microorganisms. As shown in Fig. 2, viability of Staphylococcus epidermidis was substantially decreased under the influence of 200 $\mu g/m^3$ O_3, viability of Candida albicans was decreased at higher ozone concentrations. Interestingly, Serratia marcescens revealed the greatest viability of all tested strains at extremely high ozone levels unlikely to occur outdoors (no figure).

Whereas the laboratory experiments illustrated in Fig. 2 merely revealed clear correlation between excessive ozone concentration and reduced viability, the generally lower ozone concentrations in outdoor air may indeed be deleterious due to synergistic effects with other noxious agents. This is illustrated in Fig. 3 which reveals the highest reduction rates of Staphylococcus epidermidis in the presence of concomitant increases in CO and O_3 concentration. Under the present conditions, CO should reflect increased motor vehicle pollution, whereas O_3 should reflect increased photooxidant formation; and thus the synergistic effect revealed in Fig. 3 may be due to the presence of other air-borne pollutants, such as ozonized hydrocarbons (see, e.g., DEMIK, 1976; NOVER and BOTZENHART, 1983).

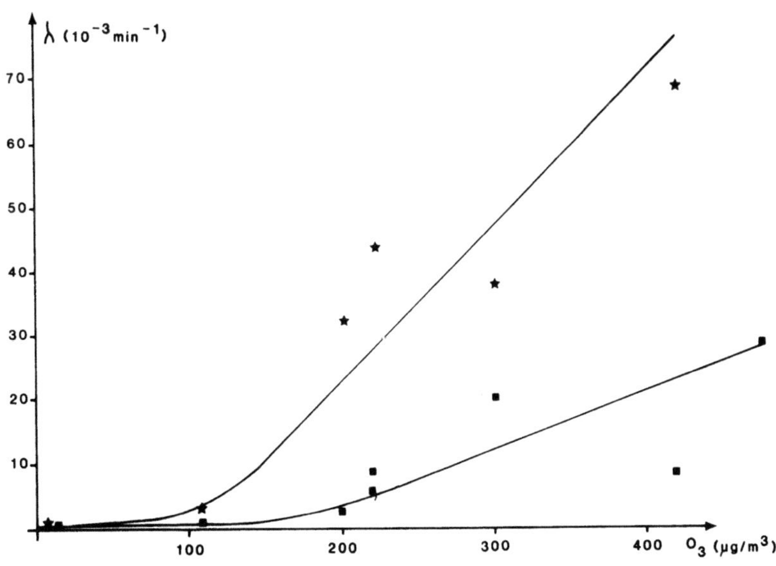

Fig. 2. Effect of ozone on viability of Staphylococcus epidermidis ★ and Candida albicans ■. Reduction rate λ is plotted against the ozone conc.

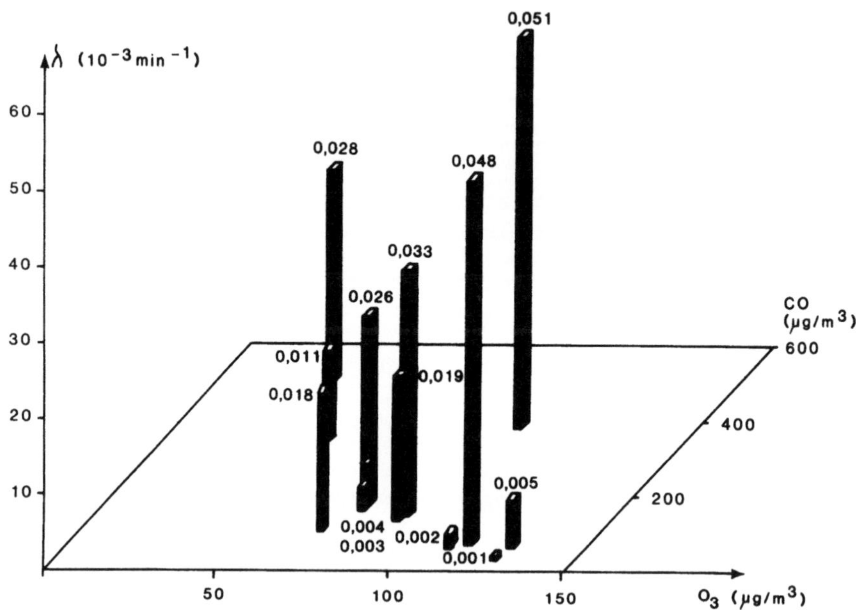

Fig. 3. Influence of Occurrence of CO and O_3 in outdoor air on viability of Staphylococcus epidermidis. λ represents the reduction rate.

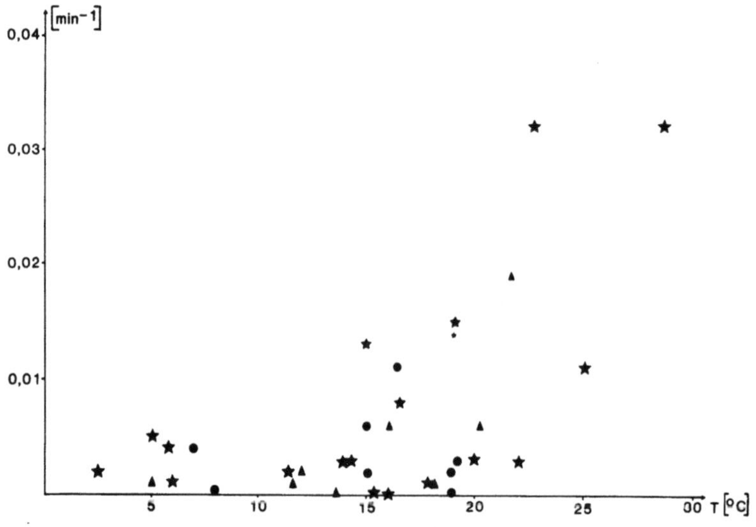

Fig. 4. Survival of Candida albicans after exposure to outdoor air at various temperatures. Ordinate axis: reduction rate λ. Exposure sites: Schönbuch ★, Welzheimer Wald ▲, Schwarzwald (Black Forest) ●.

Table 1. Comparison of exposure of Staphylococcus epidermidis and Candida albicans to outdoor air at 3 sites in 1985 in Baden Württemberg, FRG. Black Forest (B.F.), Schönbuch (Sch.), Welzheimer Wald (W.W.)

Date	Temp.	Rel. Humidity	λ (min^{-1})		Site		
1985	(°C)	(%)	Staph. epid.	Cand. alb.	B.F.	Sch.	W.W.
5/8	14	68	-0.001	-0.002			x
5/15	16	40	0.035	0.011	x		
5/19	19	43	0.022	-0.002	x		
5/22	13	91	-0.003	0.001			x
5/23	15	63	0.033	-0.001		x	
6/20	14	60	0.003	0.002		x	
6/21	16	70	0.013	0.006			x
7/3	22	47	0.002	0.019			x
7/4	18	39	0.005	0.001		x	
7/10	19	60	0.013	0.002	x		
9/4	15	–	0.050	0.006	x		
9/5	16	79	0.018	0.008		x	
9/11	20	50	0.006	0.006			x
9/17	15	57	0.027	0.002	x		
9/20	22	56	0.019	0.003		x	
9/27	22	45	0.048	0.032		x	
10/1	25	40	0.002	0.011		x	
10/2	19	55	-0.001	0.003	x		
10/4	14	50	-0.002	0.003		x	
10/17	10	80	0.028	0.040		x	
10/23	12	35	-0.002	0.002			x
11/8	5	92	0.003	0.001			x

The present method is also suited to detect the noxious effect of photoxidants presumed to occur in increased concentrations in warm, dry periods. As shown in Fig. 4, viability of Candida albicans exposed to outdoor air is reduced with increasing temperature at the present 3 monitoring sites. This deleterious effect cannot be attributed to temperature alone, as was shown in laboratory tests which ruled out a direct temperature effect on test strain viability between 10 and 30°C. It thus appears likely that noxious agents occurring in the outdoor air are responsible for the demonstrated reduction in viability in Fig. 4. Of course, it cannot be ruled out that metabolism of noxious agents is increased under increased temperature. However, the overall toxic effect of such agents is clearly revealed by the present method.

Differences in the sensitivity of various microorganisms to noxious agents in outdoor air are also clearly documented with the present method. As snown in Table 1, Candida albicans generally tolerated outdoor exposition better than Staphylococcus epidermidis. The toxic effects of relatively low pollution loads should thus be better detected using Staphylococcus epidermidis, whereas higher loads should be better reflected by the response of Candida albicans. The present results therefore demonstrate tne advantage of simultaneously using different test strains to monitor a wide range of potential toxic effects of air pollution.

The feasibility of employment of various microorganisms as bio-indicators of air-borne pollutants has been demonstrated. As mentioned above, there are differences in the sensitivity of different strains of micro-organisms to detection of toxic air pollution. Moreover, such bioindicators appear to differ in sensitivity to various types of gaseous noxa. Synergistic effects of various air constituents can also be detected using the present method. Finally, differences in the overall sensitivity of microbial bio-indicators have been shown and should permit monitoring of a wide range of toxicity resulting from air pollution. In light of the demonstrated rapidity with which the present microbial bioindicators provide information about air quality, continued testing of such methods would appear to be worthwhile.

References

Ames, B.N.; McCann, J.; Yamashi, E.: Methods for detecting carcinogens and mutagens with the Salmonella/mammalian-microsome mutagenicity test. Mutation Res. 31, 347-364 (1975)

Arndt, U.; Erhardt, W.; Keitel, A.; Michenfelder, K.; Nebel, W.; Schlüter, C.: Standardisierte Exposition von pflanzlichen Bioindikatoren. Staub - Reinhalt. Luft 45, 481-483 (1985)

Demik, G.: The open air factor. Doctoral Dissertation, Univ. of Utrecht, Faculty of Natural Science, 1976

Druett, H.A.; May, K.R.: Unstable germicidal pollutant in rural air. Nature 220, 395-396 (1968)

Druett, H.A.; Packman, L.P.: Microbiological detector for air pollution. Nature 218, 699 (1968)

Nover, H.; Botzenhart, K.: Untersuchungen zur Wirkung von photochemischem Smog aus einem Stömungsreaktor auf Bakterien. I. Mitt.: Bestimmung der bakteriziden Bestandteile des photochemischen Smogs. Zbl. Bakt. Hyg. I. Abt. Orig B 177, 298-311 (1983)

Nover, H.; Botzenhart, K.: Bactericidal effects of photochemical smog consituents produced by a flow reactor. III Comm.: Determination of mutagenic effects of photochemical smog on E. coli. K 12 343/113. Zbl. Bakt. Hyg. I. Abt. Orig B 181, 71-80 (1985)

VDI (Vereinigung Deutscher Ingeniuere, German Association of Engineers) (ed.): Richtlinie VDI 3792 (Guideline 3792): Verfahren der standardisierten Grasskultur (standardised grass culture procedures), 1978

Dr. Sylvia Waldner-Sander, Hygiene-Institut der Univ. Tübingen, Abteilung für Allgemeine Hygiene und Umwelthygiene, Silcherstr. 7, 7400 Tübingen, FRG

EXS 51:
Advances in Aerobiology
© 1987 Birkhäuser Verlag Basel

MULTIVARIATE CORRELATION OF DEPOSITION DATA OF 8 DIFFERENT AIR
POLLUTANTS TO LICHEN DATA IN A SMALL TOWN IN SWITZERLAND.

K.Ammann[*], R.Herzig, L.Liebendörfer and M.Urech
Systematisch-Geobotanisches Institut der Universität Bern, Switzerland.

1. Introduction, interdisciplinary concept

Since 1980 a group of scientists of the Universities of Berne, Lausanne,
land-use planners and politicians have been working on an interdisciplinary
study concerned with meteorology, air pollution and land-use planning in
Biel-Bienne, Switzerland ("Nationales Forschungsprogramm Nr. 14, NFP 14")
HERZIG,R. et al.1985, 1987, WANNER,H.et al.1986.
For further details regarding interdisciplinary concept of the study see
captions of Fig. 1. The study shows that the degree of air pollution in a
small town like Biel, which is surrounded by well-marked hills, differs
only slightly from that of larger towns like Basel and Zürich. It is
planned to co-operate with decision-making government agencies in Biel.

Figure captions for Fig. 1 on next page

1: Biel area: Topography, land-use, wind roses for six locations which are
   representative of conditions found within the study area (discontinuous
   line: urban area; dotted surfaces: hills or mountains.
2: Regional distribution of calibrated lichen Index of Atmospheric
   Pollution in the Biel area in five zones. Black shaded area: lichen
   desert, diagonal strips: inner struggle zone, horizontal strips: outer
   struggle zone, dotted area: transition zone, white area: normal zone.
3: Emission inventory for SO_2. All the sources have been converted to a
   500m grid. The largest circle represents 45 tons SO_2 per year. Black
   sectors show the proportion of domestic heating.
4: SO_2-concentration (half hour values), recorded in two typical stations
   (city center and southern slopes of Jura mountains).
5: Same representation as in No.3, but for NO_x. The largest circle
   represents 50 tons NO_x per year. The black sectors show the proportion
   causes by traffic.
6: Mean annual concentration of NO_2 ($\mu g/cm^3$), Gaussian plume model.
7: Nighttime fields of potential temperature (2m, thin lines) and stream-
   lines of the boundary layer air flow (10m, arrows), typical for a high
   pressure weather situation with weak synoptic winds. Cold air pools or
   heat islands are marked with a plus or minus symbol.
8: Same representation as in No.7, but for the early afternoon.

Fig. 1: next page. Wanner,H. et al. 1986

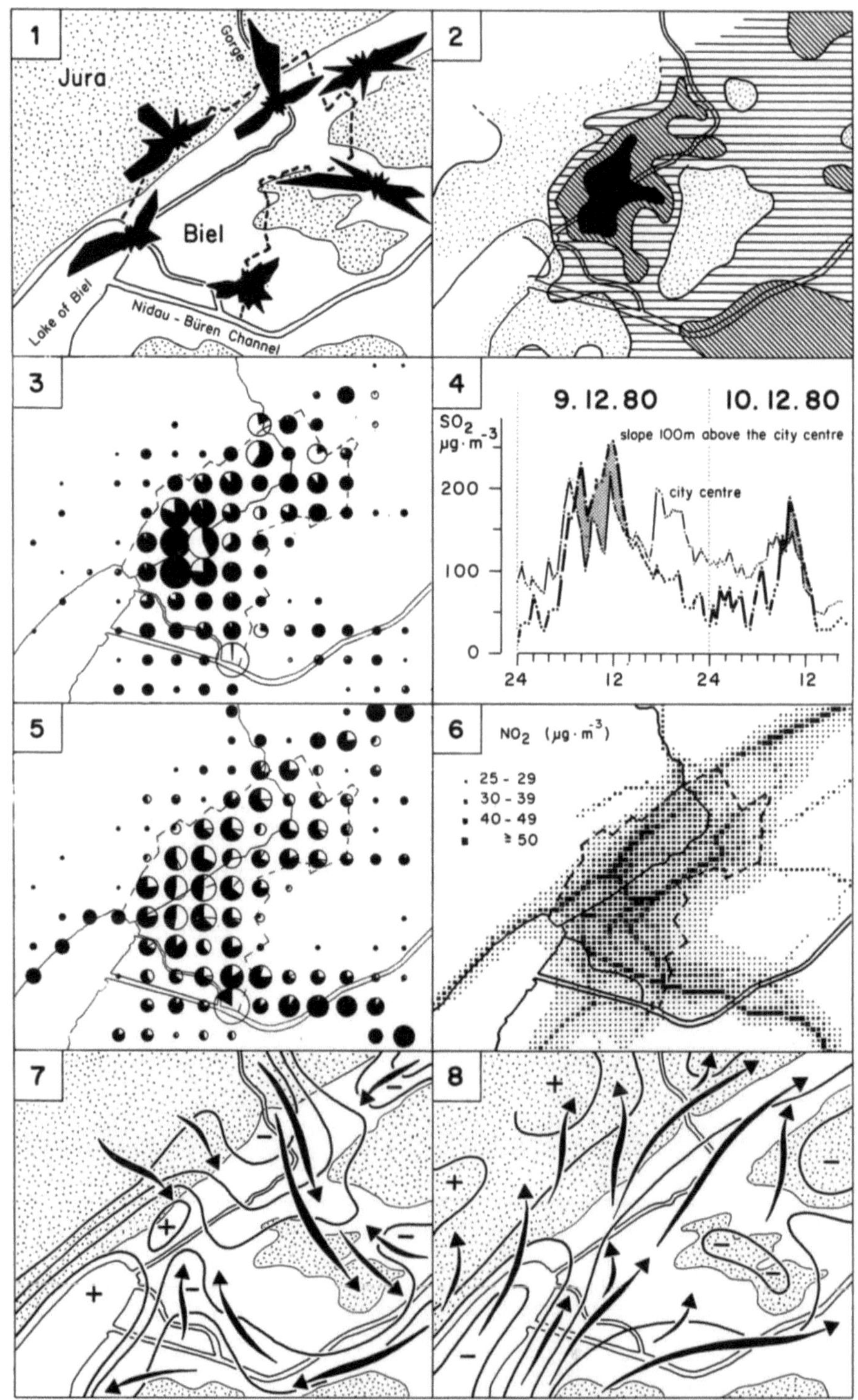
1
Jura
Gorge
Biel
Lake of Biel
Nidau - Büren Channel
2
3
4
9. 12. 80   10. 12. 80
SO$_2$
µg·m^{-3}
slope 100m above the city centre
200
city centre
100
0
24   12   24   12
5
6
NO$_2$ (µg·m^{-3})
25 - 29
30 - 39
40 - 49
≥ 50
7
+
−
+
+
−
+
−
8
+
−
+
−
−
−

2. Sampling concept of lichen data

Based on a modified grid concept after Wildi (1981), 528 sites (trees) where chosen at random, situated as well- balanced as possible within 48 grid units of 100 hectares and 52 grid units of 25 hectares within the area of Biel.
A "Fequenzleiter" (frequency grid) after Kunze 1972 has been modified, so that a grid of 10 subunits arranged in two vertical rows was attached to the trunk, the grid being delimited by a system of horizontal nylon ropes and vertical, perforated metal bars.
The following data were collected on the site:
2.1.	general data on sample and site, coordinates, detailed sketch of situation, orientation of grid.
2.2.	data on ecology of site, land-use and population density data traffic situation, hydrological situation, status of surface, topography of the nearest surrounding of site.
2.3.	tree and trunk data: tree species, trunk circumference, bark texture, illumination of analysed area, inclination.
2.4.	lichen data per each species recognized
-	frequency in 11 classes
-	coverage in 7 classes
-	vitality in 3 classes
-	thallus damage in 3 classes
-	average age of population in 3 classes
-	spectrum of species

3. Immission data

We understand biomonitoring as a method to determine an overall pollution of all components which might harm in any way the monitors of a given site. What we need consequently to meet our goal of optimization and calibration of a lichen monitor analysis is a set of deposition data on air pollutants which is as large as possible over a minimum period of one year.
An existing network on deposition data collecting sites of the town of Biel had to be expanded from 10 to 14 stations measuring not only SO_2 and dust, but also NO_3, Cl, Pb, Zn, Cd, Cu. This was made possible with the help of the government of Biel.
SO_2, NO_3^-,and Chlorine were analysed in a Liesegang instrument filter extract by HPLC (Pfenninger in Landolt,W. et al. 1985)
Dust (total annual load) was obtained gravimetrically, and Pb^{2+}, Cu^{2+}, Zn^{2+}, Cd^{2+} were measured by atomic absorption spectroscopy Bergerhoff deposition tubs.

4. Development of calibrated IAP method.

The aim of our biological study within the NFP 14 was to develop a method which would later allow to map in detail air pollution impact by means of a lichen statistics calibrated to the deposition data of eight pollutants (SO_2, NO_3, Cl, dust, Pb, Cu, Cd, Zn). An evaluation by multiple linear regression of 20 different IAP formulas (Lichen Index of Atmospheric Purity) showed that sum of frequency alone, on a moderately reduced set of 40 species, can be correlated best to pollution data.

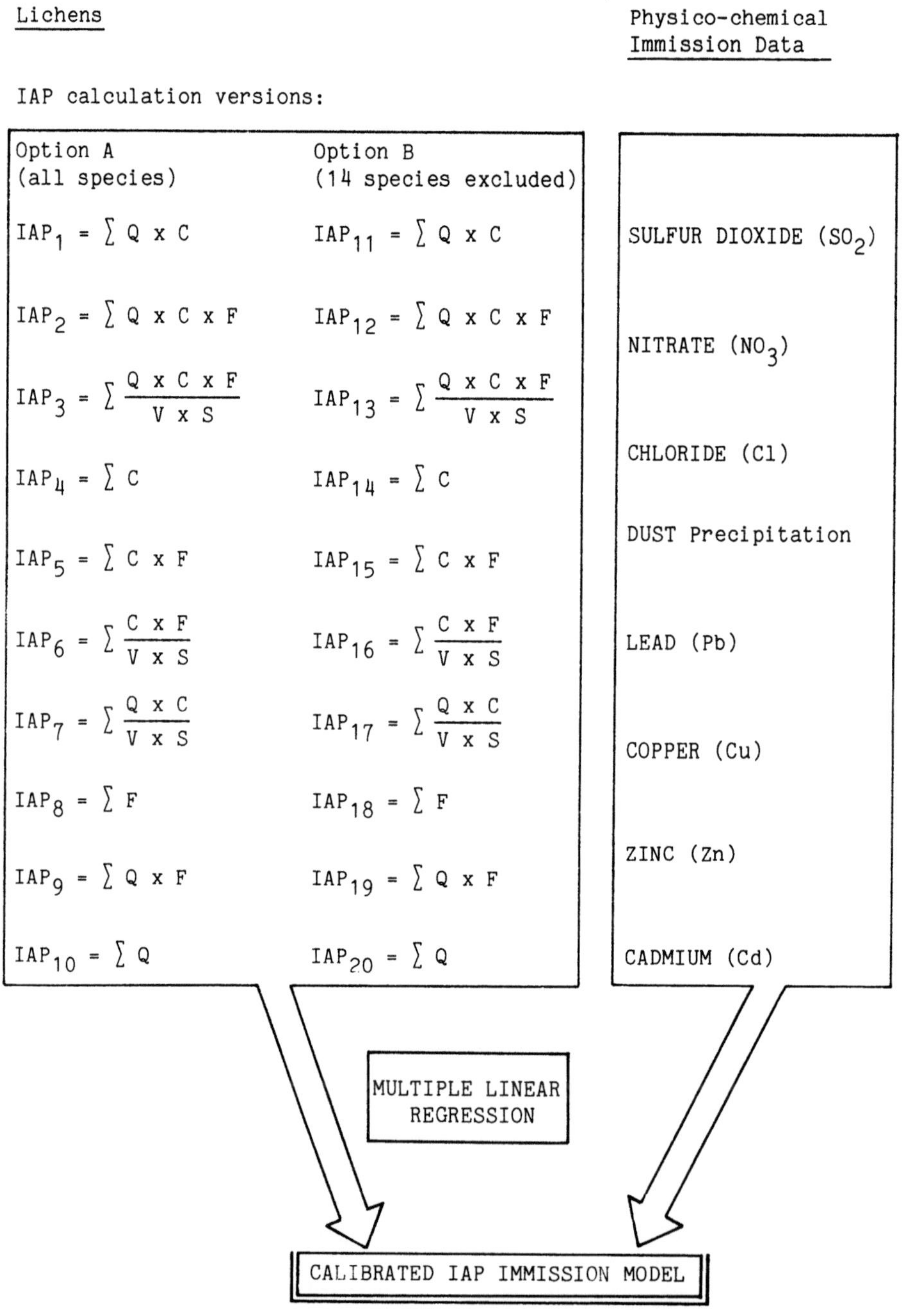

Lichens   Physico-chemical
                                                 Immission Data

IAP calculation versions:

Option A (all species)	Option B (14 species excluded)	Physico-chemical Immission Data
$IAP_1 = \sum Q \times C$	$IAP_{11} = \sum Q \times C$	SULFUR DIOXIDE (SO_2)
$IAP_2 = \sum Q \times C \times F$	$IAP_{12} = \sum Q \times C \times F$	NITRATE (NO_3)
$IAP_3 = \sum \dfrac{Q \times C \times F}{V \times S}$	$IAP_{13} = \sum \dfrac{Q \times C \times F}{V \times S}$	CHLORIDE (Cl)
$IAP_4 = \sum C$	$IAP_{14} = \sum C$	DUST Precipitation
$IAP_5 = \sum C \times F$	$IAP_{15} = \sum C \times F$	LEAD (Pb)
$IAP_6 = \sum \dfrac{C \times F}{V \times S}$	$IAP_{16} = \sum \dfrac{C \times F}{V \times S}$	COPPER (Cu)
$IAP_7 = \sum \dfrac{Q \times C}{V \times S}$	$IAP_{17} = \sum \dfrac{Q \times C}{V \times S}$	ZINC (Zn)
$IAP_8 = \sum F$	$IAP_{18} = \sum F$	CADMIUM (Cd)
$IAP_9 = \sum Q \times F$	$IAP_{19} = \sum Q \times F$	
$IAP_{10} = \sum Q$	$IAP_{20} = \sum Q$	

MULTIPLE LINEAR REGRESSION

CALIBRATED IAP IMMISSION MODEL

Fig. 2: Multivariate calibration of IAP method

With a probability of over 97 % one can predict total impact of air pollution by calculating IAP value No.18 (sum of frequency of 40 species) in the Biel area.

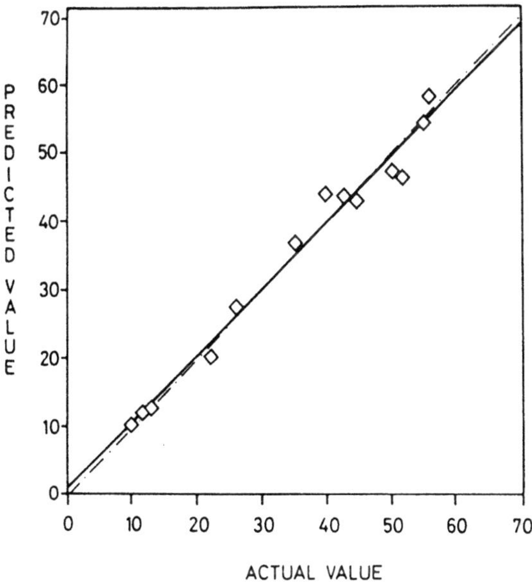

Fig. 3: Overall model obtained by multiple linear regression including SO_2, NO_3, Cl, Dust, Pb, Cu, Zn, Cd.

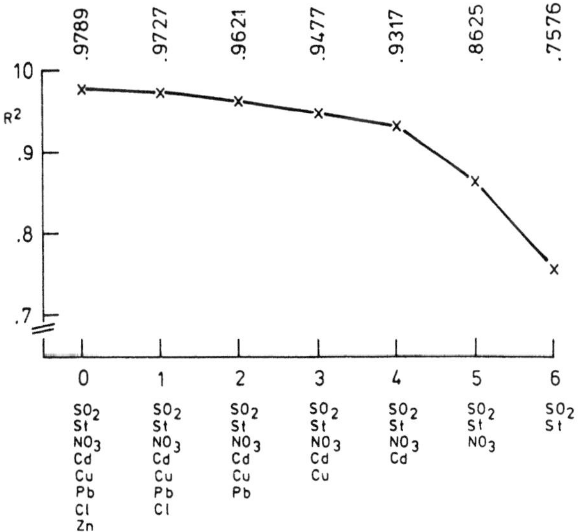

Fig. 4   Step by step elimination of air pollution parameters; related R^2 values.

Elimination of pollution parameters step by step revealed a minimum number
of 4 to 5 parameters to guarantee a good model (calibration) quality and
in addition it showed that there is no clear cut redundancy of any one
parameter.

This, and the fact one can choose almost freely any of the parameters for
elimination leads to the conclusion that it is best to keep all pollution
parameters in the model together for calibration. In our interpretation the
above mentioned 8 parameters are a good indicator collective to describe
total impact of atmospheric pollution on lichens in the region of Biel.

5. Calibrated IAP zone map

IAP zones have been constructed out of over 500 lichen analyses on trees in
the region of Biel. Further details on lichen analysis and calibration and
lichen map in colors in press: HERZIG,R. et al. 1987.
It can be clearly seen that the lowest lichen activity (lichen desert) was
registered in the center of the city. The following zone ("inner struggle
zone"), representing still a quite low lichen activity, can be seen either
in a belt surrounding the lichen desert, or far out of the city, specially
in industrial areas and village centers. (Fig. 1, No. 2).

6. Conclusion

Multivariate calibration of lichen data and pollution data has been
achieved for the first time in Biel, and, most recently, for the second
time on a larger scale on the Swiss plateau, using quantitative pollution
data of the NABEL measuring network. This makes it possible to construct
zonation maps with a statistically supported relationship to the overall
load of air pollution. Government agencies may thus be interested in this
screening method to delimit problem areas at considerably low costs.

Literature cited

Herzig,R. and L.Liebendörfer and M.Urech (1985) Flechten als biologische
Indikatoren der Luftverschmutzung in der Schweiz. Methodenentwicklung in
der Region Biel-Seeland. Lizentiatsarbeit Universität Bern, 241pp.

Herzig R., L.Liebendörfer, M.Urech und K.Ammann (1987) Evaluation und
Kalibrierung einer Flechten-Indikations-Methode mit wichtigen Luftschad-
stoffen. In: Bioindikation - Wirkungsbezogene Erhebungsverfahren für den
Immissionsschutz. Freising-Weihenstephan VDI-Kolloquium, 609, in press.

Wanner,H., K.Ammann, P.Berlincourt, P.Filliger, R.Herzig, L.Liebendörfer,
R.Rickli, M.Urech (1986) Urban meteorology and air pollution in Biel-Bienne
(Switzerland): International Symposium on Climatology Febr.1986. In
Freiburger Geografische Hefte 1986.

Dr. Klaus Ammann, Systematisch-Geobotanisches Institut, Altenbergrain 21,
CH-3013 Bern (Switzerland)

EXS 51:
Advances in Aerobiology
©1987 Birkhäuser Verlag Basel

METHODS FOR YEARLY COMPARISONS OF AEROPALYNOLOGICAL DATA

A. Buttler, M. Girard, Institute of Botany, University of
Neuchâtel (Switzerland)

Some methodological approaches to the problem of comparing
several years' aeropalynological results are discussed. The
methods include comparisons of specific and total quantities of
pollens, comparisons of specific diagrams, comparisons of
specific apparition dates and comparisons by multivariate
analysis. In the 3 last cases, the data are grouped in ten-day
periods.

It can be showed that the annual comparisons of pollen diagrams
by using the 5 classes normally employed (1 (=groundline) :
0 to 5; 2: 6 to 9; 3: 10 to 99; 4: 100 to 999; 5: $>$ 1000
pollens/10 m^3/10 days) is especially suitable for comparing
species distribution. But no overall comparison for the year can
be made. The comparison must be done species by species. In
statistical terms it means variable by variable. Therefore the
multivariate analysis described hereafter is of great interest.
Furthermore the 5 classes give a bad rendering of the quantitative
relations. When an average calendar is calculated and illustrated
by the same method, the choice of the classes can influence the
information contained in the diagram. For species which are
well represented and with close and similar distributions over
the years (i.e. _Taxus_), this choice does not have a decisive
consequence on the representation. But for species which appear
in small quantities and over a large period (i.e. _Compositae_),
the use of the first class (=ground line) alters the type of
distribution observed over the different years. For species with
variable but close distribution from year to year and normal

quantities, the use of this first class can improve the shape
of the distribution by removing low values responsible for a
wider spread distribution.

The yearly comparisons of specific apparition dates by counting
the number of ten-day periods marking the variations in the
date of apparition of species in comparison with the average is
helpful to quantify the behaviour of the species. By summarizing
the values of all species it is possible to have an idea of the
precocity of the years.

The year by year comparisons by specific and total quantities of
pollens offer absolute precision for the overall characterization
of the years (without distinction of ten-day periods). The data
can be used in different ways, i.e. annual total for each species
or annual grand total and classification of species according to
the percentage of the annual grand total (=rank). Variability
index like mean/standard deviation can be calculated with these
values.

The year by year comparison by multivariate analysis (Principal
Component Analysis) allows to bring to light the characteristic
features of each year on the basis of a simultaneous integration
of all the species (variables). The ten-day periods are then
releves (observations). Furthermore, interpretation is made
easier by its graphic illustration. This consists in the
projection of the multidimensional diagram in the most interesting
plan (axes 1+2 = maximum variance) of a reduced dimensional space.
Two methods are proposed. In the first one, the variables are
standardized with the consequence that quantitative effects are
reduced to stress distribution effects. By this way it is
possible to create groups of species for each year on an
objective basis. These groups are constructed with the species
situated in the same direction on the plan (high degree of

correlation). The profiles of releves (ten-day periods) show
peaks which correspond to a concentration of species whose
distribution at that particular moment is at its highest level.
Groups and profiles of releves can change from year to year.
In the second method the variables are not transformed, which
puts more emphasis on the species best represented quantitatively.
Consequently, the size of a peak in the profile of releves is
proportional to pollen production of the ten-day periods.

In conclusion it can be said that the use of several methods
is necessary to restore all the information contained in the
aeropalynological data. The multivariate analysis can be
mentioned as a powerful method.

Legendre, L., Legendre, P. (1979). Ecologie numérique (I + II).
Masson, Paris.

Wildi, O., Orloci, L. (1983). Management and multivariate
analysis of vegetation data. Eidg. Anst. forstl. Versuchsw.,
Ber. <u>215</u>.

A. Buttler, M. Girard, Institute of Botany, University of
Neuchâtel, Chantemerle 22, 2000 Neuchâtel 7 (CH).

AIRCRAFT SAMPLING OF THE UPPER AIRSPORA

S. N. Agashe and M. Chatterjee
Department of Botany, Bangalore University, Bangalore, India.

Introduction

The use of an aircraft to study the upper airspora was undertaken to determine the vertical profile of pollen and spores upto a height of 915m over Bangalore. Though extensive studies of the airspora at lower levels based on data collected from surface traps have been made in Bangalore and elsewhere in India, very little is known of the composition and the trend in the pollen and spore distribution pattern at higher altitudes except for the work of Mehta (1952), and Chitaley and Bajaj (1975) in India. Vertical distribution of pollen and spores in the lower layers of the troposphere are known to vary over different geographical locale, dictated by atmospheric instability. In the present paper the different meteorological phenomena which affect the dispersion and vertical distribution of pollen and spores at higher altitudes are considered.

Materials and Methods

The study of vertical profile of the airspora was carried out over the Flying Training School on the outskirts of Bangalore City situated at 13° 05'N, 77° 36'E at an elevation of 915m above sea level. The city has a rich and varied vegetation on account of salubrious climate. 12 samples were collected during the period, October 1984 to April 1985. Collections were made by Pushpak MK I aircraft which is a single engined, strut based, high winged monoplane. The flight plan during this investigation followed a spiral ascent over the airfield. Sampling was done during ascent at heights of 305m, 610m and 915m at a cruising speed of

115 kmph. Flights were made between 1100 hrs and 1200 hrs at about fortnightly intervals. At the specified altitudes samples were collected by exposing nutrient petriplates and glycerine coated glass slides held by hand outside the window for 30 seconds, each facing the wind to ensure impaction on its adhesive coat. For comparison with airspora near ground level, one sample was taken at the airfield following each flight from the top of a tower about 8m above ground level where the duration of exposure was proportionately for long periods. The results of pollen and spore catches were expressed as the number deposited on the total area of exposed sticky surface viz. 4.84 sq.cm. on slides and the total colony counts with respect to plates. Attempts were made to correlate weather conditions with spore profiles.

Results

Results of culture plate exposure studies

The study revealed the abundance of viable spores at all heights. The maximum colony counts of 866 for 12 samples were recorded at 305m, the number declining to 610 at 610m and 653 at 915m. A similar pattern emerged when day to day spore profiles were examined with the highest concentration at 305m except nominally lower counts on 4 days when the counts at higher altitudes increased. All samples were dominated by a few mould spores viz. Penicillium, Cladosporium, Alternaria, Aspergillus, and Fusarium (Table I).

Table I. Total colony counts of major fungal spore types at different heights

Spore types	Spore counts at different heights above ground level			
	8m	305m	610m	915m
Cladosporium	134	414	220	280
Penicillium	270	264	243	259
Alternaria	25	34	19	21
Aspergillus	12	29	24	11
Fusarium	6	11	7	3

Results of exposed slides for fungal spores

This method enables the study of important pathogenic forms such as smuts and rusts, viable and nonviable spores and those which do not sporulate easily in cultures. 467 and 630 spores were trapped at 305m and 610m respectively, with lower counts of 172 spores at 915m. The principle spore types were <u>Penicillium, Cladosporium</u>, Smut, <u>Alternaria</u>, <u>Aspergillus</u>, Uredospore, <u>Helminthosporium</u> and <u>Curvularia</u>.

Results of exposed slides for pollen grains

Pollen trapped on exposed slides were relatively less compared to fungal spores. The total value for the 12 samples were highest at 8m and lowest at 915m. The middle altitudes showed a moderate catch with a marginal difference between the two levels. Day to day profiles show a higher count in 5 samples at 8m and low counts at 915m throughout. The major pollen types were <u>Parthenium hysterophorus</u>, Poaceae, <u>Eucalyptus</u> sp., <u>Casuarina equisetifolia</u> and <u>Ricinus communis</u> (Table II).

Effects of meteorological factors

Flights for this study were restricted to days when favourable weather conditions were prevalent. It was observed that on fine days fungal spore profiles showed relative uniformity. Spore counts were also found to decrease above cloud levels.

Table II. Variations of major pollen types at different heights above ground level

Pollen types	Pollen counts at different heights			
	8m	305m	610m	915m
<u>Parthenium hysterophorus</u>	142	15	16	7
Poaceae	27	22	29	4
<u>Eucalyptus</u> sp.	16	4	18	5
<u>Casuarina equisetifolia</u>	11	19	7	3
<u>Ricinus communis</u>	9	4	7	3

Conditionally unstable conditions in the lower strata of the atmosphere were responsible for moderate thermal turbulance and increased convection, mixing the spores in the lower levels was also observed by Hirst et al., (1967). Though, pronounced effects of any weather factors were not evident on spore profiles, they appeared to be affected by tempe- rature profiles, wind shears and the abundance of spores in the air at the particular time.

It is intended to continue the above investigation for further period of one year by undertaking additional flights at higher altitudes and by employing improved pollen and spore sampling device.

References

Chitaley, S.D. and Bajaj, A. (1975) Airspora of Nagpur at high altitudes - III. Botanique, 6, 59 - 68.

Hirst, J.M., Stedman, O.J. and Hogg, W.H. (1967) Long distance spore transport: Methods of measurement, vertical spore profiles and the detection of immigrant spores. J.Gen. Microbiol. 48, 329-355.

Mehta, K.C. (1952) Further studies on cereal rusts in India,The Indian Council of Agricultural Research Sci. Monog. no. 18, 1-368.

Author's address

Prof. Shripad N. Agashe, Department of Botany, Bangalore University, Bangalore-560056, India.

METHODS TO REDUCE ALLERGIC EFFECTS: ELIMINATION OF ALLERGENES
(POLLEN, MITES, INDOOR DUST, BACTERIA, GASES ETC.) FROM INDOOR
AIR

Prof. Antonius Kettrup, Arnsberg
Peter R. Schmidt, Bonair, Gröbenzell

The sensibility of human organism against environmental influences leads to an always increasing number of allergic diseases. These allergic diseases are a severe problem for the individual and due to their number for the community as well as for the national economy.

It can be considered as proved that one of the main reasons of allergic reactions is environmental burden of the human organism. This environmental burden leads to different kinds of allergies as for instance obstructive diseases of respiratory tract, skin allergies, contact allergies etc.

It is out of discussion, that medical treatment of the patient is basis of any success against such diseases.

Industry however can considerably support medical treatment by developping filtration units reducing air-pollution in the surroundings of the patient decisively. So human organism can be relieved from excess of environmental burdens in a way that allergic reactions can be considerably reduced or even avoided.

We have concentrated our efforts actually on developping filtration units as a support for medical treatment of obstructive diseases of respiratory tract. Basic investigations gave the result that pollution of air with harmful substances like pollen, mites, mushroom-shaped spores, gases and bacteria

most probably plays a very important role for the existance of
such diseases. Pollen, mites, mushroom-shaped, spores and
bacteria are particles with a diameter larger than 0,1 μm.

The process efffected in our units, protected by gran-
ted patents in all industrial nations of the world, secures a
safe filtration of particles larger than 0,1 micron diameter in
indoor air by help of electrostatic charging of microporous fil-
ter-elements. Filtration is effected by circulating air inside
the room without any constructural changes become necessary.

Continuous controls of tailor made units for patients
suffering from flour-dust allergy, obstructive disease of respiratory tract in a barber-shop and from pollen allergy clearly
showed that allergic reactions either were considerably reduced
or disappeared completely as soon as pollution in surrounding
air was reduced. Patients could even continue their professional
work almost without restrictions. Allergic reactions against
pollen, registered in unprotected areas with normal pollution
of the surrounding air disappeared within a few minutes after
the patient stayed in a "protected" room equipped with our filter unit. The long term test in the barber shop arranged by the
responsible trade association for health service and social welfare showed the result that the efficiency of the unit remained
unchanged at high level within so far 14 months and the patient
continues working without restriction and even medical treatment
could be stopped. So these tests can be considered as positively
finalized.

These fields tests have been effected as follows:
First for each case air pollution under existing local circumstances has been measured. Basing upon these results the details
of the filter unit have been determined assuring that the air
blown back into the room after passing the filter is almost
100 % free from particles, resulting to two major effects:
- the total level of pollution in the room was reduced to
  approx. 25 % although constantly new pollution came up
- the allergic reactions disappeared within a very short time,
  but came back after the unit has been stopped and air pollu-

tion passed a certain level of pollution.

The same effects have been registered from the patient suffering from pollen-allergy.

For scientific control of these results a filter unit has been installed in the Institute of applied Chemistry at University of Paderborn (Prof. Dr. A. Kettrup) to control its efficiency for filtration of allergenes.

Figure 1. shows the principle of the system. 1) aerosol generator, 2) ventilator, 3) ionisation unit, 4) electrostatically charged activated carbon filter element.

The unit consists of 2 ventilators connected in series (2) sucking a defined quantity of indoor air passing the ionisation unit (3) and the filter element (4).

The test solution for allergy diagnosis is introduced into the aerosol generator. So a test aerosol is produced, which is injected into the indoor air stream.

The sampling is done before and after the filter-element by help of pumps, sucking a defined airstream through a fibreglass-filter. The occupancy with particles is examined with respect to form, size and number by help of an electron-grid-microscope.

Alternatively concentration of particles has been measured before and after the filter-element with a core-condensation-counter type TSI 3020. This unit allows to detect particles of a diameter larger than 0,01 micron.

The following test solutions for diagnosis of allergy have been used for our test:
- Indoor dust 150 % GEM; (3201) V 1026
- Dog hairs / dog flakes 60 % GEM; (3205) T 2106
- Birch pollen 6 % GEM; (4204) T 2550
- Penicillum notatum 10 % GW; (1603) T 1369

Summarising the obtained results and calculating the efficiency by counting occupancy of the filter sampling before and after the electrostatically charged activated carbon filter the efficiency of the system is as follows, shown in the next picture.

Table 1. shows the efficiency for provocation test aerosoles.

In the same chart the particle concentrations measured with the core-condensate-counter are shown. It is obvious that an efficiency of better than 99,99 % is obtained.

It is now the question, how to come to solutions for the individual case which are practicable also moneywhise.

Here I actually can give the following general statements:

1. With respect to allergies caused by professional work a solution should be arranged as follows:
   It should suck the air at places where the main pollution arises directly and in a direction away from the breathing organs. This means in horizontal direction not upwards to avoid passing nose and mouth of the patient. In some cases it might be as well considered to suck downwards. The contaminated air should be brought to a central filter unit having an efficiency of at least 99 % for particles larger than 0,3 microns in size to assure that the filtered air reentering the room is almost free of particles. The mechanical installations possibly necessary have to be determined for each place individually.

2. For the private life of patients within their residence an area has to be found, possibly the bedroom where a unit with the same efficiency as stated before can be installed. A 10-times air exchange should be calculated as a maximum. This would allow the patient to go to a protected area for a certain time as soon as allergic reactions in unprotected living areas are registered. Additional measures of air filtration in these living areas are certainly to be recommended but will often fail because of financial problems. These additional measures should base upon 3-times air-exchange.

The peculiarity of our system is that these effects can be achieved on long terms with stable results without special installations. The installation of a unit for a

<u>Tab. 1</u>   Efficiency of precipitation for allergens,
          numbers of particles, permeability

Aerosol	impure air part./cm^3	purified air part./cm^3	efficiency %	part. μm
Penicillum not.	$1,8 \cdot 10^4$	8	99,999556	< 0,1
Dog hair	$3,6 \cdot 10^4$	6	99,999833	< 0,1
Birch pollen	$3,6 \cdot 10^4$	3	99,999917	< 0,1
House dust	$3,4 \cdot 10^4$	6	99,999824	< 0,1

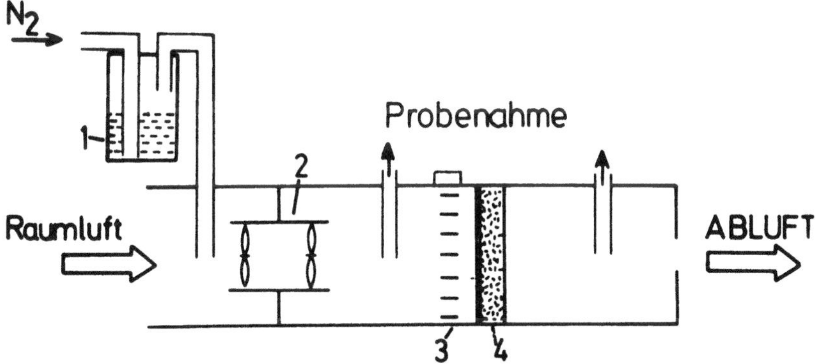

<u>Fig.1. Construction of the air filter device</u>

Probenahme = sampling of specimens
Raumluft   = indoor air
Abluft   = purified air

1. Vorlage mit Zerstäuber = aerosolgenerator with atomizer
2. Lüfter = ventilators
3. Elektrofilter = electrostatic filter
4. Aktivkohlefilter = filter with activated carbon

private living area is comparable to the installation of a ceiling mounted lamp.

Additionally this system allows rapid elimination of suddenly appearing allergenes, so that the well-feeling of the patient even in such - rarely foreseeable - cases is not considerably influenced.

Finally let me say that the effectivity of the system is actually under evaluation for application in cases of skin allergies in close cooperation with the responsible trade association. First results lead to the hope that positive results may be obtained.

Prof. Antonius Kettrup, Rumbecker Höhe 10, 5760 Arnsberg 2
Peter R. Schmidt, Messrs. BONAIR Luftreinigungsgeräte GmbH,
Starenweg 15, 8038 Gröbenzell

Subject index

Compiled by *G. Boehm,* Basel

Acidic fog, containing H_2O_2, effects on the sensitivity of agricultural crops to important fungal diseases 325

Acidic inputs; ozone and acidic inputs into forests in Northwestern USA 319

Aeroallergens (Coakroach, mites, fungus spores etc.) in the New York inner-city apartments of asthmatics 133

Aerobiological investigations; correlation between aerobiological and phytogeographical investigations in the Florence area 43

Aerobiology – its past and its future 3

«Aerobiology» term created by *Fred Campbell Meier,* v. Aerobiology – its past and its future

Aeropalynological data; methods for yearly comparisons of aeropalynological data 407

«Aeroplankton»: term created by *Hans Molisch,* v. Aerobiology – its past and its future

Aeroplankton; microalgae as aeroplankton and allergens 171

Aerosols; fine particle aerosols in experimental legionnaires disease: their role in infection and treatment 357

-; v. Bioaerosols

Aerosols, microbiological; collection efficiency of two samplers for microbiological aerosols (Cyclone sampler and modified personal impinger) 391

Agricultural crops; effects of acidic fog containing H_2O_2 on the sensitivity of agricultural crops to important fungal diseases 325

Air ambient; the quantitative significance of asbestos fibres in the ambient air 213

Air cleaning, v. Apparatus

Air conditioned and conventional heated buildings: the building illness syndrome. Comparative studies 287

Airconditioned buildings, v. Building (s); Building, ventilation

Airconditioning installation; filters of an airconditioning installation as disseminators of fungal spores 283

Air, indoor, v. Indoor air

Air pollutant, new, v. Calcium sulphate

Air pollution; influence of meteorological and air pollution factors on acute diseases of the airways in children as illustrated by the Biel (Switzerland) region 255

-; microbial and dust pollution in non-industrial work places 279

-; microorganisms as biological indicators of air pollution 395

-; statistical correlation of deposition data on 8 air pollutants to lichen data in a smal town - Biel - in Switzerland 401

-; v. Bacterial atmospheric contamination, Wasts domestic

Airborne Botrytis conidia, v. Botrytis conidia, airborne

Airborne crystals of anhydrous calcium sulphate - a new air pollutant 261

Airborne Gram negative bacteria, associated with the handling of domestic waste 371

Airborne particles, environmental influences on their deposition 13

Airborne particulate matter, v. Burkard sampler; Particles, unanimate

Airborne pollen, v. Pollen, airborne

Airborne pollutant new, v. airborne crystals

Airborne spores; clinical significance of airborne Alternaria tenuis spores: Seasonal symptomes positive skin and bronchial challenge tests with Alternaria in subjects with asthma and rhinitis 147

Airborne spread of footh-and-mouth disease; improvement of mathematical models for predicting 351

Airborne spread of microorganisms affecting animals 347

Airborne spread, v. Barley mildow

Aircraft sampling of airspora in the upper atmosphere 411

Airspora concentration, uniformity 313

Airways in children; influence of meteorological and air pollution factors on acute diseases of the airways in children as illustrated by the Biel (Switzerland) region 255

Alexandria; study of airborne pollen grains of Alexandria (Egypt) 55

Algae; microalgae as aeroplankton and allergen 171

Allergen-containing dust samples from the interior of the house 189

-; v. Guanine, House dust, Mites

Allergen standardization; intrinsic variability in airborne fungi: implications for allergen standardization 143

Allergen type, unrecognized; allergic reactivity to Parthenium hysterophorus pollen: an unrecognized type I allergen 125

Allergenic pollen; air borne pollen types of allergenic significance in India 61

Allergens, indoor allergens; apparatus for elimination of indoor allergens (pollen, spores, bacteria, dusts, gases etc.) 415

Allergens, inhalative 71

Allergens; microalgae as aeroplankton and allergens 171

Allergens, mite allergens; guanine dosage in house dust samples and quantification of mite allergens 203

Allergic attacks; influence of inversion layers on the daily pollen count and on the allergic attacks of patients 19

Allergic patients; airborne pollen and symptoms in allergic patients undergoing immunotherapy 89

Allergic sensitization, v. Botrytis extract

Allergic symptomes, seasonal, v. Alternaria tenuis

Alternaria tenuis; clinical significance of airborne Alternaria tenuis spores: Seasonal symptomes positive skin and bronchial challenge tests with Alternaria in subjects with asthma and rhinitis 147

Alternaria Nees ex Fr., occurence in indoor and outdoor habitats in Cordoba (Spain) 157

Altitudes (high altitudes); house dust mites in different altitudes of Grisons (Switzerland) 197

Ambrosia, v. Ragweed

Animals; airborne spread of microorganisms affecting animals 347

Annual and daily variation of pollen from Olea europaea, v. Pollen, airborne, Olive pollen (atmosphere of Cordoba) 109

Antarctica; coniometric dust measurements during a sea voyage from the northern to the southern hemisphere, along the Antarctic coast, and a coast station in Antarctica 267

Antigen analysis; efficiency of a new bioaerosol sampler in sampling Betula pollen for antigen analysis 383

Apparatus for elimination of indoor allergens (pollen, spores, mites, bacteria, dust, gases etc.) from indoor air 415

Artemisia, v. Mugwort

Asbestos fibres; quantitative significance of asbestos fibres in the ambient air 213

Asbestos indoor problem; the indoor asbestos problem: facts and questions 221

Asthma, v. Alternaria tenuis

Asthmatics; aeroallergens in New York inner-city apartements of asthmatics 133

Atmospheric conditions; annual and daily variation of pollen from Olea europaea L. in the atmosphere of Cordoba (Spain) along two year of sampling 109

-; monitoring of atmospheric conditions and forecast of Olive pollen season 95

Atmospheric pollen, v. Pollen, airborne

Bacteria; airborne Gram negative
bacteria associated with the
handling oof domestic waste 371

-; the size distribution of whirlpool-
generated droplets, their ability
to contain bacteria and their
disposition potential in the
human respiratory tract 361

-; v. Indoor allergens (apparatus for
elimination)

Bacterial atmospheric contamination in wastewater treatment
plants 365

Barley mildew; spread of barley
mildew by wind and its significance for phytopathology, aerobiology and for barley cultivation
in Europe 331

Basel (Switzerland), v. Burkard-sampler (radioactive dust), Inversion
layers

Beetle as an inhalative allergen v.
Gibbium psylloides, Tribolium
confusum

Berlin (West) (Germany), v. Pollution of the air

Betula pollen; efficiency of a new
bioaerosol sampler in sampling
betula pollen for antigen analysis
383

Biel (Switzerland), v. Airways in
children, Biological indicators

Bioaerosol sampler, new, efficiency
in sampling Betula pollen for
antigen analysis 383

Bioaerosols and office building ventilation system 297

Biological indicator; statistical correlation of deposition data on 8
air pollutants to lichen data in a
small town - Biel - in Switzerland 401

-; Microorganisms as biological
indicators of air pollution 385

Biometeorology and its relation to
pollen count 37

Botrytis conidia, airborne; concentrations of airborne Botrytis conidia, and frequency of allergic sensitization to Botrytis extract 165

Bronchial challenge tests, v. Alternaria tenuis

Bronchial reactivity; effects of
short-term exposure to ambient
nitrogen dioxide concentrations
on human bronchial reactivity
and lung function 227

-; v. Pulmonary defense

Building illness syndrom; comparative studies in air conditioned
and conventional heated buildings 287

-; v. Filters of an airconditioning
installation

Building, ventilation; bioaerosols
and office building ventilation
system 297

Buildings, indoor air; inquiries
received by the California indoor
air quality program on biological
contaminants in buildings 275

Buildings, v. Radon

Burkard-sampler for collecting airborne pollen and spores, special applications

I.: Semiquantitative continuous determination of dust-particles, soot, flue ash etc. in the air 377

II.: Record of radioactive dust in the air of Basel (Switzerland) (after Chernobyl) 381

Burkard samples, v. Bioaerosol sampler

Calcium sulphate: airborne crystals of anhydrous calcium sulphate – a new air pollutant 261

California; inquiries received by the California indoor air quality program on biological contaminants in buildings 275

California, Southern; indoor mold spore exposure: characteristics of 127 homes in Southern California with endogenous mold problems 139

Carcinogenic and mutagenic effects of airborne particulate matter from polluted areas on human and rodent tissue cultures 231

Castanea, v. Chestnut

Cell cultures; lysosome response and cytoskeleton alteration in cell cultures exposed airborne lead 243

-; v. Tissue cultures

Chernobyl (U.S.S.R.), atomic reactor accident, v. Radioactive dust

Chestnut (Castanea) pollen counts related to patients pollinosis in Paris 113

Children, diseases of the airways, v. Airways in children, Respiratory tract diseases

Climate, high altitude climate, v. Pollen, airborne and airborne spores (Davos/Switzerland)

Coackroachs, v. New York

Coniometric dust measurements during a sea voyage from the northern to the southern hemisphere. along the Antarctic coast, and at a coast station in Antarctica 267

Contaminants, biological, in buildings; inquiries received by the California indoor air quality program on biological contaminants in buildings 275

Cordoba (Spain); annual and daily variation of pollen from Olea europaea L. in the atmosphere of Cordoba (Spain) along two years of sampling 109

-; v. Alternaria tenuis

Cour collector, v. Ragweed (Rhône Bassin)

Crops, agricultural crops; effects of acidic fog containing H_2O_2 on the sensitivity of agricultural crops to important fungal diseases 325

Cross-reactions; clinical cross-reactions between various moulds with special reference to Fusarium 147

Cultivation of barley in Europe, v.
Barley mildew

Cytoskeleton; lysosome response
and cytoskeleton alteration in
cell cultures exposed to airborne
lead 243

Daily and annual variation of pollen
from Olea europaea L., v. Pollen
airborne. Olive pollen (Cordoba)

Davos (Switzerland); relationship of
airborne pollen and spores to
symptoms on the skin and
mucous membranes of patients
in the high altitude climate in
Davos 81

Desensibilization, v. Immunotherapy

Droplets, whirlpool-generated, size
distribution, v. Whirlpool

Dust; coniometric dust measurements during a sea voyage from
the northern to the southern
hemisphere, along the Antarctic
coast, and at a cost station in
Antarctica 267

Dust and microbial air pollution in
non-industrial work places 279

Dust particles in the air: special
applications of the Burkard sampler for collection airborne pollen
and spores

I . Semiquantitative continuous
determination of dust particles
377

II. Record of radioactive dust particles in the air of Basel (Switzerland) (after Chernobyl) 381

Dust, respirable; health risks from
respirable dusts produced during
the operation of big harvester 251

Dust samples from the interior of
the house, investigations of allergen containing 189

-; v. Guanine, House dust, Mites

Dust, v. Indoor allergens (apparatus
for elimination)

Egypt, v. Alexandria (airborne pollen)

Environmental influences on deposition of airborne particles 13

Erysiphe, v. Mildew

Europe, v. Mildew

Exposure, acute, to nitrogen dioxide, modulation of pulmonary
defense mechanism 235

Exposure, short-term, to ambient
nitrogen dioxide concentrations
on human bronchial reactivity
and lung function 227

Filters of an airconditioning installation as disseminators of fungal
spores 283

Fine particles aerosols, v. Legionnaires' disease

Florence area; correlation between
aerobiological and phytogeographical investigations in the Florence area (Italy) 43

Fog, acidic, containing H_2O_2, effects
on the sensitivity of agricultural
crops to important fungal diseases 325

Foot-and-mouth disease; improvement of mathematical models for predicting the airborne spread of footh-and-mooth disease 351

Forests; ozone and acidic inputs into forests in Northwestern USA 319

Fruite-fly; falling behaviour of sterilized and slept fruit-fly and melonfly 337

Fungal diseases; effects of acidic fog containing H_2O_2 on the sensitivity of agricultural crops to important fungal diseases 325

-; v. Barley mildew

Fungi, airborne; intrinsic variability in airborne fungi: implications for allergen standardization 143

Fungus spores; filters of an airconditioning installation as disseminators of fungal spores 283

-; indoor mold spores exposure: characteristics of 127 homes in Southern California with endogenous mold spores 139

-; relationship of airborne pollen and spores to symptoms on the skin and mucous membranes of patients in the high altitude in Davos (Switzerland) 81

-; v. Alternaria; Botrytis; Filters; Fungal...; Fungi...; Fusarium; Spores

Fusarium; clinical cross-reactions between various moulds with special reference to Fusarium 147

Gases, v. Indoor allergens (apparatus for elimination)

Gibbium psylloides («mite beetle») as an inhalative allergen 183

Grisons (Switzerland); house dust mites in different altitudes of Grisons 197

Gregory, Philipp Herries (1907–1986) 9

Guanine dosage in house dust samples and quantification of mite allergens 203

Guanin dosage, v. Dust samples, House dust, House dust mites, Mites

Harvesters; health risks from respirable dusts produced during the operation of big harvesters 251

Heated buildings, conventional, and air conditioned: the building illness syndrome. Comparative studies 287

Hemisphere, northern, to hemisphere southern, sea voyage, v. Coniometric dust measurement

High altitude climate; relationship of airborne pollen and spores to symptoms on the skin and mucous membranes of patients in the high altitude climate in Davos (Switzerland) 81

House dust; samples from the interior of the house, investigations of allergen containing 189

House dust mites in different altitudes of Grisons (Switzerland) 197

-; v., Guanine, Mites

Human bronchial reactivity, v.
Bronchial reactivity

Human tissue cultures, v. Tissue
cultures

Hyposensibilization, v. Immunotherapy

Immunotherapy; airborne pollen
and symptoms in allergic patients
undergoing immunotherapy 89

India; airborne pollen types of allergenic significance in India 61

Indicators, biological; microorganisms as biological indicators of
air pollution 395

-; Statistical correlation of deposition data on 8 air pollutants to
lichen data in a small town –
Biel – in Switzerland 401

Individual pollen collector; experiences with the «Individual Pollen Collector» developed by G.
Boehm 87

Indoor air; inquiries received by the
California indoor air quality program on biological contaminants
in buildings 275

Indoor allergens; apparatus for elimination of indoor allergens (pollen, spores, bacteria, dusts, gases
etc.) from indoor air 415

Indoor asbestos problem: facts and
questions 221

Indoor environment: radon and its
decay products in the indoor
environment: radiation exposure
and risk estimation 303

Indoor habitats; occurence of Alternaria Nees ex Fr. in indoor and
outdoor habitats in Cordoba
(Spain) 157

Indoor mold spore exposure: characteristics of 127 homes in Southern California with endogenous
mold problems 133

Inhalative allergens 71

Inhalative allergens, v. Gibbium
psyloides

Interior of houses; investigations of
allergen-containing dust samples
from the interior of the house 197

Inversion layers in Basel (Switzerland), influence on the daily pollen count and on the allergic
attacks of patients 19

Italy, v. Florence area

Legionnaires' disease; fine particle
aerosols in experimental legionnaires' disease: their role in
infection and treatment 357

Legionella pneumophila, v. Legionnaires' disease

Lead, airborne; lysosome response
and cytoskeleton alteration in
cell cultures exposed to airborne
lead 243

Lichen data; statistical correlation
of deposition data on 8 air pollutants to lichen data in a small
town – Biel – in Switzerland 401

Lung function; effects of short-term
exposure to ambient nitrogen
dioxide concentrations on human
bronchial reactivity and lung
function 227

Lung, v. Bronchial reactivity, Pulmonary defense, Respiratory tract

Lysosome response and cytoskeleton alteration in cell cultures exposed to airborne lead 245

Mathematical models, v. Airborne spread of footh-and-mouth disease

Meier, Fred Campbell, creator of the term 'Aerobiology', v. Aerobiology – its past and its future

Melonfly; falling behaviour of sterilized and slept fruite-fly and melonfly 337

Meteorological factors; influence of meteorological and air pollution factors on acute diseases of the airways in children as illustrated by the Biel (Switzerland) region 255

-; obstructive respiratory tract diseases of children in Berlin (West) (Germany), influences of pollution of the air and wheater 25

Meteorology, v. Biometeorology

Method to reduce indoor allergens; apparatus for elimination of indoor allergens (pollen, spores, bacteria, dusts, gases etc.) from indoor air 415

Method, v. Individual pollen collector

Methods for yearly comparisons of aeropalynological data 407

Microalgae as aeroplankton and allergens 171

Microbial aerosols; collection efficiency of two samplers for microbial aerosols (Cyclone sampler and midified personal impinger) 391

Microbial and dust air pollution in non-industrial work places 279

Microorganisms affecting animals, airborne spread 347

Microorganisms as biological indicators of air pollution 395

-; v. Lichen data

Mildew; spread of barley mildew by wind and its significance for phytopathology, aerobiology and for barley cultivation in Europe 331

Mite allergens; guanine dosage in house dust samples and quantification of mite allergens 203

Mite beetle, v. Gibbium psylloides

Mites; house dust mites in different altitudes of Grisons (Switzerland) 197

-; investigations of allergen containing dust samples from the interior of the house 189

Mold spores; indoor mold spore exposure: characteristics of 127 homes in southern California with endogenous mold problems 139

Molisch, Hans, creator of the term 'Aeroplankton', v. Aerobiology – its past and its future

Moulds, various; clinical cross-reactions between various moulds with special reference to Fusarium 147

Mugwort; pollen counts of ragweed (Ambrosia) and mugwort (Artemisia) (Cour collector) in 1984 measured at 12 meteorological centers in the Rhône Bassin and surrounding regions (France) 119

Mutagenic and carcinogenic effects of airborne particulate matter from polluted areas on human and tissue cultures 231

New York inner-city apartments of asthmatics, aeroallergens 133

Nitrogen dioxide; effects of short-term exposure to ambient nitrogen dioxide concentrations on human bronchial reactivity and lung function 227

-; modulation of pulmonary defense mechanisms by acute exposure of nitrogen dioxide 235

Obituary: *Philipp Herries Gregory* (1907–1986) 9

Obstructive respiratory tract diseases of children in Berlin (West), influences of pollution of the air and wheater 25

Occupational sensitization, v. Tribolium confusum (beetle)

Olea europaea, v. Olive pollen

Olive pollen; annual and daily variation of pollen from Olea europaea L. in the atmosphere of Cordoba (Spain) along two year of sampling 109

-; pollen yield of two cvs. of Olea europaea L. ('Manzanillo' and 'Swan hill') (Tucson/Arizona, USA) 101

Olive pollen season; monitoring of atmospheric conditions and forecast of Olive pollen season (Bari/Italy) 95

Outdoor and indoor habitats; occurence of Alternaria Nees ex Fr. in indoor and outdoor habitats in Cordoba (Spain) 157

Ozone and acidic inputs into forests in Northwest USA 319

Palynology, v. Aeropalynological data

Paris; chestnut (Castanea) pollen counts related to patients pollinosis in Paris (France) 113

Parthenium hysterophorus; allergic reactivity to Parthenium hysterophorus pollen: an unrecognized type I allergen 125

Particles, airborne, environmental influences on their deposition 13

Particles, unanimate; mutagenic and carcinogenic effects of airborne particulate matter from polluted areas on human and rodent tissue cultures 231

Particles, unanimate, in the air, v. Asbestos (fibres), Burkard-sampler, Calcium sulphate, Dust, Radioactive dust

Phytogeographical investigations; correlations between aerobiological and phytolgeographical investigations in the Florence area 43

Phytopathology, v. Barley mildew, Crops, agricultural

Plants; bacterial atmospheric contamination in wastewater treatment plants 365

Pneumonia, v. Legionnaires' disease

Pollen, airborne; airborne pollen and symptoms in allergic patients undergoing immunotherapy 89

Pollen, airborne; airborne pollen types of allergenic significance in India 61

Pollen, airborne; air sampling studies in a tropical area (Venezuela). Four year results 49

Pollen, airborne, Olive pollen; annual and daily variation of pollen from Olea europaea L. in the atmosphere of Cordoba (Spain) along two years of sampling 109

Pollen airborne, Olive pollen; monitoring of atmospheric conditions and forecast of Olive pollen season in Bari (Italy) 95

Pollen airborne, Olive pollen; pollen yields of two cvs. Olea europaea L. ('Manzanillo' and 'Swan hill') (Tucson/Arizona, USA) 101

Pollen, airborne; study of airborne pollen grains of Alexandria, (Egypt) 55

Pollen, airborne; survey of atmospheric pollen in various provinces of Thailand 65

Pollen airborne and symptoms in allergic patients undergoing immunotherapy 89

Pollen, airborne and airborne spores, relationship to symptoms on the skin and mucous membranes of patients in the high altitude climate in Davos (Switzerland) 81

Pollen, airborne, v. Indoor allergens (apparatus for elimination), Recent pollen

Pollen collector, individual; experiences with the 'Individual Pollen Collector' developed by G. Boehm 87

Pollen count; biometeorology and its relation to pollen count 37

Pollen counts of ragweed (Ambrosia) and mugwort (Artemisia) in 1984 measured with the Cour collector at 12 meteorological centers in the Rhône Bassin (France) and surrounding regions 119

Pollen counts, v. Inversion layers

Pollen data; comparison of recent and quarternary pollen data 31

-; methode for yearly comparisons of aeropalynological data 407

Pollen, Olive pollen, v. Pollen airborne, Olive pollen

Pollen of Parthenium hysterophorus; allergic reactivity to Parthenium hysterophorus pollen: An unrecognized type I allergen 125

Pollinosis; chestnut (Castanea) pollen counts related to pollinosis patients in Paris 113

Pollutant of the air, new, v. Calcium sulphate

Pollution of the air; influences of pollution of the air and wheater on respiratory tract diseases of children in Berlin (West)-(Germany) 25

Pulmonary defense; modulation of pulmonary defense mechanisms by acute exposures to nitrogen dioxide 235

Quarternary pollen; comparison of recent and quarternary pollen data 31

Radiation exposure; radon and its decay products in the indoor environment: Radiation exposure and risk estimation 303

Radioactive dust particles in the air of Basel (Switzerland) (after Chernobyl); record with the Burkard-sampler for pollen, spores etc. 381

Radon and its decay products in the indoor environment: radiation exposure and risk estimation 303

Ragweed; pollen counts of ragweed (Ambrosia) and mugwort (Artemisia) (Cour collector) in 1984 measured at 12 meteorological centers in the Rhône Bassin and surrounding regions (France) 119

Recent pollen; comparison of recent and quaternary pollen data 31

Respiratory diseases, v. Inhalative allergens

Respiratory tract; the size distribution of whirlpool-generated droplets, their ability to contain bacteria and their disposition potential in the human respiratory tract 361

-; v. Airways in children, Bronchial reactivity, Dust particles, Legionnaires' disease, Obstructive respiratory tract disease, Pulmonary defense

Rhinitis, v. Alternaria tenuis

Rodent tissue cultures, v. Tissue cultures

Rhône Bassin; pollen counts of ragweed (Ambrosia) and mugwort (Artemisia) (Cour collector) in 1984 measured at 12 meteorological centers in the Rhône Bassin and surrounding regions (France) 119

Sampler, new construction, for sampling of Betula pollen for antigen analyses 383

Sampler for pollen, spores, algae and dusts particles, v. Burkardsampler, special applications

Samplers; collection efficiency of two samplers for microbial aerosols (Cyclone sampler and modified personal impinger) 391

-; v. Individual pollen collector

Seasonal symptoms positive skin and bronchial challenge tests with Alternaria in subjects with asthma and rhinitis- Clinical significance of airborne Alternaria tenuis spores 153

Spores, airborne, v. Davos

Spores; aircraft sampling in the upper atmosphere 411

-; airspora concentrations, uniformity 313

-; v. Alternaria; Botrytis; Filters; Fungal...; Fungi...; Fusarium...; Mildcw

Standardization, v. Allergen standardization

Tests, v. Alternaria tenuis, Fusarium

Thailand; survey of atmospheric pollen in various provinces of Thailand 65

Therapy, v. Immunotherapy, Inhalative allergene

Tissue cultures (human and rodent); mutagenic and carcinogenic effects of airborne particulate matter from polluted areas on human and rodent tissue cultures 231

-; v. Cell cultures

Tribolium confusum DuVal; sensitizations against Tribolium confusum DuVal in patients with occupational and non-occupational exposure 177

Tropical area, pollen; air sampling studies in a tropical area (Venezuela). Four year results 49

Tucson/Arizona, v. Pollen yield of two cvs. of Olea europaea...

Upper atmosphere; aircraft sampling of airspora in the upper atmosphere 411

USA, Northern USA; ozone and acidic inputs into forests in Northern USA 112

Variability, intrinsic, v. Fungi, airborne

Venezuela; air sampling studies in a tropical area. Four year results 49

Viral disease, v. Footh-and-mouth disease

Waste, domestic; airborne Gram negative bacteria associated with the handling of domestic waste 371

Wastewater; bacterial atmospheric contamination in wastewater treatment plants 365

Wheater, influence; obstuctive respiratory tract diseases of children in Berlin (West), influences of pollution of the air and wheater 25

Whirlpool; the size distribution of whirlpool -generated droplets, their ability to contain bacteria and their disposition potential in the human respiratory tract 361

Wind, spread of spores, v. Barley mildew

Work places; microbial and dust pollution in non-industrial work places 279

Index of authors

Agashe, S. N. and Chatterjee, M. 411

Aliani, M. vide Macchia, L.

Ammann, K.; Herzig, R.; Liebendörfer, L. and Urech, M. 401

Arrigoni, P. V. vide Zerboni, R.

Azuma, A.; Onda, Y. and Ichikawa,
R.-h. 337

Baron, P. A. vide Willeke, K.

Basabe, F. A. vide Edmonds, R. L.

Baskerville, A. vide Fitzgeorge,
R. B.

Baturay, O. F. vide Orsi, E. V.

Bavlsik, M. vide Orsi, E. V.

Beaumont, F. vide Spieksma,
F. Th. M.

Benninghoff, W. S. 13

Bergmann, E. M. vide Schultze-Werninghaus, G.

Bessot, J. C. vide Pauli, G.

Bischoff, E. and Schirmacher, W.
189

Boehm, G. 3

Boehm, G. 377

Boehm, G. 381

Boehm, G. and Leuschner, R. M. 87

Boehm, G. vide Leuschner, R. M.

Boissinot, E. vide Boutin, P.

Borelli, S. vide Kneist, W.

Bossert, J. vide Fuchs, E.

Botzenhart, K. vide Waldner-Sander, S.

Boutin, P.; Torre, M.; Moline J. and
Boissinot, E. 365

Brombacher, Chr. vide Leuschner,
R. M.

Brown, H. M. and Jackson, F. A. 261

Buchmann, St. L. vide O'Rourke,
M. K.

Buffat, Ph. vide Guillemin, M.

Bunnag, C. vide Dhorranintra, B.

Burge, H. A.; Simmons, E. G.;
Muilenberg M.; Hoyer, M.;
Gallup, J. and Solomon, W. 143

Burkart, W. 303

Buttler, A. and Girard, M. 407

Bylin, G.; Lindvall, T.; Rehn, T. and
Sundin, B. 227

Caiaffa, M. F. vide Macchia, L.

Campi, P. vide Zerboni, R.

Carbonara, A. M. vide Macchia, L.

Casella, G. vide Macchia, L.

Chatterjee, M. vide Agashe, S. N.

Coetzee, J. A. 31

Cour, P. vide Déchamp, C.

Crook, B.; Higgins, S. and Lacey, J.
371

Cummins, L. vide Gallup, J.

David, B. vide Sutra, J. P.

Déchamp, C. and Cour, P. 119

De Luca, H. vide Sutra, J. P.

Dhorranintra, B.; Bunnag, Ch. and
Limsuvan, S. 65

Doll, Sir Richard 213

Donaldson, A. I.; Lee, M. and Gibson, C. F. 351

Düngemann, H. vide Kneist, W.

Eberhard, K. vide Wedler, E.

Edmonds, R. L. and Basabe, F. A.
319

El-Ghazaly, G., and Fawzy, M. 55

Elixmann, J. H.; Jorde, W. and Linskens, H. F. 283

Eversmeyer, M. G. vide Kramer,
Chs., L.

Fängmark, I. vide Henningson, E.

Fasani, F. and Gorini, M. 89

Fawzy, M. vide El-Ghazaly, G.

Featherstone, A. S. R. vide Fitzgeorge, R. B.

Fegeler, U. vide Wedler, E.

Fitzgeorge, R. B.; Baskerville, A.
and Featherstone, A. S. R. 357

Frankland, A. W. 147

Frinking, H. D. 9

Fuchs, E. 71

Fuchs, E.; Bossert, J.; Honomichl,
K.; Maasch, H.-J.; Risler, H. and
Wahl, R. 183

Galán, C.,; Ruiz de Clavijo, E.;
Infante, F. and Gallego, G. 109

Galán, C. vide Infante, F.

Gallego, G. vide Galán, C.

Gallego, G. vide Infante, F.

Gallup, J.; Kozak, P.; Cummins, L.
and Gillman, S. 139
Gallup, J. vide Burge, H. A.
Gatti, E. vide Macchia, L.
Gehrken, H. vide Kneist, W.
Gibson, C. F. vide Donaldson, A. I.
Gillman, S. vide Gallup, J.
Girard, M. vide Buttler, A.
Goldstein, I. F.; Reed C. E.; Swan-
son, M. C. and Jacobson, J. S. 133
Gorini, M. vide Fasani, F.
Gravesen, S. 279
Guillemin, M.; Litzistorf, G.; Made-
laine, P.; Iselin, F. and Buffat, Ph.
221
Hadnagy, W. vide Seemayer, N. H.
Haupthof, M. vide Rudolph, E.
Henningson, E. and Fängmark, I.
391
Herzig, R. vide Ammann, K.
Higgins, S. vide Crook, B.
Honomichl, K. vide Fuchs, E.
Hoyer, M. vide Burge, H.
Hoyet, C. vide Pauli, G.
Hurtado, I. and Riegler-Goihman,
M. 49
Iacobelli, A. vide Macchia, L.
Ichikawa, R.-h. vide Azuma, A.
Ickovic, M.-R. vide Sutra, J. P.
Infante, F.; Ruiz de Clavijo, E.;
Galán, C. and Gallego, G. 157
Infante, F. vide Galán, C.
Iselin, F. vide Guillemin, M.
Jackson, F. A. vide Brown, H. M.
Jacobson, J. S. vide Goldstein, I. F.
Jakab, G. J. 235
Jantunen, M. J. vide Pitkänen, E.
Jorde, W. vide Elixmann, J. H.
Kalliokoski, P. vide Pitkänen, E.
Kappos, A. D. vide Schultze-
Werninghaus, G.
Kauppinen, E. vide Rantio-Lehti-
mäki, A.
Kettrup, A. and Schmidt, P. R. 415
Kneist, W.; Düngemann, H.; Gehr-
ken, H. and Borelli, S. 81
Koivikko, A. vide Rantio-Lehtimäki,
A.

Kozak, P. vide Gallup, J.
Kramer, Chs. L. and Eversmeyer,
M. G. 313
Kroeling, P. 287
Kundel, M. vide Stalder, K.
Lacey, J. vide Crook, B.
Lee, M. vide Donaldson, A. I.
Leuschner, R. M.; Boehm, G. and
Brombacher, Chr. 19
Leuschner, R. M. vide Boehm, G.
Lévy, J. vide Schultze-Werninghaus,
G.
Lewis, W. H. vide Wedner, H. J.
Liebendörfer, L. vide Ammann, K.
Limpert, E. 331
Limsuvan, S. vide Dhorranintra, B.
Lind, P. vide Menz, G.
Lindvall, T. vide Bylin, G.
Linskens, H. F. vide Elixmann, J. H.
Litzistorf, G. vide Guillemin, M.
Maasch, H.-J. vide Fuchs, E.
Macchia, L.; Aliani, M.; Caiaffa,
M. F.; Carbonara, A. M.; Gatti,
E.; Iacobelli, A.; Strada, S.;
Casella, G. and Tursi, A. 95
Macher, J. M. 275
Madelaine, P. vide Guillemin, M.
Mallant, R. K. A. M. vide Masuch,
G.
Mandrioli, P. 37
Manfredi, M. vide Zerboni, R.
Marty, H. 255
Masuch, G.; Paul, V. H. and Mal-
lant, R. K. A. M. 325
Meier-Sydow J. vide Schultze-Wer-
ninghaus, G.
Menz. G.; Petri, E.; Lind, P. and
Virchow, Chr. 197
Moline, J. vide Boutin, P.
Moyzes, R. vide Wedler, E.
Muilenberg M. vide Burge, H. A.
Nolard, N. vide Spieksma, F. Th. M.
Onda, Y. vide Azuma, A.
O'Rourke, M. K. and Buchmann,
St. L. 101
Orsi, E. V.; Bavlsik, M.; Viera,
R.; Petersheim, M. and Baturay,
O. F. 243

Paul, V H. vide Masuch, G.
Pauli, G; Tenabene, A.; Bessot,
    J. C. and Hoyet, C. 203
Pellikka, M. vide Pitkänen, E.
Peltre, G. vide Sutra, J. P.
Petersheim, M. vide Orsi, E. V.
Petri, E. vide Menz, G.
Pitkänen, E.; Pellikka, M.; Kallio-
    koski, P. and Jantunen, M. J. 297
Rantio-Lehtimäki, A.; Kauppinen,
    E. and Koivikko, A. 383
Reed, C. E. vide Goldstein, I. F.
Rehn, T. vide Bylin, G.
Riegler-Goihman, M. vide Hurtado,
    I.
Risler, H. vide Fuchs, E.
Rudolph, E.; Stresemann, E.; Stre-
    semann B. and Haupthof M. 177
Ruiz de Clavijo, E. vide Galán, C.
Ruiz de Clavijo, E. vide Infante, F.
Schirmacher, W. vide Bischoff, E.
Schmidt, P. R. vide Kettrup., A.
Schrader, G. 267
Schultze-Werninghaus, G.; Lévy, J.;
    Bergmann, E. M.; Kappos, A. D.
    and Meier-Sydow, J. 153
Seemayer, N. H.; Hadnagy, W. and
    Tomingas, R. 231
Sellers, R. F. 347
Simmons, E. G. vide Burge, H. A.
Singh, A. B. 61
Solomon, W. vide Burge H. A.
Spieksma, F. Th. M.; Nolard, N.;
    Beaumont, F. and Vooren P. H.
    165
Stalder, K. and Kundel, M. 251
Strada, S. vide Macchia, L.
Stresemann, B. vide Rudolph, E.
Stresemann, E. vide Rudolph, E.
Sundin, B. vide Bylin, G.
Sutra, J. P.; Ickovic, M.-R.; De Luca,
    H.; Peltre, G. and David, B. 113
Swanson, M. C. vide Goldstein, I. F.
Tenabene, A. vide Pauli, G.
Tiberg, E. 171
Tomingas, R. vide Seemayer, N. H.
Torre, M. vide Boutin, P.
Tursi, A. vide Macchia, L.

Urech, M. vide Ammann, K.
Viera, R. vide Orsi, E. V.
Virchow, Chr. vide Menz, G.
Vooren, P. H. vide Spieksma,
    F. Th. M.
Wahl, R. vide Fuchs, E.
Waldner-Sander, S. and Botzenhart,
    K. 395
Wedler, E.; Fegeler, U.; Moyzes, R.
    and Eberhard, K. 25
Wedner, H. J.; Zenger, V. E. and
    Lewis, W. H. 125
Willeke, K. and Baron, P. A. 361
Zenger, V. E. vide Wedner, H. J.
Zerboni, R.; Manfredi, M.; Campi,
    P. and Arrigoni, P. V. 43

MIX
Papier aus verantwortungsvollen Quellen
Paper from responsible sources
FSC® C105338

If you have any concerns about our products,
you can contact us on
ProductSafety@springernature.com

In case Publisher is established outside the EU,
the EU authorized representative is:
Springer Nature Customer Service Center GmbH
Europaplatz 3, 69115 Heidelberg, Germany

Printed by Libri Plureos GmbH
in Hamburg, Germany